Defense Industries: Science and Technology Related to Security: Impact of Conventional Munitions on Environment and Population

NATO Science Series

A Series presenting the results of scientific meetings supported under the NATO Science Programme.

The Series is published by IOS Press, Amsterdam, and Kluwer Academic Publishers in conjunction with the NATO Scientific Affairs Division

Sub-Series

I. Life and Behavioural Sciences	IOS Press
II. Mathematics, Physics and Chemistry	Kluwer Academic Publishers
III. Computer and Systems Science	IOS Press
IV. Earth and Environmental Sciences	Kluwer Academic Publishers
V. Science and Technology Policy	IOS Press

The NATO Science Series continues the series of books published formerly as the NATO ASI Series.

The NATO Science Programme offers support for collaboration in civil science between scientists of countries of the Euro-Atlantic Partnership Council. The types of scientific meeting generally supported are "Advanced Study Institutes" and "Advanced Research Workshops", although other types of meeting are supported from time to time. The NATO Science Series collects together the results of these meetings. The meetings are co-organized bij scientists from NATO countries and scientists from NATO's Partner countries – countries of the CIS and Central and Eastern Europe.

Advanced Study Institutes are high-level tutorial courses offering in-depth study of latest advances in a field.
Advanced Research Workshops are expert meetings aimed at critical assessment of a field, and identification of directions for future action.

As a consequence of the restructuring of the NATO Science Programme in 1999, the NATO Science Series has been re-organised and there are currently five sub-series as noted above. Please consult the following web sites for information on previous volumes published in the Series, as well as details of earlier sub-series.

http://www.nato.int/science
http://www.wkap.nl
http://www.iospress.nl
http://www.wtv-books.de/nato-pco.htm

Series IV: Earth and Environmental Sciences – Vol. 44

Defense Industries: Science and Technology Related to Security: Impact of Conventional Munitions on Environment and Population

edited by

Pelagio Castelo Branco
Danotec Association,
Lisbon, Portugal

Hiltmar Schubert
Fraunhofer Institute of Chemical Technology,
(ICT) Pfinztal, Germany

and

Jose Campos
Laboratory of Energetics and Detonics (LEDAP),
Coimbra, Portugal

Kluwer Academic Publishers

Dordrecht / Boston / London

Published in cooperation with NATO Scientific Affairs Division

Proceedings of the NATO Advanced Research Workshop on
Defense Industries: Science and Technology Related to Security: Impact of Conventional Munitions on Environment and Population
Porto, Portugal
28–31 October 2001

A C.I.P. Catalogue record for this book is available from the Library of Congress.

ISBN 1-4020-2794-X (PB)
ISBN 1-4020-2793-1 (HB)
ISBN 1-4020-2795-8 (e-book)

Published by Kluwer Academic Publishers,
P.O. Box 17, 3300 AA Dordrecht, The Netherlands.

Sold and distributed in North, Central and South America
by Kluwer Academic Publishers,
101 Philip Drive, Norwell, MA 02061, U.S.A.

In all other countries, sold and distributed
by Kluwer Academic Publishers,
P.O. Box 322, 3300 AH Dordrecht, The Netherlands.

Printed on acid-free paper

Printed in the Netherlands.

Table of Contents

PREFACE

The state of the art in the area of security related civil science and technology concerning energetic materials and munitions is of pressing importance at the present time, because 10 years after the end of the cold war, in many countries in east and west, huge stockpiles of munitions still await destruction.

Military munitions with their high concentration of energetic materials, special and sophisticated hardware and toxic agents, develop varied dangerous impacts on environment and population.

Scientific research and development have the aim to increase performance of munitions on one side and to improve safety in relation to unpredictable events on the other side. The impact on the environment by the production and normal use of munitions and their residues after use will acquire an increasing importance in environmental awareness.

The impact of munitions on environment is caused during the whole life cycle of the item, such as:

- Manufacturing, storage and transportation, usage and misfires, residues and undetected ordnance, recycling and destruction.

The exchange of results of R&D activities of NATO and Atlantic Partnership Countries, the general approach to scientific understanding of explosive reactions, and their initiation and chemical behaviour are important to increase safety with regard to environmental protection.

Hence the justification for an Advanced Research Workshop focused on 2 key areas of the environmental impact of the munitions, which are the safety effects of munitions themselves and the possible damage to environment at the end of their life cycle:

- The safety behavior of munitions and their energetic material is the determining factor for all precautionary measures for the protection of environment and population. The knowledge about the probability of and the effects during an incident, and therefore the development of insensitive munitions and "green energetic materials", are areas subject to special research activity and should be dealt with in the workshop.
- The probability of a catastrophic reaction of munitions and their energetic components will be so more likely, the closer the munitions are to the end of their life time. Residues and undetected ordnance, recycling, destruction of ammunition and depletion or reuse of energetic materials will either cause more or less damage to environment.

This NATO Advanced Research Workshop, under the theme *"Defence Industries: science and technology related to security, and the impact of conventional munitions on environment and population"*, was held in Porto, Portugal, from 27 October to 1 November 2001. Delegations from western and eastern countries (Czech Republic, Georgia, Germany, Netherlands, Poland, Portugal, Romania, Russia, Sweden, United Kingdom, United States of America) brought to the meeting a significant number of Guest Speakers, Romania deserving a particular mention.

The excellent technical presentations were classified in five Groups, according to the technical programme. The *"Introduction to the subject; Methods, tests and application; Insensitive and "green" munitions; Storage and transportation; and Demilitarisation and recycling"* sessions, were closed by a final and general interesting discussion among participants.

The first session started with an opening lecture on "*The impact of munitions on the environment in a general view*", and defined the point of this workshop. This lecture, followed by two other contributions, introduced the presentation of several technical tests, predicting methods and results, related to the main evaluation of performance, life time and environmental impact of existing munitions. The possibility of using predictor-corrector computer codes and of the miniaturisation of destructive samples, significantly reduces the mass of test samples, their costs and environmental impact. The insensitive and "green" munitions were largely discussed during the third session. Extensive communications were presented focussing mainly on the problems of environmental impact of munitions on military proving grounds. A large number of military institutions and laboratories related thereto, especially the Naval Surface Warfare Center, Indian Head, USA, concluded the existence of comprehensive information on insensitive munitions, relatively wider than that related to "green" munitions. These data and criteria were also associated to the issues of storage, transportation and disposal. The last session concerned demilitarisation and recycling and included presentations from both western and eastern countries. The problems concern not only the environmental research of demilitarisation impact, but also the technical proposals for demolition and recycling of hazardous explosive materials and components.

The final and general discussion started from a presentation by the formally Chairman of the WEAG, CEPA 14 "*Energetic materials*", where he defined a cost effective approach to the common problems of disposal and cleaning up. Different environments cause different problems, which, all the same, need similar risk reduction and management. Effective risk assessment also implies risk quantification, including the whole life time assessment and costing. The discussion was immediately extended to all the participants, questioning the use of explosive components in future munitions. Reality shows that explosives and propellants mean less than three percent of the munitions total cost, but they also represent their energetic power, and there is no other alternative to this power source.

The solution to the problem at the present NATO workshop implies a collaborative approach, with shared experience and shared studies/assessment between countries, in order to reduce capability gaps between nations and develop generally acceptable methods for costing offset, through recycling of both existing and future munitions.

A great many people deserve thanks for having contributed to a truly successful workshop.

First of all, let me express my appreciation to my Romanian colleague and co-director of the event, Mrs. Catalina Teodorescu.

Speakers and Participants are to be commended for their efforts and commitment.

A reference is to be made to the presence of the Portuguese Minister of Defence, who in turn was hosting the Secretary of State for Defence of Romania, in attendance of the last day's Discussion Session of the workshop.

Last but not least, my gratitude to the Scientific Affairs Division of NATO for financial support, as well as for invaluable advice and encouragement.

Pelágio Castelo Branco

Hiltmar Schubert

José Campos

Introduction into the Problematics

THE IMPACT OF MUNITION ON THE ENVIRONMENT IN A GLOBAL VIEW

Hiltmar Schubert
Fraunhofer Institute of Chemical Technology (ICT)
Pfinztal , Germany

ABSTRACT

The general impact of munition on the environment is caused by the original purpose of these energetic materials to fight against an object. By this impact also the environment will be in most ases badly injured.
Beside this original object, the environment is threatened by the munition and its components in princible more or less during the whole life-cycle of the munition because of its energetic content, which may cause chemical pollution. During application or by an accidents, physical stresses by pressure waves, flames and also air- and ground pollution may occur.
These impacts may appear during the whole life-cycle of the munition, that means during:

- *manufactoring*
- *transportation*
- *storage*
- *application and*
- *demilitarisation.*

In an overview the possible impact of munition on the environment will be explained by means of these steps of the energetic material's life-cycle.

To reduce these impacts , efforts are known to avoid or decrease toxic materials or components of the munition and to reduce the sensitivity of the energetic materials without decreasing the performance
and other properties of the munition. We have to learn, especially insensitivity is a kind of performance and- of course in practice we must come to compromises like we do it in budgetplannings.
The general impact of munition on the environment is caused by the original purposes of these energetic materials to fight against an object. By this impact also the environment will be in most cases badly injured.
Beside this original object, the environment is threatened by the munition and its components in priciple more or less during the whole lifer-cycle of the munition because of ist energetic content, which may cause chemical pollution. During application or by an accident, physical stresses by pressure waves, flames and also air- and ground pollution may occur. These impacts may appear during the whole life-cycle of the munition and/or its components.
In an overview the possible impact of energetic materials on the environment will be explained by means of the life-cycle-steps.
To reduce these impacts, efforts are known to avoid or decrease toxic materials and to reduce the sensitivity of the energetic materials without decreasing the performance and other necessary properties of the munition. We have to learn, insensitivity and environmental compatibility is also a kind of performance and – of course in practice – compromises have to be done.

1. Introduction

Environmental qualification of technical products has come up during the 70ies of the last century and since that time an increasing realization of ideas for environmental protection

P. Castelo Branco et al. (eds.),
Defense Industries: Science and Technology Related to Security, 3–8.

came round. Along with general commercial products also military materials has come into focus, because public feeling estimates munition and its components as special dangerious materials.Nowadays expressions like „Environment Friendly Behavior",Green Explosives",„Life-Time-Properties", „Recycling" etc are equally entitled concerning properties like performance, cost, reproducebility, stability etc.

If one considers the various impacts of munition on the environment and human being, the following whole life-cyle- profil of energetic materials is to be considered:

- Manufactoring
- Tranportation
- Storage
- Application and
- Recycling and Demilitarization
- Destruction and Waste

The objectives for the decreasing of impacts to the environment and human being two possibilities should be mentioned:

1. For the prevailing energetic materials : Treatment during the whole life-cycle by means of environmental compatible processes.
2. Using munition with compositions or combinations which have a lower risk by nature to the environment and human being. Expressions like „Insensitive- or Less Sensitive Munition"," Low Vulnerability","Green Explosives",etc. are known for these developments.
3. Development of amunition for" Non lethal weapons" with minimal undesired damage or impact on the environment (personnel and equipment).

2. Manufactoring

Concerning safety aspects the explosive industry has developed different procedures for manufactoring of material with high energy content to decrease the impact in case of an accident.
The aim in principle is to decrease the volume of material during the dangerous phase of production. More and more batch processes are changed into continious processing to decrease the amount of explosive material during manufactoring. Steps of the production-chain will be local seperated and protected from each other if possible.
A remotely controlled processing during the dangerous steps will be the rule. The number of workers are general limited. Special hardware is used to avoid friction, sparking or electric loading.
Strong regulation are established for every process etc.

Concerning the environmental protection same duties are used as in the commercial Industry to minimize the impact on the environment. (Waste water,Air- and Ground-pollution).

3. Tranportation

Because of the high energy-content tranportation of munition is dangerous for the environment and human being. Therefore very tight and extensive regulations in both fields of commercial and military application are existing more or less in every country including also the tests for permission. All these regulations are based on extensive practical knowledge. To my experiences about 8o% of the accidents during processing and tranportation are caused by human failures, that means regulations are not enough considered in the daily routine.
Stresses to the explosive items which courses disasters do not cover the test procedure for permission because of different reasons:

- Specific stresses on the material was not perceived mostly because of the lack of experiences and knowlege.
- The explosive object is exposed to more than one stress or condition which are not dangerous for its own but in a cumulative manner the overall impact is high enough for a disaster.
- According to the sensitivity of the munition during a dangerous situation the impact is too large, therefore the material react. A transition from deflegration to detonation are very often observed.

4. Storage

Concerning handling of explosive materials and munition we have the same conditions like during transportation.
Long-term advance planning of the life-time of defense equipment in general is a factor of basic importance to the economy of the defense expenditure of a country. I have opened with this sentence an ICT Conference on lifetime of explosive materials exactly 30 years ago and nowadays this sentence is more valid than ever before.

We differ from

- a life-time, in which all functions of the munition show all dates within the prescribed tolerances,
- a life-time, in which the munition can be used with some restiction (for inst.temperature)
- a life-time, in which storage or tranportation will be not dangerous, but a misfire may be possible.

Life-time prediction of energetic material and munition are in this area a very important item of research and not yet solved in all cases, because
Life-time will be influenced be environmental conditions:
The main influences are temperatur, humidity and mechanical stresses and the compatibility of the different components.
The aim will be the life-time prediction.

Concerning temperature, the Arrhenius equation is used as a routine. But the factor may change over distinct temperatur ranges and other influences like temperatur-cycling, stresses, humidity or compatibility with other materials have additional influences.

For a tight life-time prediction a set up of a life-cycle-profil is necessary, that means to calculate all proposed stresses over all phases of the life cycle. For this objective a test-tailoring process is used to decrease the scope of tests.
The solution will be in future for larger systems to register the stresses of the different life-cycle phases. Research is working to develop small and cheap sensors to register all the environmental influences at the single munition.

5. Application

The original purpose of energetic materials are to create gases in more or less short time with the object of propulsion or blasting. The output of such reactions are hot gases and solid residues from the explosive charge itself and its confinement. Dependent on the reaction time pressure waves are produced.
The whole output of such charges are more or less hurtful to the environment and the question is, in which way these outputs may be reduces without decreasing the performance of the munition, because all these outputs were also produced by testing and proving the munition and training the armed forces.
Another object will be to develope energetic material and munition which are less sensitive to impacts coused by accidents.

During the last 25 years research and development were very active to improve the insensitivity of energetic materials without decreasing the performance of the munition. These objectives have animated very much the research-activities with questions like the investigation of the microstructure of energetic materials, the understanding of initiation-processes, investigation of the mechanical properties and understanding damping processes in the material etc. Also new energetic compounds were synthesized to reach this aim.
Practice has shown, that insensitivity will be also some kind of performance. To decrease the amount of toxic gases during combustion or detonation and also to avoid toxic residues of non-reacted materials, investigations were done during the past 5 –10 years to use environmental compatible energetic compounds.

Examples are to replace Tetryl by HMX or to replace leadcompounds in igniters or detonaters by other primary explosives.“ Green explosives“ should avoid all aromatic compounds. In this connection Ammoniumnitrate as an oxidizer has become interest again.

To increase performance of penetraters heavy metalls are used which may be toxic.
For instance large discussions have taken place in the past years about environmental problems replacing tungsten by unriched uranium.
For training purposes a good solution was found to use plastic munition instead of metall rounds. Also in the field of „Green Munition“ we have the same situation like in the area of Insensitive Munition“, that means improving insensitivity or environmental friendlyness without descreasing performance. To reach this aim intensive researchwork must be done in future.

The general purpose of conventional munition is to fight against a target and the aim will be to destroy or to disable the object at least.

During the last years local conflicts and civil wars have increased and military forces must act in many situations no longer as compatants, but also as conflict managers or peace keepers in dangerous situations. To protect solidiers during their functions and to put out of action single opponents or groups „Non-Lethal Weapons“ were developed.

„Non Lethal Weapons“ are according to NLW-Concept of NATO:
„Non-Lethal Weapons are weapons which are explicitly designed and primarily employed so as to incapacitate personnel or equipment, with a very low probability of death or permanent injury and with minimal undesired damage or impact on the environment.
Unlike conventional lethal weapons which destroy their targets principally through blast, penetration and fragmentation, Non-Lethal-Weapons employ means other than gross physical destruction to prevent the target from functioning.“

In general a large number of different methods and ammunitions may be used based on Lasers, Microwaves, Electroshocks and optical-, acustic-,chemical- and bioligical means, which are directed against men or facilities.
The fight against terrorism nowadays will increase the importance of these new developments.

6. Recycling and Demilitarization

After the end of the cold war Germany was confronted to be responsible for 3oo ooo t of munition. (With about 17 ooo t of TNT) of the former peoples army of the DDR. The question was to deplete the muniton environmental friendly or to recycle for different civil or military applications.
The ICT was very much engaged in this task and much research was done in those days.
We collect the following experiences of recycling energetic materials:
The amount of military explosives are - comparing with commercial explosives -not so high, to investigate a new process for converting the explosive material for civil use. There may be a chance to recristalize relative expensive explosives like HMX. Some pyrotechnic material with expensive metals like Mercuryfulminate are of interest.

Of course the sheaper the manpower will be the better is the possibility for recycling or usage for commercial means.
There may be a chance, if there is a possibility to use energetic materials for civil purposes in a simple process without expensive investigations. Because of the explosive behaviour of the material the processes has to be done environmental friendly und under safety aspects until the explosive properties are still present. These conditions will increase the expenses. There may be examples, where the production of energetic material will be as high as the recycling or depletion under environment friendly conditions.

7. Destruction and Waste

Under common circumstances energetic materials will be incinerated. This process impact the environment, because the combustion gases contain the toxic Nitrogenoxid. Therefore chemical or physical means are necessary to convert this substance to Nitrogen. Because of the explosive behaviour of the material safety precautions are necessary to Prevent deflagration or detonation of the material. Numerous accidents have occured in the past by incineration.

It is known, that –because of the expenses for wasting - the munition is sunk into lakes or into the open sea.
Without regard to the general impact to the environment this process is dangerous as examples have shown. Sometimes the munition will wash against the shore again or unstability will cause unterwater explosions(Picatinny Arsenal) or the granates will get leaky and the material will come into contact with the water. A very dangerious example is the sunk munition with chemical agents in the baltic sea or the transport and incineration of the chemical munition itself.

SAFETY AND COMPATIBILITY IN SERVICE OF MUNITIONS AND EXPLOSIVES IN ROMANIA

Lt. Col. Eng. Adrian COMAN, PhD. *

Dip. Eng. Cătălina TEODORESCU**

ABSTRACT

The article refers to the importance of the security concept in Romania and the compatibility for service of the munitions and explosives materials, and presents the disposals taken already and those to be taken/ to come/ to reach/ in the future by the Romanian Government in order to fulfil the NATO requirements in this domain (STANAG 4170, AOP-7), by elaborating/issuing/ improving/ the national military standards.

After presenting the general legal bounds of the explosive material regiment/system/ in Romania (law no. 126/1995 and 17/1996), the article is focused on the validation procedures of the military explosive materials and ammunition. Thus, there are presented the responsibilities of the National Authority which has the responsibility/duty/ which is in charge/ to validation the explosive materials and ammunition in Romania and the article also presents the strategy of National Authority to put in accord/ to meet the requirements of/ the NATO standards and the national standards and their compatibility in the ammunition and explosive materials system, as a part of the general strategy in order to (fully) integrate in NATO.

Finally, there are briefly presented the procedures , methods and evaluating tests which are valid in Romania for ammunition, explosive materials, explosive powders and other pyrotechnic composites in order to ensure the security and the compatibility in the system(in charge).

* *Senior lecturer*, Military Technical Academy, Bucharest, Romania

** Deputy director, Ministry of National Defence, Bucharest, Romania

P. Castelo Branco et al. (eds.),

Defense Industries: Science and Technology Related to Security, 9–19.

1. GENERAL CONSIDERATIONS

Part of the security and protection concept, safety and compatibility in service of ammunition and explosives requires special attention by the Romanian authorities and specialists who are making efforts to comply with the regulations adopted by the developed countries, especially NATO members .

In Romania, actions for ensuring the safety and compatibility in service of ammunition and explosives are very serious, involving even penal responsibility for responsible, irrespective of their hierarchical level.

1.1 The laws in force

- Law no. 126 / 1995 on the legal status of explosives, amended with Law no. 464/2001;
- SMT 40269-2002 on safety and compatibility in service of ammunition (complying with STANAG 4297 and AOP-15);
- SMT 40268/1-2002 on introducing, assessing and testing IM (insensitive munitions) policy in Romania;
- SMT 40268/2-2002 - Guide for implementing the policy in the IM field;
- SMT 40362-2001 on principles and methodology of qualifying military explosives (complying with STANAG 4170 and AOP-7).

1.2 General conditions for evaluation the security of ammunition

In Romania, for a correct evaluation the safety of ammunition are considered the following general conditions:

- The assessment of safety and suitability for service of a new munitions needs to be addressed within the concept of system safety, meaning that while a munitions may be viewed as a system in and of itself, it should also be seen as part of a larger weapon system when addressing system safety. Thus, system safety efforts traditionally encompass thorough assessment of the intended use of all elements of the larger weapon system over its entire life- from design concept to development and manufacturing to use a disposal- as well as an assessment of any external influences that could impact safety.
- The program for ensure system safety will work in harmony with the activities of project management and of conception and also with concurrent efforts of other disciplines as reliability, maintainability, human factors, health and

environment, storage and transportation, to achieve an integrated and balanced approach throughout the entire system life cycle. Early identification and elimination or control of hazards that may arise from design flaws, incorrect assembly, inadequate procedures, and general misuses is intended to ensure that the highest possible degree of safety consistent with mission effectiveness requirements and cost constraints is designed and built into each munitions and weapon system. That integrated and co-ordinated approach will support management decisions, assist in minimising risks, and should preclude wasted effort in the proving of a design and the associated cost and time penalties of late design changes. The end products of this process should include ample evidence to enable a positive assessment that the munitions and the larger weapon system are acceptably safe and suitable for service use, consistent with mission requirements and cost.

- For designing munitions it hat to be used materials and constructive solutions to give munitions a very law sensitiveness of the external stimuli, in order to reduce the risk of an incidental initiating by stimuli which normally exist in the environment, or could appear in unordinary situations.
- If the resulting munitions meets the requirement goals given in SMT 40268/1-2002 and SMT 40268/2-2002, then the munitions may have ascribe some descriptor signed MURAT in order to indicate the assessed degree of compliance with those standards..
- Industrial process and technologies for munitions destruction pursuant to demilitarisation and disposal of it, must be environmentally acceptable, free of health hazards and also to maximise the recycling of component materials.

1.3 Priorities for safety munitions

In order to obtain a high safety level for a munitions system, applying the eliminate and control measures must be done conform following priorities:

a) Aim to avoid hazards in the design concept phase.

b) Design to eliminate the hazard.

c) Design to control the hazard that cannot be eliminated through design.

d) Use safety devices when elimination or design control is not possible.

e) Use warning devices to advise of a hazardous condition that cannot otherwise be eliminated or controlled.

f) Use procedures and training when it is impossible to eliminate or adequately control a hazard through design selection or use of safety and warning devices.

Because it is unanimously known that human intervention is the poorest safety method to avoid an accident, reversing the order of priorities from bottom to top is not recommended, excepting only the cases when all danger-avoiding design alternatives have been examined and they were not feasible. Such exceptions must be previously written and approved.

1.4 Safety design criteria

To minimise mishaps resulting from the inadvertent, premature, or incorrect operation of materiel, the following system safety criteria are recommended for the design and development of munitions systems and components:

a) In its design mode, no single design defect, component failure, environmental condition, personnel error, or other mishap shall result in a Level 1 or Level 2 risk.

b) In general, in the absence of the designed firing stimulus being applied to the explosive train, no two independent component failures, no two human errors, nor a combination of a component failure and a human error shall result in a Level 1 or Level 2 risk.

c) No single mishap shall result in a Level 3 risk unless trade studies show that design alternatives capable of reducing the risk would also seriously compromise system effectiveness.

2. SAFETY AND SUITABILITY ASSESSMENT PROCESS

Munitions development plan should include a methodical process to assess safety and suitability for service. Those plan should include analyses and testing regimes aimed at ensuring that the explosive components as well as the complete munitions will function as designed with a degree of safety that has been achieved as a balance with system effectiveness, in service schedule and life cycle cost.

That methodical safety assessment process should commence as soon as possible after inception of the development project and should actively progress so that all relevant safety analyses and tests are completed in sufficient time to allow the safety evaluation organisation to advise on the safety and suitability of the munitions prior to any required in service date.

Figure 1 depicts the safety and suitability assessment logical process.

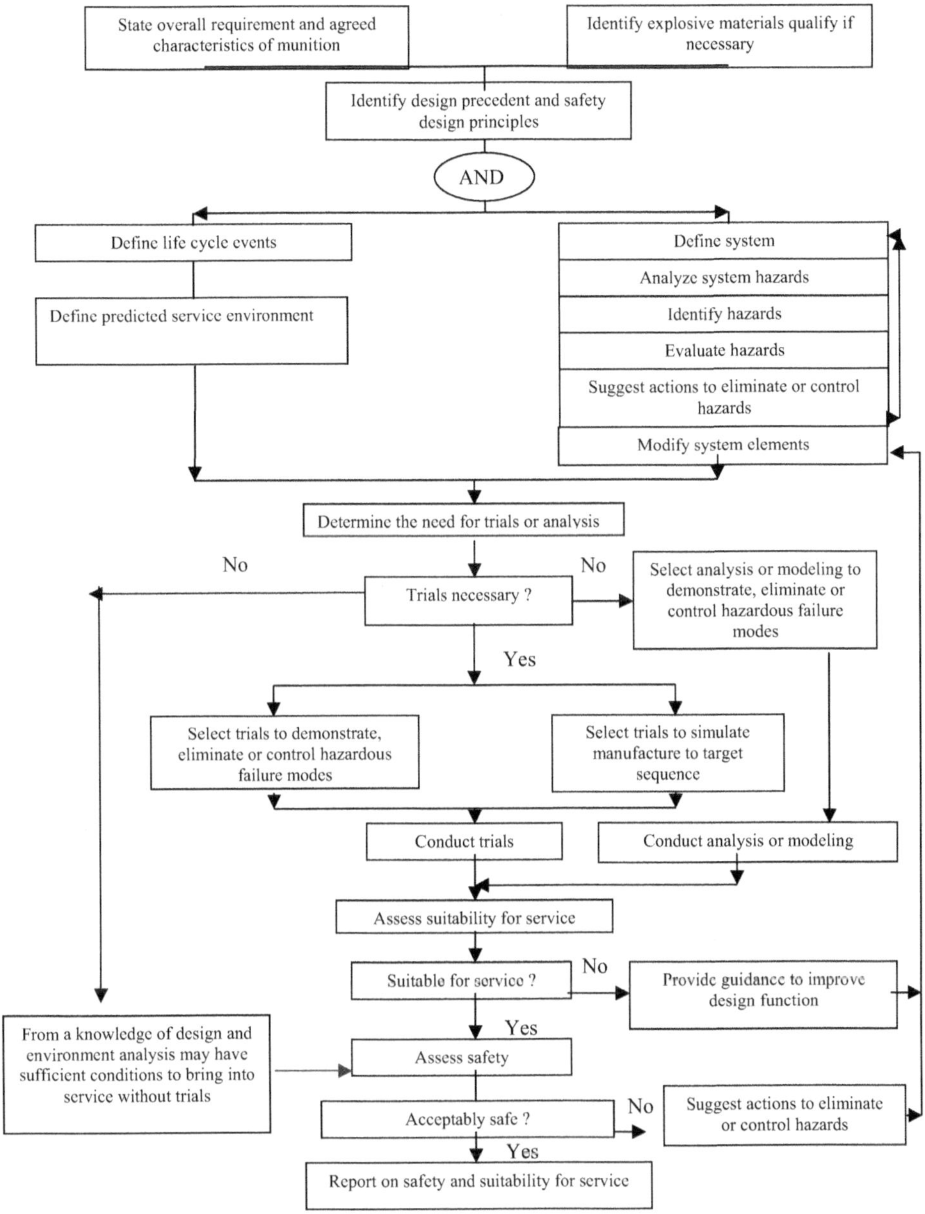

Fig. 1 *Safety and suitability assessment process*

3. ACTIONS FOR REDUCING RISKS OF MUNITIONS IN ROMANIA

3.1 General frame

Progress in explosives technology made possible the development of a large range of munitions, generally called *munitions with attenuate risk.* (Munitions a Risques Attenues – MURAT) in France language countries and *insensitive munitions* (Insensitive Munitions – IM) in English language countries, which are less vulnerable at incidental stimulation than classic munitions.

Such munitions remain very effective for their assigned missions, but they are less sensitive at extreme environmental conditions (hot, froze, mechanical sock or impact) than their predecessors.

To introduce IM in service means both development of logistic and strategy system efficacy by reducing risks of serving people, reducing the price and efficiencing the transportation, storage and operating with munitions

The policy for introducing, assessing and testing IM in Romania is the object of SMT 40268/1-2002. A guide for implementing this policy is SMT 40268/2-2002, which stipulates provisions and procedures to be considered in munitions design, in order to reduce their vulnerability regarding the real threats.

IM evaluations does not replace evaluations carried out in accordance with SMT 40269-2002 for settling the security and compatibility in service (S^3) and law requirements concerning munitions production, transport, storage and drawing out of service. MURAT evaluations completes all those rules and laws, aiming to:

a) identify specific threats to which munitions can be exposed, during their life cycle;

b) establish the test regime which have to provide information containing munitions response when subjected at representative stimuli;

c) comparing those answers with requirements of MURAT as stated in SMT 40268/1-2002 .

3.2 Energetic materials and design techniques

In order to reduce the probability of an accidental initiating and producing collateral damages, that for designing a munitions is necessary to consider using appropriate energetic materials and/or to be used proper design techniques, unconcerned if the munitions is new, modernised, repaired or it is a part of an acquisition process.

The design techniques utilised which have to be considered:

a) reducing thermal and mechanical sensitivity of the explosives;
b) reducing the detonation of explosives;
c) reducing the risks by selecting the proper configuration of explosive cargo;
d) protection against thermal stimulation;
e) protection against shocked and sympathy reaction;
f) preventing the disastrous growing pressure, unconcerned of initiating stimulation;
g) preventing accidental ignition of the initiating systems.

The measures in order to reducing the risk in designing the munitions are:

- partial or entire use the less sensitive explosives;
- protection internal energetically cargo with an external composition less sensitive;
- using the polymers as isolation material;
- using comps material or thermal initialised aeration system on designing munitions body;
- using foil or based on laser explosives initiators;

When all those measures are insufficient, it has to be calculate some modifies of wrapper/cabinet design such as:

a) protection by armours and shields;
b) shock absorber for reducing or eliminating the probability of sympathy reaction;
c) attenuate machine:

At the munitions design its very important to pursue the cumulative effects of all those measures, not one by one effect

4. VALIDATION PROCEDURES FOR MUNITIONS AND EXPLOSIVES UTILISED IN ROMANIA

Closely connected whit assuring safety and compatibility in service of munitions and explosives, in Romania has been conceived the SMT 40362-2001 standard, which settled the *principles and methodology for military explosives validation,* corresponding to STANAG 4170 and AOP-7 standards.

4.1 Military departments responsible for validation

In Romania, the responsibility for military munitions and explosives validation is assigned to the National Ministry of Defence, by the Research and Development Direction (RDC) of the Department of Armaments (DA).

The Program Directors and Testing & Evaluation Service are implementing the policy of RDC for munitions and explosives validation.

4.2 National authority

National authority in military explosives field is the Research and Development Direction. It is responsible for validation of all explosives and munitions used by Romanian Army Forces.

In explosives and munitions validation process, the Research and Development Direction collaborates whit the representatives of the following institutions: Military Techniques and Technology Research Agency, National Company ROMARM S.A., Army Artillery Laboratory, Military Technical Academy.

RDC is collaborating whit the representatives of Ground Troops, Aviation & Air Defence, and Navy General Headquarters, both in requirement establishment and final validation phases, in order to ensure that munitions and explosives are safe in service, storage and transport.

National authority accomplishing following services:

- initiating the validation procedures, evaluate preliminary dates, select necessary dates for validation of military explosives used by Romanian army forces, according with SMT 40362;

-establishing the accepting or rejecting requirements for validation of explosives

-receiving from the abilities country or foreign institutions dates, concerning tests effects which have to be done for validation, thus validate materials to be safety and compatible in service for application which they have been made for;

-publishing tests and evaluation results which leaded to validation with or without limits dictating by different situations;

-synchronising both the implementation of STANAG 4170, AOP-7 and international and NATO standards in explosives area in Romania;

-collaborating with Standardising Agency for maintaining and improving a national and international standards collection from safety and compatibility in service of munitions and explosives area;

-maintaining and improving data base with validate explosives and munitions in Romania;

-assuring a contact between Romanian National Authority and National Authorities from other countries, for realising a good information exchange regarding to safety and compatibility in service of munitions and explosives;

-collaborating with abilities governmental authorities to assure environmental protection regarding to the impact of munitions and explosives military activities on it, assuring the respect of regulation in this area;

-preparing and improving Romanian section in AOP-7.

4.3 Validation procedures

Romania has the capability to produce and to load large range of explosives: primer explosives, gunpowder, rocket powder, main disruptive and detonator explosives, pyrotechnics compositions. In the past, explosive validating process was displayed according Varsovia Treatise and Romanian military standards. Those standards allowed both the international requirements and the own researcher results. Performed tests and testing procedures were established by military standards, task bucks or internal rolls. After 1990, Romania started to be preoccupied about implementation of international standards, especially the NATO and MIL-STD. That is why the Romanian technical military standards were and will be modified and the process of estimating and testing provided to be made for validation will be displayed according to STANAG 4170 and AOP-7.

SMT 40362-2001 regulating the activity of validation and represent the implementation of STANAG 4170 in Romania. Other NATO standards established as indispensable for explosives validation will be adopted as SMT and will be implement in the nearest future, depending of the responsible institutions in safety and compatibility in service of munitions and explosives possibilities.

The majority scientific researches in Romania, regarding to military explosives are realised in research, designed and output institutions belonging Armament Department or National Company ROMARM.

Creation and validation of an explosive is addressed to National Authority which undertaken all legal actions for starting the validation process.

Formal validation of an explosive is not possible until it successfully passed all scientific research and development stages: lab model, prototype lot, punctual series of industrial production. The explosive must be produced in prototype lot or punctual series, using the equipment, conditions and methods used on large-scale. Nevertheless, the estimation and validation process is started from the evaluation and development phase, to assure that the security and compatibility performances and characteristics in service can be accomplish in final production phases without discovering unknown or unsolved obstacles.

Before validate the explosive, subjected on evaluation process according to STANAG 4123 *"Methods of munitions classification from the risk point of view"*, it is necessary to determine the risk class for transportation and storage activities. This activity is made by Armament Department from Minister of National Defence.

4.4 Tests and methods of validation

National Authority haven't the evaluation and testing facilities for explosives. The tests are done normally by the Research Agency for Military Technology and Techniques laboratories, Army Artillerist Laboratory, National Company ROMARM S.A. and Technique Military Academy. In certain events, tests could be realised also in state or private laboratories and institutions accredited by the National Authority. The institution or the laboratory which made tests, send obtained results and observations to National Authority.

National Authority decide:

- the obtained results are sufficient for validation the explosives;
- there are necessary other tests;
- the explosive doesn't answer with obligated requirements and can not be validate;

When the National Authority consider that the explosive accomplish all safety and compatibility in service requirements, the permanently secretariat prepared necessary documents for approving by the Technique Commission and after the event to National Authority.

5. CONCLUSIONS

As a Partnership for Peace member, Romania decided to undertake all necessary legal measures for its NATO integration.

Regarding the safety and compatibility in service of munitions and explosives and their validation, Romania is at the beginning of the process of implementation the requirements of AOP-15, STANAG 4297, STANAG 4170.

So, the information presented in this article must be considered preliminary and can be modified: - within the frame of the restructuring process of the Ministry of Defence and the up-mentioned institutions for military technique research, development and production, - and also within the frame of the process of lining up to the international standards

BIBLIOGRAPHY

1. Lee, P.R. –*Insensitive Munitions: Background And Technology*, Military Technology MILTECH 9/99, page. 49-58.
2. AOP-15, ed. a 2-a.
3. AOP-39, ed. a 2-a.
4. STANAG 4297.
5. Parlamentul României- *Legea nr. 17/1996 privind regimul materiilor şi materialelor explozive.*
6. SMT 40269-2002- *Evaluarea siguranţei şi compatibilităţii în exploatare ale muniţiilor în România.*
7. SMT 40268/1-2002 - P*olitica pentru introducerea, evaluarea şi testarea MURAT în România.*
8. SMT 40268/2-*Ghidul pentru implementarea politicii şi cerinţelor SMT 40268/1-2002;*
9. SMT 40362-2001-P*rincipiile şi metodologia omologării materialelor explozive cu destinaţie militară în România.*

Methodes, Tests and Application

ENERGETIC MATERIALS – MODELS, MICROMEASUREMENTS AND RESULTS

J. Campos, A. Portugal, I. Plaksin, J. Gois, R. Mendes, P. Simões and J. Ribeiro

Laboratory of Energetics and Detonics, Faculty of Sciences and Technology of University of Coimbra, 3030 Pinhal de Marrocos, Coimbra, PORTUGAL

ABSTRACT

The development and evaluation of new classes of energetic materials requires new metrology methods able to test small samples of theses materials. The same problem appears when it is necessary to check time life of explosives and propellants from existing warheads. They are usually formed by crystals, surrounded by binder, and their detonation properties are very dependent of the type of nitration, of the used solvents for their crystallization and its granulometric size. Theoretic phenomenological models, more and more, take in account the transmission of shock wave between crystal and binder, the delay of its initiation, the cavitation phenomena between crystal and binder and the hot spot formation. However, it is very important to optimise these theoretical models with experimental results. In the presented particular case study, detonation PBX micro-samples, based in HMX with an inert (HTPB, epoxy) or energetic (GAP) binder, was performed on the meso-scale level, using the multifiber optical probes of 50 μm of maximum resolution, connected directly to a fast electronic streak camera with 0.6 ns resolution. The direct 2D observation of detonation progression and the particle to particle successive transition, of transmitted shock wave, through the binder, allows to analyse and to discuss, not only the cooperative formation of a multihead detonation front (DF), in the collection of particles surrounded by binder, but also the synenergetic effect, behind the DF, by the appearing of dissipative structures drawing spatial and temporal DF oscillations.

P. Castelo Branco et al. (eds.),
Defense Industries: Science and Technology Related to Security, 23–35.

Advanced Research Workshop
Defence Industries: Science and Technology Related Security
- Impact of Conventional Munition on Environment and Population -
29-31 October 2001, Porto, Portugal

Energetic Materials – models, micro-measurements and results

J. Campos, A. Portugal, I. Plaksin, J. Gois, L. Durães*, R. Mendes , P. Simões* and J. Ribeiro*

LEDAP - Laboratory of Energetics and Detonics
*Mech. and *Chem. Eng. Dept. – Fac. of Sciences and Technology*

University of Coimbra

PORTUGAL

Introduction I

- The development of a new classes of molecules of energetic materials (EM) needs a new metrology technique for the experimental test of small amounts.
- **The influence of particle size and shape is a very important parameter in these new EM.**
- **The same problem appears when it is necessary to check time life of explosives and propellants, from existing warheads, extracting from them samples and avoiding its destruction.**
- The developed methods allow the simultaneous registration, **from very small samples**, of two/three parameters such as, D, P_d, DF local curvature, sizes of the DZ, SDT zone and the processes of DW collisions (Mach Wave formation/attenuation).
- Sampling technique will be required.

Introduction II

- **Objectives**

  To study the detonation properties of crystals of energetic materials and their derivated PBX, not only from new PBX but also from PBX samples, collected in booster fixing zones, in existing warheads and ammunitions.

- **Motivations**

  Development of data for phenomenology computer codes from the development of an accurate optical technique for detonation metrology.

The classical technique using the gauges can not be applied because their sizes exceed the particle size.

Existing computer codes take in account the properties of energetic particle materials and shock transmission through the inter-particle binder

Introduction III

- **Mechanisms of DW propagation**
  - → *associated with macrostructure of PBX as a composite material:*
  - → *course EM crystals + fine EM crystals and binder (successively energetic and inert)*

- **Resolution of DW fine structure in PBX**

- **Unstable behaviour of DW in PBX:**
  - → *heterogeneity of DW*
  - → *local oscillations of Detonation Front (DF)*

- **Estimation of the energetic binder (EB) influence to the shock burning of EM crystals**

Mechanisms of know-how progression

Starting assumption:

PBX is a heterogeneous explosive formed by pure explosive crystals and pure homogeneous inert or reactive binder (and no voids between them).

Consequently it is required:

→ *1. The knowledge of reacting properties of explosive material and its detonation products.*

→ *2. The knowledge of reacting and pyrolysis properties and its reaction products of binder material.*

→ *3. The shock behaviour of both materials.*

→ *4. The shock interaction between particle and binder.*

→ *5. The shock interaction between clusters of particle/binder zones*

→ *6. The SDT between these clusters…* … … …

… … … … … … … …

THOR Code structure

THOR CODE

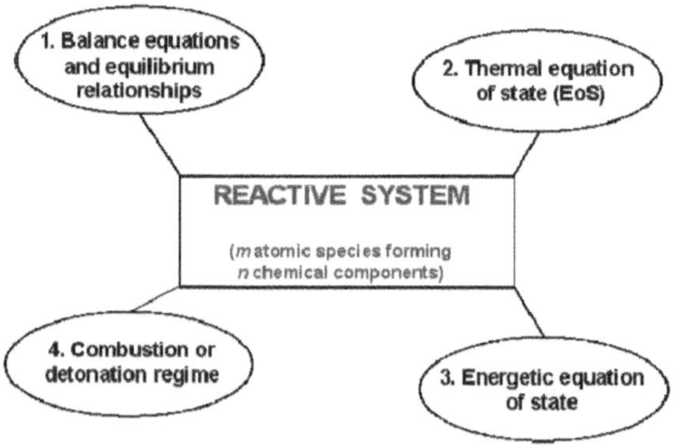

Calculation of combustion or detonation products

ADIABATIC COMBUSTION AND DETONATION CONDITIONS

The combustion regimes of energetic materials are assumed as global adiabatic processes generating reaction products (global validation)

- ■ the isobar adiabatic combustion verifies
- equal initial and final total enthalpy $Hb^{Tb}=Ho^{To}$;
- ■ - the isochor adiabatic combustion verifies
- equal initial and final internal energy $Eb^{Tb}=Eo^{To}$;
- ■ - the Chapman-Jouguet detonation condition

(mass, momentum and energy balances and $dp/dV]_S=((P-Po)/(V-Vo))$ based on the assumption that the detonation velocity D is obtained adding the sound velocity a_o with particular velocity u_p ($D=a_o+u_p$).

Building EoS for computer code

➤ **Results of our previous work (1995-1997) have been compared with other codes and EoS, proving the validity of THOR Code and H_L EoS.**

HOW TO BUILD A USEFUL EoS OF EASY INTEGRATION IN A THERMOCHEMICAL COMPUTER CODE ?

❖ **Take a simple polynomial equation form, which can be easily integrated in a computer code;**

❖ **Reduce to the minimum the "fitting" from experimental data and consider physical intermolecular potential information;**

❖ **Select critical thermodynamic parameters (Γ ; ΣX_i).**

↳**THOR Thermochemical Code uses modified H_L EoS, based on a Boltzmann EoS type, after review of theoretical work of Heuzé et al. (1985) and Durães et al. (1995).**

Exp. and theoretical D deviation

Numerical calculation for 49 different energetic molecules, using H_L EoS with θ=1 and θ=1.4 values.

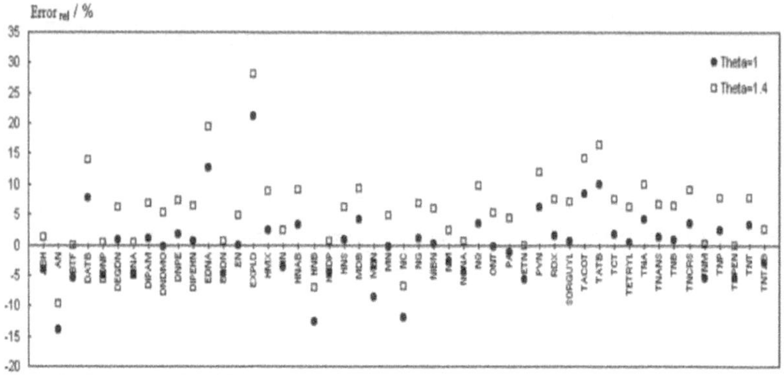

Real experimental study needs the analysis of interaction particle vs. binder under shock

- **Experimental study, on the meso-scale level, of DW propagation, can be performed with:**
  - → *Single HMX crystals, within a binder, energetic or inert*
  - → *Collection of a few HMX coarse crystals*
  - → *Microsamples of PBX = Coarse HMX crystals + Fine HMX crystals + Binder*
- **Resolution of DW fine structure in PBX**
- **Unstable behaviour of DW in PBX:**
  - → *Irregularities of DW*
  - → *Local oscillations of Detonation Front (DF)*
- **Estimation of the energetic binder (EB) influence to the shock burning of HMX crystals**

Standard Basic Experimental Tests

- Long charge tests
- μ-gap tests
- Mini-gap tests for SDT characterization
- Light witness tests

Witness plate test

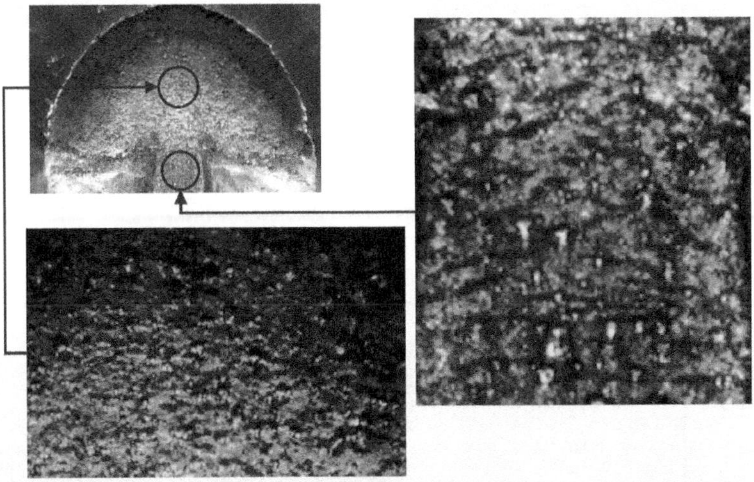

Corner Turning - Witness Cu Plate

HMX - HMX - GAP

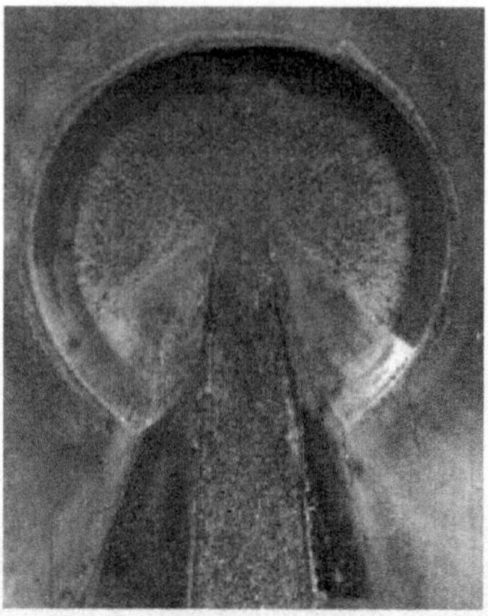

HMX - DNAM - GAP

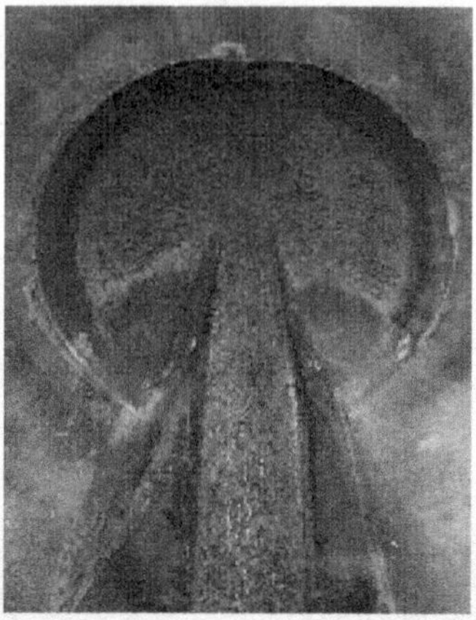

Optical Recording Method

- The light associated to the DW or Shock is recorded by tha Electronic Streak Camera (THOMSON TSN 506N), using optic probes, technique developed in LEDAP.

  → ***The main element of this method is a Polymeric 64 optic fiber ribbon***

  → ***64 Independent channels***

  → ***Fiber's diameter is 250 μm***

  → ***No intermediate optics ↖ Low losses of light***

  → ***Easy connection between experimental set-up and Electronic Streak Camera.***

  → ***Spatial resolution is not less than 250 μm for Direct View Registration and 60 μm with the collimation***

  → ***Maximum temporal resolution 600 ps (at 200 ns sweeping time)***

Explosive μ-samples and PBX Compositions

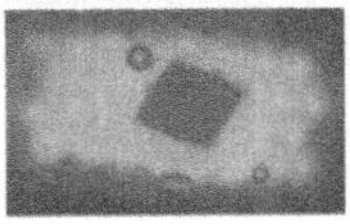
Single Crystal

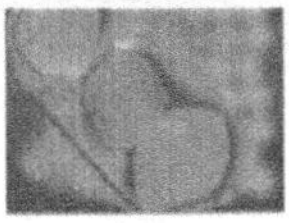
Collection of Crystals

- PBX with energetic binder - GAP

65.6% HMX class 3 (d_{50}=203.7 μm)
16.4% HMX class 2 (d_{50}=16.5 μm)
13% GAP
5% BNDA/F
ρ_o=1.745 g.cm^{-3}
D = 8.52 mm.μs^{-1}

- PBX with inert binder - HTPB

65.6% HMX class 3 (d_{50}=203.7 μm)
16.4% HMX class 2 (d_{50}=16.5 μm)
15.85% HTPB
2.15% DOS
ρ_o=1.575 g.cm^{-3}
D = 7.89 mm.μs^{-1}

- PBX = HMX coarse crystals + water
- PBX = HMX fine crystals + water
- PBX = HMX coarse crystals + epoxy (18%)

μ-gap test with a single HMX coarse crystal

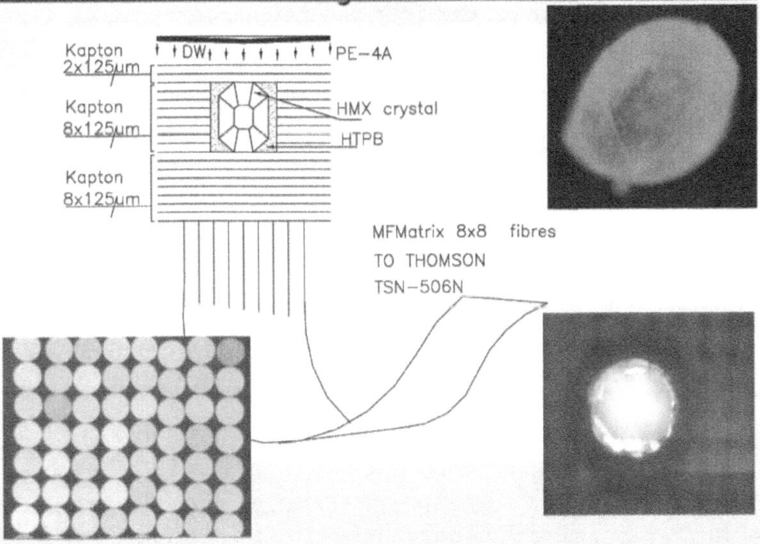

Experiment with HMX crystal, surrounded by water

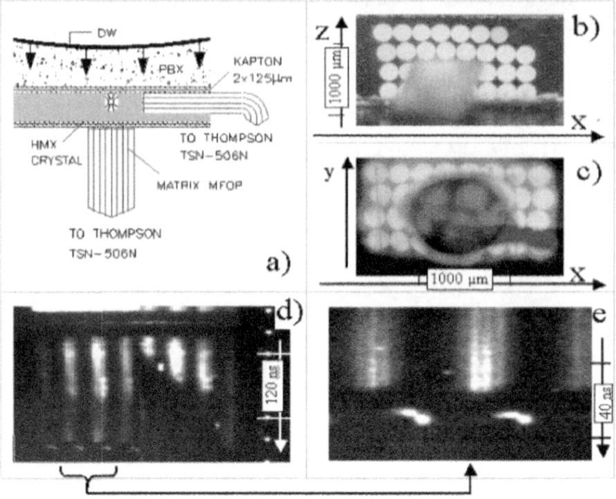

a) experimental set-up, b) HMX crystal and side MFOP, c) HMX crystal and the bottom MFOP, top view
d) photo-chronogram showing the pulsing SW propagation inside the crystal, e) detail of photo-chronogram

μ-gap test with a single HMX coarse crystal

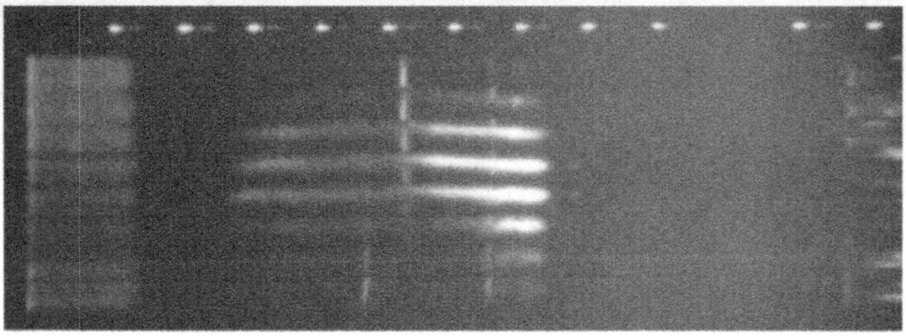

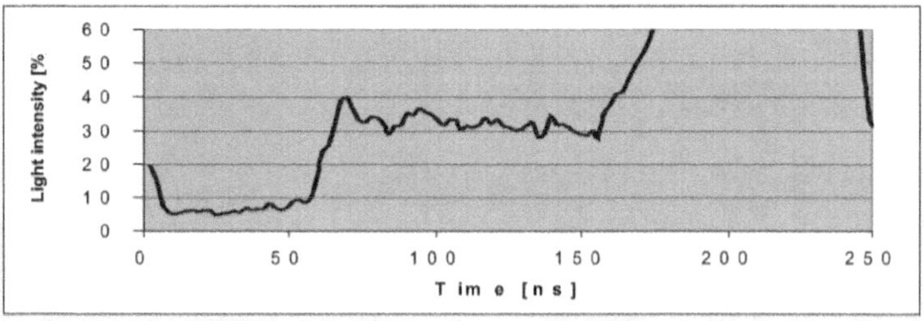

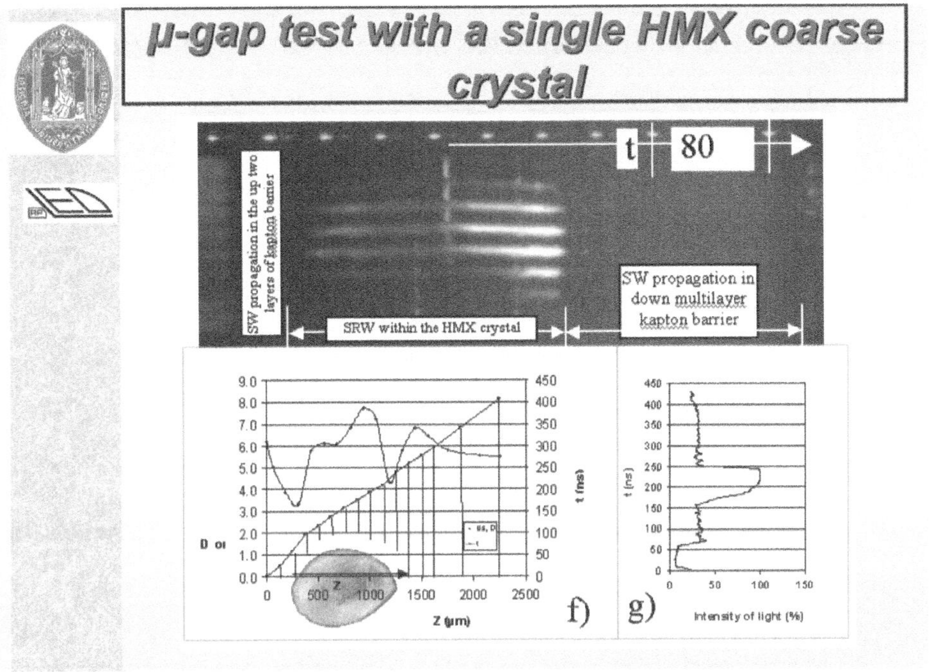
μ-gap test with a single HMX coarse crystal
t
80
SW propagation in the up two layers of kapton barrier
SRW within the HMX crystal
SW propagation in down multilayer kapton barrier
D
t (ns)
Z (μm)
f)
g)
t (ns)
Intensity of light (%)

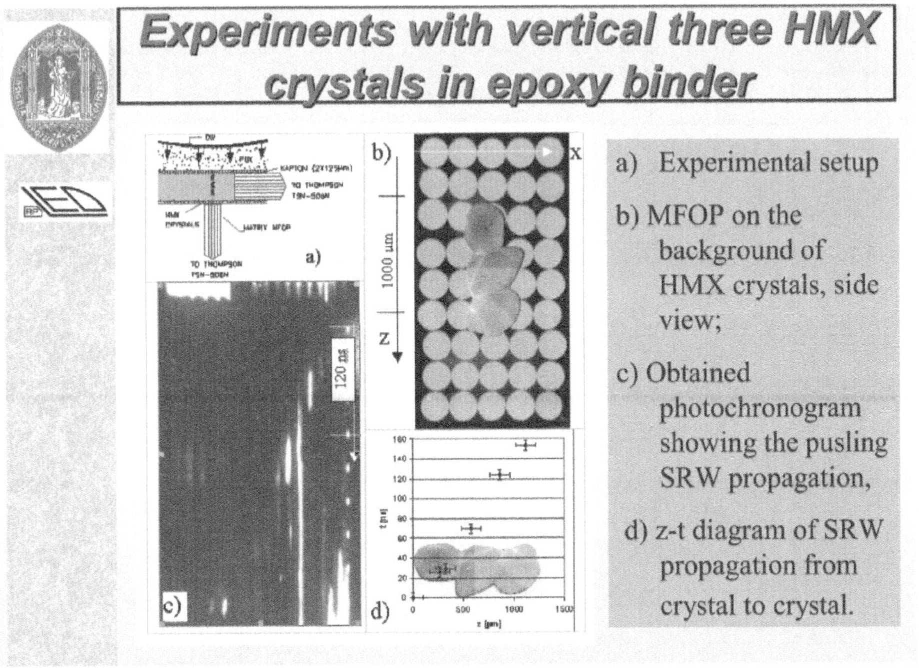
Experiments with vertical three HMX crystals in epoxy binder
a)
b)
c)
d)
x
1000 μm
z
120 ns
a) Experimental setup
b) MFOP on the background of HMX crystals, side view;
c) Obtained photochronogram showing the pusling SRW propagation,
d) z-t diagram of SRW propagation from crystal to crystal.

Characteristic tests of Energetic micro-samples and mini-samples

- μ-Gap test of new crytals, particles and their clusters
  - ➔ *Characteristic sample volume and mass:* $V_0 \leq 2\ mm^3$; $m_0 \leq 4\ mg$
- μ-Conical/μ-Prism test ⇖ $V_0 \leq 4\ mm^3$; $m_0 \leq 8\ mg$
- Coaxial double charge test ⇖ $V_0 \leq 0.05\ cm^3$; $m_0 \leq 100\ mg$
- Long Channel test with crystal scale resolution: $V_0 = 0.5 \div 1\ cm^3$
- Conical Tube test ⇖ $V_0 = 0.3 \div 1\ cm^3$
- Gap test with crystal scale resolution ⇖ $V_0 = 0.5 \div 1\ cm^3$
- Cornet turning tests ⇖ $V_0 \approx 1.5\ cm^3$
- Colliding test ⇖ $V_0 \approx 1.5\ cm^3$

Validation of method – discussion of results from typical and micro samples

Experimental results of typical scale samples (L=100 mm)

Compositions	ρ_0 (g/cm^3)	D (mm/μs)	P_{det} (GPa)
PE-4A	1.56	7.87	24.2
HMX(82%)-GAP	1.75	8.56	38.6
HMX(82%)-EPOXY	1.82	8.30	35.4
RDX(70%)-EPOXY	1.60	7.52	21.0

Experimental results of μ-samples (L=1 mm)

Compositions	ρ_0 (g/cm^3)	Curvature	P_{det} (GPa)
HMX(82%)-EPOXY	1.82	(+)	35.3
RDX(70%)-EPOXY	1.60	(-)	20.2

Conclusions

- The developed methods allow the simultaneous registration of two/three parameters such as, D, P_d, DF local curvature, sizes of the DZ, SDT zone and the processes of DW collisions (Mach Wave formation/attenuation) from micro samples that can be extracted from existing warheads.
- The detonation study of PBX on the meso-scale level has been carried out in the developed originally characteristic testes by the application of the high-resolution multi-fiber optical technique.
- The obtained results show the phenomena of the kinetic instability in shock reaction of HMX coarse crystals interacting between themselves and with binder, followed by the effects of self-organization in PBX detonation generating products.

SAFETY ASPECTS of COMBUSTIBLE GASES and VAPOURS

Anatoly A.Vasil'ev
Lavrentyev Institute of Hydrodynamics,
Siberian Branch of Russian Academy of Sciences,
630090 Novosibirsk, Russia
gasdet@hydro.nsc.ru

Abstract

The critical energy E_* of direct initiation of detonation wave (DW) is accepted as the main criterion parameter of detonation hazards of gaseous mixtures.

The blast explosion approximation is proposed to use at experimental definition of E_*. This technique is identically suitable both to various mixtures and various initiators with of large distinctions in the spatial-temporary energy-release characteristics (flame igniters, electrical or laser spark, exploding wire, explosive, high-speed bullet...).

The priority at theoretical definition of critical detonation energy is given up to «multi-points» initiation model (MPI) based on DW transverse waves collisions as the micro-initiators of a detonation («hot» points).

The hazard parameters were calculated with the help of Computer Code «SAFETY».

The basic parameters of detonation and explosive hazards of some typical fuels (interesting from scientific and practical points of view) are presented in a wide range of its concentration in mixtures with oxygen and air, and at dilution of inert gases. The dependence of critical initiation energy E_* on percentage of fuel in a mixture allows to find out comparative detonation hazards of various combustible systems at plane, cylindrical and spherical initiation.

Calculated results are in a good agreement with experimental data.

Keywords: *detonation, explosive, combustion, catastrophes, ecology, safety, initiation*

Introduction.

A great number of gas-dynamics, chemical and physical parameters of chemically active systems are important in hazard problem. The demonstration of correct experimental and calculation methods for estimation of explosive and detonation hazards is the main purpose of this paper.

The consequences of large-scale accidents at emergency explosions of gaseous fuels directly depend on characteristic modes of a mixture burning: from low-speed (around m/s) laminar or turbulent flame up to high-speed (km/s) detonation. Of course, there are many other important factors such as geometric form of fuel-air charge, mixture homogeneity, different obstacles, etc.

Some hazard parameters can be calculated using classical detonation theory based on conservation laws of mass, impulse and energy

P. Castelo Branco et al. (eds.),
Defense Industries: Science and Technology Related to Security, 37–63.

$$\rho_0 u_0 = \rho u$$
$$P_0 + \rho_0 u_0^2 = P + \rho u^2$$
$$H_0 + u_0^2/2 = H + u^2/2 + Q$$

(ρ - density, u - flow velocity, P - pressure, H - thermodynamic part of enthalpy, Q - chemical energy-release of a mixture, the index 0 relates to initial condition).

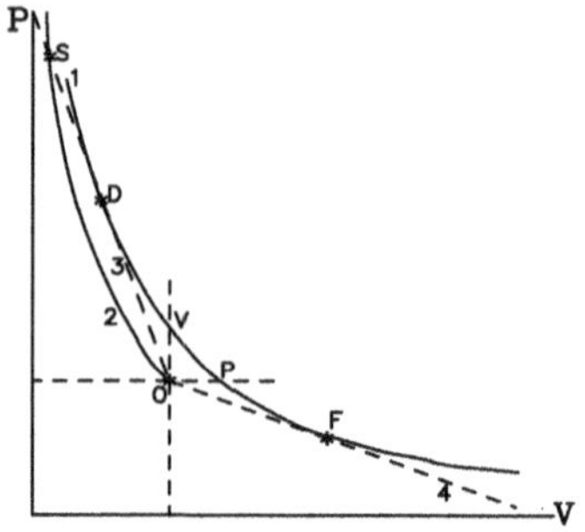

Fig. 1

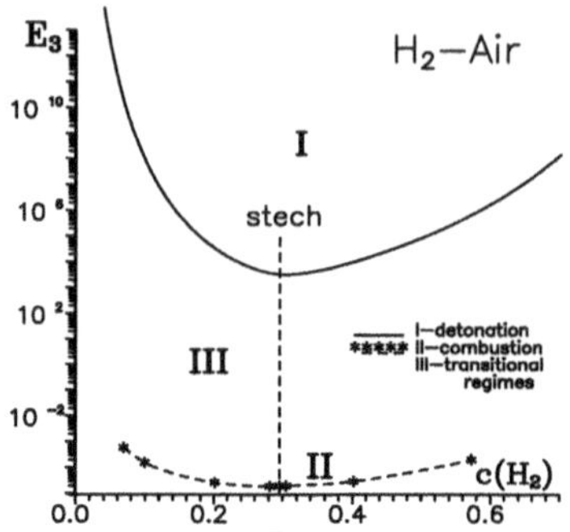

Fig. 2

The final states of reacting system (with chemical energy-release Q) are represented by adiabatic curve 1. Two types of stationary (self-sustaining) solutions are known for this equations, namely, subsonic deflagration (point F in Fig.1) and supersonic detonation (point D). The F and D are the points of a contact of straight lines 3 and 4 with adiabatic curve 1 (the lines start from the point O corresponding to initial condition). On adiabatic curve 1 there are two characteristic points, which are important also at estimations of fire- and explosion hazard: P - the combustion at constant pressure, V - instant explosion in a volume. The distinction in values of basic gas-dynamics parameters of reaction products in points F, P, V and D determines the character of accident consequences with gaseous fuels. At estimations of detonation hazard it is necessary to remember that the detonation wave (DW) except a point D for detonation products is characterised also by a point S - condition at DW front prior to the beginning of chemical reaction (shock adiabatic curve 2 for DW chemical spike).

According to modern classification the excitation of fast chemical reaction can be carried out by three basic ways: 1) - the weak initiation when the initiator only ignites a mixture and low-speed combustion (deflagration) occurs; 2) - the intermediate regimes when mixture is ignited on initial stage only and then due to natural or artificial acceleration of a flame front the subsequent transition from deflagration to detonation (DDT) proceeds; 3) the strong initiation when self-sustaining DW is formed in immediate proximity from a initiation point. In the first case the critical flame initiation energy E_{flame} is used as the main hazard parameter traditionally (Fig.2 - dashed line II with the minimal energy of ignition E_{flame} = 1.7 10--5 J, as example, for a hydrogen-air mixture at spherical initiation). For the second case the quantitative criterion of DDT conditions still has not been formulated because of multi-parameter nature of DDT phenomenon. The initiation of detonation wave is multi-parameter process too, but for «ideal» (see below) initiators only one parameter - the critical initiation energy E* for detonation may be considered as a measure of explosion hazard: the less E*, the more hazardous a combustible mixture (see Fig.2 - solid line with the minimal energy of DW initiation E* = 3.35 103 J). The tasks of determination of critical energy for burning and critical energy for detonation are important in hazard problem. In this paper only critical energy for detonation initiation is discussed.

Correct experimental measurement of critical initiation energy.

The range of critical energies for various fuel-oxygen (FOE) and fuel-air (FAE) explosive mixtures is very large. Therefore the only individual type of initiator cannot be used for the DW process triggering for all cases. As a rule, each initiator has its specific features of energy-release. Consequently a creation of simple and universal technique for determining a source effective energy (which must be suitable for various initiators as well as for various FOE and FAE) becomes a prime task.

At initiation the explosive mixture absorbs during some finite time interval t_0 in finite space domain V_0 some quantity of energy E_v released by the initiator (which is only η-portion of the E_{00} energy originally stored in the initiator):

$$E_v = \int_0^{t_0}\int_0^{V_0} \varepsilon(t,V)\cdot dt\cdot dV = \eta E_{00},$$

where $\varepsilon(V, t)$ is a function describing spatial - temporary law of energy-release.

From the formal mathematical point of view this integrate equality represents typical functional of two variable in a variation task about optimisation of energy E_v at movable limits of integration. At mutual influence of the temporal and spatial energy-release factors the requirement of minimisation of initiation energy up to value E_{min} is reduced to minimisation of power density up to $\varepsilon_{min} = \varepsilon(r_*, t_*)$. At the account only of temporal factor (V_0 = const) it is required simultaneously with E_{min} to optimise the initiator power $\varepsilon_{min} = \varepsilon(t_*)$. Last conclusion is confirmed experimentally by [1] at variation only of the electrical discharge duration (at the fixed pressure). If time of discharge $t_0 \leq t_*$ then the required for initiation energy E_t = const $\cong E_{min}$ while at $t_0 > t_*$ («stretching» discharge) E_t exceeds E_{min} and grows with increase t_0 (Fig. 3). The value E_{min} is accepted as the critical energy E_* of a detonation initiation. The temporal parameter t_* is individual for each mixture and depends on initial pressure P_0. The condition $t_0 \leq t_*$ represents for the given mixture a criterion for instantaneous initiator or criterion of its «ideality». In this case excitation of multifront DW is determined only by energy.

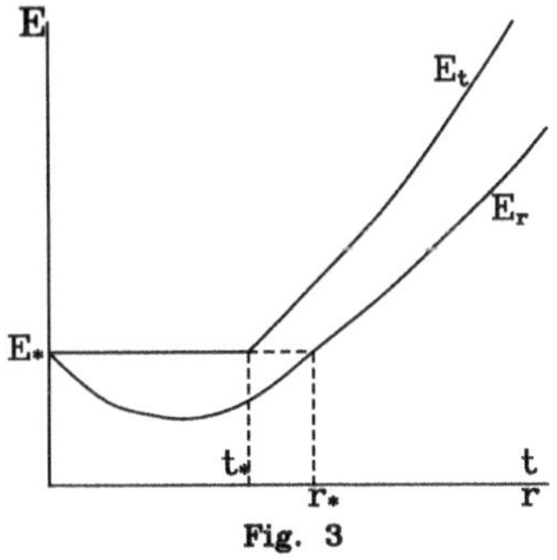

Fig. 3

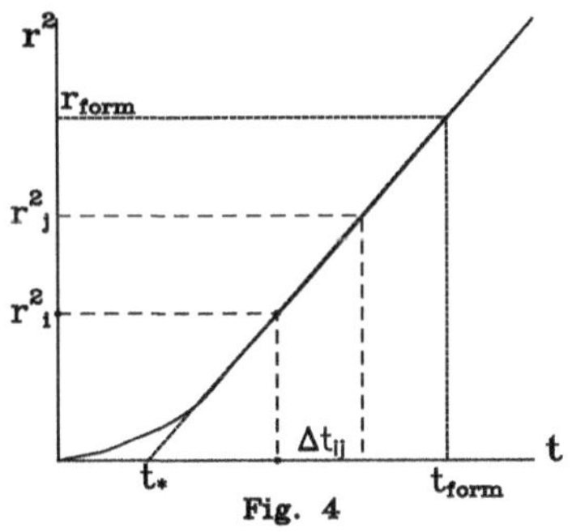

Fig. 4

The influence of spatial components of energy-release E_r is reduced to optimisation of density of supplied energy $\varepsilon_{min} = \varepsilon(r_*)$ simultaneously with E_{min} - Fig. 3 [2]. In other words, the optimality of each mixture is characterised by some temporary t_* and spatial r_* scales of initiation.

Thus, correct experimental determining E_* for each individual initiator requires the control of spatial - temporary characteristics of energy-release. It appreciably complicates measurements, however without information about E_t and E_r the error in E_* can achieve several orders of magnitude.

In order to elaborate an universal approach the experimental method to determine the energy absorbed by mixture was developed and used in practice [2]. The method is based on processing the trajectory r(t) of an explosive wave excited by initiator, from the point of view of strong explosion model (at an initial stage of a movement) [3-4]:

$$r(t) = \left(E_0 / \alpha_\nu \rho_0\right)^{1/(\nu+2)} t^{2/(\nu+2)}$$

Here E_0 - energy of explosion, ρ_0 - initial density, ν=1,2,3 for plane, cylindrical and spherical symmetry, α_ν - parameter dependent from ν and an adiabatic index γ. The effective initiator energy E_0 is given by

$$E_0 = \alpha_\nu \rho_0 \left[\left(r_i^{(\nu+2)/2} - r_j^{(\nu+2)/2}\right) / \Delta t_{ij}\right]^2,$$

where r_i and r_j are the co-ordinates of a shock wave in the moments divided by time interval Δt_{ij} (Fig.4). If for an explosive mixture $2H_2+O_2$ a critical initiation regime (simultaneous influence of initiator and internal chemical energy-release) is observed at some pressure P_* the effective energy only of the initiator at given P_* is defined on a trajectory r(t) in an inert mixture $2H_2+N_2$. It ensures similarity of densities ρ_0 and adiabatic indexes γ_0, that is condition of similarity of mixture parameters and their profiles in the DW formation area. The E_0 value, determined in such inert mixture, represents the critical initiation energy of a multifront detonation E_*.

The approach described allows to determine adequately the effective energy of the various initiators and to establish the power equivalents between them. The validity of the approach has been checked up for such initiators as the electrical discharge, exploding wire, explosive charges. It was found that under identical conditions (mixture composition and initial pressure, initiation geometry) the results of experiments correlate satisfactory with the authentic data of other researchers [5-7].

It is necessary to remind that the real DW in gaseous mixtures essentially differs from that predicted by the plane classical model. Real DW represents multifront pulsing system which consists of shock waves, rarefaction waves, contact discontinuous, local zones of chemical reaction, etc. The most important role at propagation of such multifront detonation belongs to transverse waves as the main elements of the detonation front structure. The motion of transverse waves has a periodic character, and their trajectories form a cellular structure with a characteristic linear scale *a* - so-called size of an elementary cell. The collisions of transverse waves play the important role at initiation process also.

Initiation models

The three types of initiation models may be emphasized:

I) numerical models within the framework of one-dimensional DW model (*Chernij-Korobejnikov-Levin-Markov-Osinkin, He-Klavin...*) [8-11];

II) approximate models for DW with one-dimensional smooth front (*Zeldovich-Kogarko-Simonov, Lee-Ramamurthi-Matsui-Bach-Knystautas-Guirao-Benedick-Sulmistras, Edwards-Hooper-Morgan, Sichel,*

Nicholls, Urtiew-Westbrook, Zhdan-Mitrofanov, Uljanitsky, Shulenin-Bohon, Borisov-Zamansky-Lisjansky-Skachkov-Troshin...) [12-24];
III) approximate models for real multifront DW (*Vasil'ev-Grigor'ev-Nikolaev-Uljanitsky*) [25-29,5].

I) Numerical models are based on gas-dynamic and kinetic equations for reactive mixture. For example, at Task about Explosion in Combustible Mixture (*Chernij-Korobejnikov-Levin-Markov*) [8] the next equations and initial-boundary conditions are used:

Gasdynamic equations

$$\frac{\partial \rho r^{\nu-1}}{\partial t}+\frac{\partial}{\partial r}\rho u r^{\nu-1}=0$$

$$\frac{\partial \rho u r^{\nu-1}}{\partial t}+\frac{\partial}{\partial r}[(P+\rho u^2)r^{\nu-1}]=(\nu-1)r^{\nu-2}P$$

$$\frac{\partial}{\partial t}[r^{\nu-1}(\rho u^2/2+\rho h-P)]+\frac{\partial}{\partial r}[\rho u r^{\nu-1}(u^2/2+h)]=0$$

ν=1,2,3 for plane, cylindrical and spherical cases; P - pressure, ρ - density, u – mass velocity, h – specific entalphy.

Equations for induction and reaction zones

$$\frac{dc}{dt}=-\frac{1}{\tau_i}=-k_1P^{n_1}\rho^{l_1}\exp\left(-\frac{E_1\rho}{\mu P}\right)$$

$$\frac{d\beta}{dt}=-k_2\beta^{m_1}P^{n_2}\rho^{l_2}\exp\left(-\frac{E_2\rho}{\mu P}\right)+k_3(1-\beta)^{m_2}P^{n_3}\rho^{l_3}\exp\left(-\frac{E_3\rho}{\mu P}\right)$$

c=1 behind shock wave and c=0 when induction time is finished; β - portion of nonreacting mixture, β=1 up to starting of chemical reaction, then β tends to zero during chemical energy release. Entalphy $H=\gamma/(\gamma-1)\,P/\rho+\beta Q$, Q – energy-release on unit mass.

Initial and boundary conditions

a) u(r,0)=0; $P(r,0)=P_0$=const; $\rho(r,0)=\rho_0$=const; $r_{SW}(0)=0$. Energy E_0 is released in symmetry center at t=0 instantaneously. At this case on initial stage the blast wave model (*Sedov, Korobejnikov*...[3-4]) for inert gases was used because at this stage an addition of chemical energy release was much less then initiator energy.
b) u(0,t)=0 in symmetry center; on Shock Wave (SW) – parameters are in accordance with the conservation laws for SW.

As a results a qualitative picture of one-dimensional sub-critical ($E<E_*$), critical and supper-critical ($E>E_*$) regimes of detonation initiation were calculated. The critical initiation energy E_* can be determined if correct data about induction and reaction parameters are known. The longitudinal instability of DW with a great velocity pulsation is typical for this calculations.

The full scheme of elementary reactions can be used instead of simplest equations of Arrhenius type for induction and reaction zones.

For non-stationary regime the induction time τ may be defined from the next integral correlation (u - particle velocity, r_1 and r_2 - DW radiuses at moment t_*, when a particle crosses the DW front, and at moment, when for this particle the induction period is finished):

$$1=\int_{t_*}^{t_*+\tau}\frac{dt}{\tau_{st}}=\int_{r_1}^{r_2}\frac{dr}{D\tau_{st}}=\int_{r_1}^{r_2-\lambda}\frac{dr}{u\tau_{st}}=\varphi(E,r_1)-\varphi(E,r_2)\,.$$

II). Up to day about 20 approximate models of detonation initiation are known too. Practically all initiation models are based on one-dimensional conception of DW as a complex from shock wave and zones of induction and chemical reaction with characteristic times τ_{ind} and τ_{ch} (or characteristic sizes ΔR_{ind} and ΔR_{ch}). Nonzero value of any τ (or ΔR) leads to delay of energy release which must be compensated by initiator for successful forming of DW. Insufficient quantity of initiator engaging energy at a great delay of mixture chemical energy release is a main physical reason for DW failure.

1. The functional correlation for initiation criteria was formulated firstly by *Zeldovich-Kogarko-Simonov* [12] for spherical case:

$$E_{in}\sim(\Delta R)^3,$$

where ΔR is a size of induction zone in expanding blast wave. *The ΔR -value and proportionality coefficient are indeterminate.* Qualitatively the idea of [12] was formulated by double variants: a) the blast wave from initiator must be «sufficient» duration and intensity (with pressure $P>P_{CJ}$) for triggering of chemical reaction behind the expanding wave; b) the period of decaying blast wave influence on mixture up to a moment when $P=P_{CJ}$ must be exceed the induction time. P is the detonation pressure for ideal CJ DW.

2.The ideas of *Zeldovich-Kogarko-Simonov* [12] in one or another form and blast wave model *Sedov-Korobejnikov...* [3-4] were used in the greater part of approximate initiation models. The law of blast wave decaying in inert mixture may be transformed to some modifications: *r=f(E,t), D=g(E,t), D=h(E,r), P=p(E,t)...* and any combination of parameters may be considered as critical. For example, in *Chernij-Korobejnikov-Levin-Markov* [8] its supposed that the blast wave must be a sufficient intensity up to moment (or radius) when a chemical energy of mixture in expanding wave equals with the initiator energy.

3. In accord *Edwards-Hooper-Morgan* [13] the decay time of blast wave up to CJ-velocity must exceed the duration of particle displacement in DW from the front to CJ-plane.

4. The next formula is proposed by *Sichel* [14]

$$E_*=\alpha_\nu(\nu+2)^2\rho_0c_0^2M_*^2(\Delta_*/\delta)^\nu/4,$$

where c_0 is initial sound speed, M_* and Δ_* are Mach number and induction size in critical point, δ is some criterial constant.

5. In *Bach-Knystautas-Lee* [15] the next step was made for initiation problem: the chemical energy release on initial stage was take into account in some model form (up to auto-ignition limit, $M_s^*\sim2.8$) and the next approximate formula was suggested

$$E_{0c}=k_\nu\rho_0c_0^2(\Delta/\delta_c)^\nu,$$

where δ_c is some critical value which define in calculations.

6. In model *Lee-Ramamurthi* [7] the auto-ignition limit was increased up to $M_s^*=4$. The initiator must create the blast wave with $M>M_s^*$ up to some critical radius R_s^*, when the initiator energy is equal to chemical ones

$$R_s^*=\frac{c_0M_s^*\tau(M_s^*)}{1-[0.5(M_s^*/M_{CJ})^2]^{1/\nu}},$$

$$E_c=0.5\rho_0c_0^2(M_s^2)^2k_\nu(R_s^*)^\nu J,$$

$$J \cong Q/(\nu D_{CJ}^2),$$

7. The model of *Uljanitsky* [16] postulates the longitudinal pulsation DW at its propagation because of frustration effects of chemical energy release. The main hypothesis is that at any stage the auto-ignition limit is not achieved. The integral correlation for induction time was used and some function E(r) with maximum value was constructed. Then with the help of criteria $\partial E / \partial r = 0$, and after some transformation the next formulas are proposed for initiation energy (10 corresponds to CJ shock wave parameters)

$$\gamma = 0.5(\gamma_0 + \gamma_{10}),$$

$$\beta_{CJ} = \frac{1}{1 + \dfrac{2}{(\gamma - 1)M_0^2}},$$

$$\varepsilon_* = [\nu \beta_{CJ} E_{act} / RT_{10} - 1]^{-1/\ell},$$

$$D_* = D_0(1 + \varepsilon_*^\ell / 2\ell), \qquad T_* = T_{10}(1 + \beta_{CJ}\varepsilon_*^\ell / \ell),$$

$$r_* = \nu D_* \beta_{CJ} \cdot A(\rho_a / \rho_{10})^n E_{act} / RT_* \cdot \exp(E_{act} / RT_*),$$

$$E_* = (\nu + 2)^2 \alpha \varepsilon_* \rho_0 D_0^2 r_*^\nu / 4,$$

8-9. In *Shulenin-Bochon* [17-18] the experimental data was approximated by formulas:

$$E \cong 10\mathrm{a}^3, \qquad E \cong 17\mathrm{b}^3$$

a, b are transversal and longitudinal sizes of detonation cell for multifront DW.

10. In *Lee-Knystautas-Guirao* [19] the characteristic time was equal to the cell period -

$$E_c = 62.5\pi\gamma_0 P_0 (c_0 / c_{CJ})^3 J M_{CJ}^5 \mathrm{a}^3,$$

11-13. In *Benedick-Guirao-Knystautas-Lee* [20] the some modifications of approximate model of *Lee-Knystautas-Guirao* [19] for spherical initiation are analyzed

$$E_c = K P_0 (c_0 / c_{CJ})^3 \mathrm{a}^3,$$

$$E_c = 395.5\pi\gamma_0 P_0 J M_{CJ}^2 \mathrm{a}^3,$$

but the next formula had the main preference in accordance with conception about the leading role of «surface energy» in initiation process

$$E_c = \frac{2197}{16}\pi\gamma_0 P_0 J M_{CJ}^2 \mathrm{a}^3,$$

where P_0 - initial pressure, M_{CJ} - Mach number of ideal DW (velocity D_0), $J \cong Q/(\nu D_0)$.

14. In *Zhdan-Mitrofanov* [21] at approximation of calculating results for gaseous and heterogeneous mixtures the next formula are proposed

$$R_{cr} \cong 4(\nu - 1)\mathrm{k}^2(\sigma_{10} + 1/\sigma_{10} - 2)\lambda_{10} \cdot E_{act} / RT_{10},$$

$$E_{*\nu} = \alpha_\nu P_0 (8R_{cr} / \nu)^\nu,$$

where $k = I/U$, - ratio of thermodynamic entalpy I and internal energy U, $\lambda_{10} = (D_0 - u_{10})\tau_{10}$, σ - density ratio on CJ shock wave. For estimation the next formula are proposed (P_0 - N/m², b - m):

$$E_2 \cong 250 P_0 b^2,$$

$$E_3 \cong 6000 P_0 b^3,$$

15. In *Westbrook-Urtiew* [22] the length λ_{10} of induction zone may be calculated with the help of model of detailing kinetics. Then from the experimental data for E_* the B-value may be defined. The next assumption - B=const for any mixtures:

$$E_{*3} = B\lambda_{10}^3,$$

16. In *Troshin* [23] the forming radius DW r_0 was considered as critical

$$r_0 \geq 40(\nu - 1)\lambda_{10} \cong 8(\nu - 1)b,$$

$$E_{*3} \approx \left[\left(0.31\frac{\rho_3 u_3^2}{2} + 0.59\frac{P_3}{\gamma - 1}\right) - \left(\rho_0 Q + \frac{P_0}{\gamma - 1}\right)\right]\frac{4\pi r_0^3}{3},$$

$$E_{*2} \approx \left[\left(0.25\frac{\rho_3 u_3^2}{2} + 0.57\frac{P_3}{\gamma - 1}\right) - \left(\rho_0 Q + \frac{P_0}{\gamma - 1}\right)\right]\pi r_0^2,$$

index 3 corresponds to CJ product parameters, Q - chemical energy.

17-18. In *Borisov-Zamansky-Lisyanski-Skachkov-Troshin* [24] another formula was proposed for spherical case

$$E_{*3} \approx \left[\left(0.31\frac{\rho_3 u_3^2}{2} + 0.59\frac{P_3}{\gamma - 1}\right) - \left(\rho_0 Q + \frac{P_0}{\gamma - 1}\right)\right]\left(\frac{40\lambda^{10}}{\gamma_0 - 1}\right)^3,$$

or approximately (E - kg TNT, τ - sec):

$$E_{*3} \approx 10^{14}\tau_{10}^3.$$

III). All one-dimensional models postulate that at any moment the DW may be characterized by some induction size which is equal for any front elements. But really in all gaseous mixtures DW is non-dimensional (multifront) and the sizes of induction zone in different points of DW fronts may differ by two order. The ignition occurance is caused by local temperature in the most «hot» point instead of the average temperature for all front elements (as in one-dimensional models). Such points in real DW are the domains of collision of transverse waves and its are similar to 'hot spots' or local microexplosions. Such micro-explosions play an important role in propagation of multifront DW and promote an initiation process. Taking into account similar non-one-dimensional collisions of shock-wave configurations of real detonation front it is possible to lower appreciably a level of critical energy (in comparison with one-dimensional models). The model of multipoints initiation (MPI) of DW was proposed firstly by *Vasil'ev* in [25]. In accordance with MPI-model the critical energy $E_{*\nu}$ for initiation of multifront DW is proportional to the energy $E_{0\nu}$ in individual area of TW collision -

$$E_\nu = n_{*\nu} E_{0\nu},$$

n - coefficient of collective influence.

The E value may be calculated with the model of detonation cell *Vasil'ev-Nikolaev* [26]:

$$E_2^c = 4\varepsilon^2\alpha_2\rho_0 D_0^2 b^2,$$

The MPI-model was developed in *Vasil'ev-Nikolaev-Uljanitsky-Grigor'ev* [27-28]. According to the most completed version (for ν= 1,2,3) - in *Vasil'ev-95* [5]:

$$E_1 = \frac{16\varepsilon^2\alpha\sqrt{\pi}}{\gamma_0 - 1}\cdot\frac{E_{act}}{Q}\cdot\rho_0 D_0^2 b = A_1\cdot\rho_0 D_0^2 b,$$

$$E_2 = \frac{16\varepsilon^2\alpha\sqrt{\pi}}{\gamma_0 - 1}\cdot\frac{E_{act}}{Q}\cdot\rho_0 D_0^2 b^2 = A_2\cdot\rho_0 D_0^2 b^2,$$

$$E_3 = \frac{512\varepsilon^2\alpha\cdot tg\varphi}{(\gamma_0 - 1)^2}\cdot\left(\frac{E_{act}}{Q}\right)^2\cdot\rho_0 D_0^2 b^3 = A_3\cdot\rho_0 D_0^2 b^3.$$

where E_{act} - effective activation energy of an induction period (within the framework of the average description based on the Arrhenius equation), Q- chemical energy-release in DW, tg $\varphi = a/b$, α - parameter of strong explosion model [7].

Diffracting estimation of critical energy for initiation of multifront detonation was proposed by *Vasil'ev* [29]. At this the formulas for critical energy are similar to formulas of MPI-model.

Detonation cell models.

At present some cell models are known and can be used for estimation of detonation cell sizes and construction of hazard raw of combustible mixtures.

First type - numerical modelling (for example - Computer Codes of Naval Research Laboratory (USA) for two-dimensional channel with constant cross-section, *Kailasanah-Oran-Boris-...*[30]).

Second type - "cell" models for ideal DW. For example, in [22] (*Westbrook-Urtiew*) the induction zone length $l_{10} = (D_0 - u_{10})\tau_{10}$ is calculated for DW. After that cell size a is calculated by $a = k\, l_{10} = 29\, l_{10}$ (τ - induction period, index 10 corresponds to induction zone parameters for ideal Chapman-Jouguet DW). Constancy of a/l_{10} for different mixtures and pressures is the main assumption of [22], and it may be subjected to critics. In [31] (*Gelfand-Frolov-Nettleton*) for rough tube the minimal diameter d_{limit} is calculated for DW propagation and a is defined as $a = d_{limit}/\pi$.

Third type - cell model *Vasil'ev-Nikolaev* [26] for real multifront DW, where longitudinal cell size b is calculated from the integral relationship

$$b^{-1} = \int_x^{1+y} \frac{d\zeta}{D(\zeta)\cdot\tau_{st}(\zeta)},$$

where $\zeta = r/b$, $x = r_*/b$... (see [26]). It must be mentioned that the last relation may be used in reverse problem for determination of τ_{st} and average kinetic constants for induction zone.

Recomendations about initiation models

The MPI-model [25,27-28,5] and diffracting model [29] may be recommended as the basic models for estimation of critical initiation energy.

From other approximate models it is possible to recommend for estimations only model [21], model of «surface» energy [20] and «model» [22]. In last case at first stage an induction zone length l_{10} is calculated with the help of model of detailed kinetics. The critical energy of spherical initiation was considered proportional l_{10}:

$$E_{*3} = B\lambda_{10}^3,$$

where coefficient B was determined from experimental data of E_{*3} for some particular mixture composition and then was considered constant for any other mixtures with given fuel.

The formulas of other models give a much more divergence with experimental data [32-46].

Basic calculation idea

In terms of the cell size *a*, the remaining basic parameters of multifront detonation (with the length dimension) can be determined: the critical diameter d_{**} of DW-«exit» from the tube into a volume, the geometrical characteristics of channels for limiting DW-propagation d_S, l_{lim}, and δ_{lim} (the tube diameter, and the channel width and depth, respectively), the critical diameters of free gaseous charges d_*, the linear sizes of gaseous charges for DW-formation L_{form}, the sizes of obstacles and the law of their space-orientation in turbulent devices for artificial transformation of the burning process to detonation, etc.

In terms of *a* and the gas energy $E_{0\nu}$ in the region of transverse wave collision, the critical initiation energies $E_{*\nu}$ can be determined for various symmetry ν (ν=1, 2, 3 for plane, cylindrical and spherical cases), as well as the diameters d_w of a high-velocity bullet for excitation of the detonation mode in a combustible mixture, etc...

In terms of *a,* the kinetic parameters of an induction zone such as the effective activation energy, the pre-exponent factor, and the effective reaction order can be determined (for the induction time in the Arrhenius equation form).

The technique of calculation of the main parameters of multifront detonation (DW velocity, pressure, temperature, density, mass and sound speed of products, their equilibrium composition, parameters of an induction zone, cell size, critical energy of DW initiation for plane, cylindrical and spherical cases, critical parameters of diffraction re-initiation, HVB parameters, geometrical limits for DW propagation in channels of arbitrary cross-section, etc...) is described in much details in [5] and serves as a basis of the Computer Code «SAFETY» (Detonation Hazards of Combustible Systems) [47].

Computer Code «SAFETY» for estimation of hazards.

The program of calculation of detonation and explosion hazards of combustible systems includes:

a) - thermodynamic data base for individual substances (chemical formula, sign of state, molecular weight, polynomial coefficients for entalphy and entropy, standard formation heat...);

b) - kinetic data base for various mixtures - coefficients of the Arrhenius formula for the induction period τ (within the framework of the «generalized» kinetic model named as the «average» model):

$$\tau = \frac{A \cdot \exp(E/RT)}{[f]^{k_1}[o]^{k_2}[in]^{k_3}},$$

or its logarithmic analogue

$$\lg\{[f]^{k_1}[o]^{k_2}[in]^{k_3} \cdot \tau\} = A + B/T,$$

where E is the activation energy of an «average» («limiting») reaction of the induction period, R is the universal gas constant, T is the mixture temperature in the induction zone, the square brackets indicate the concentration of a mixture component (f - fuel, o - oxidizer, in - inert additive), A and k_i are numerical factors;

c) - subroutine of calculation of the parameters of instantaneous explosion (V_0 = const);
d) - subroutine of calculation of combustion at P_0 = const;
e) - subroutine of calculation of the Chapman-Jouguet detonation;
f) - subroutine of calculation of shock-wave parameters (including DW chemical spike);
g) - subroutine of calculation of normal and oblique collisions of DW;
h) - subroutine of calculation of an elementary cell *a* and its parameters;
i) - subroutine of calculation of critical initiation energies for various cases of symmetry;
j) - subroutine of calculation of major geometrical parameters of multifront detonation (see above);
k) - subroutine of calculation of kinetic constants («inverse» problem of a cell);
l) - experimental data base for detonation parameters.

The calculations are conducted within the framework of the ideal gas model and chemical equilibrium of the products (including ions and condensed components). The vapor form of fuel in initial state takes into account too (with own vaporization energy). For hazard evaluation task the main detonation parameters - DW velocity, pressure and temperature of products,... - are calculated by traditional methods (for ideal gas model with chemical equilibrium of products).

By analogy with low-velocity burning we used the critical initiation energy E_* as the basic parameter at comparison of explosive and detonation hazards of combustible mixtures. Energy E_* is defined as the minimal energy of an initiator which ensures a propagation of ones or another process in a researched mixture. The lower the critical energy, the more dangerous the mixture. The multipoints initiation model and the diffracting initiation model are used as the basic for determination of critical initiation energy for plane, cylindrical and spherical cases.

More 150 substances containing C, H, N, O atoms take into account at calculations: saturated C_nH_{2n+2} (alkanes, n=1÷20) and unsaturated hydrocarbons C_nH_{2n} (alkenes, n=1÷20), alkadienes C_nH_{2n-2} (n=3÷5), alkynes C_nH_{2n-2} (n=2÷20), raws of cyclic hydrocarbons (cycloalkanes C_nH_{2n} n=3÷8, cycloalkenes C_nH_{2n-2} n=4÷6, alkylcyclopentanes C_nH_{2n} n=6÷21, alkylcyclopentenes C_nH_{2n-2} n=6, alkylcyclogexane C_nH_{2n} n=7÷22), raws of benzenes, styroles, ethers, spirits, aldehydes, ketones, amines, nitriles, nitrites, nitrates and many others...

The effectiveness of «SAFETY» Code was demonstrated in previous papers [48-51].

Calculated results.

Only the part of great information about detonation and flame hazards is demonstrated in a graphical form for typical fuels - hydrogen, hydrogen peroxide, hydrazine, methane, acetylene, cyanogen, ethylene... $P_0 = 10^5$ Pa (1.0 atm) and T_0 = 298 K are chosen as standard conditions. The basic parameters of various gaseous mixtures are presented for cases of propagation of DW and at mixture instantaneous combustion (named as «explosion in constant volume») for fuel-oxygen and fuel-air explosions (FOE and FAE).

Hydrogen. In Fig.5 the values of detonation velocities D_0 (m/sec) and temperatures T (in K degree) in DW products are presented. The DW velocity D_0 monotonously grows with increasing of hydrogen molar concentration *c* and only near to the upper limit it begins to decrease. For FOE maximal $D_0 \sim 3800$ m/sec is observed in a mixture with *c* = 0.9, for FAE the maximum $D_0 \sim 2345$ m/sec is located with *c* = 0.8. For stoichiomertic FOE and FAE D_0

are equal 2837 and 1966 m/sec respectively. Maximal temperature of products T ~ 3682 K for FOE is reached at stoichiometric ratio. The maximum T ~ 2956 K for FAE is reached with c = 0.35 and its displaced slightly from a stoichiometry, where T = 2947 K. In Fig.6 the pressure ratio $p = P/P_0$ and the dimensionless chemical energy-release $q = Q/c_0^2$ (Q is mixture chemical energy release at detonation mode, c_0 is a sound speed of an initial mixture) are presented. For stoichiometric mixtures p = 18.8 and q = 22.9 (FOE) and p = 15.6 and q = 16.8 (FAE). The dependence of the cell size a of a multi-front detonation on $c(H_2)$ has characteristic U - form - Fig.7. The minimal a ~ 1.0 mm for FOE and a ~ 11.0 mm for FAE also are close to a-values for stoichiometric mixtures. The minimal calculated energies of a detonation initiation for plane (E_1 - J/cm^2), cylindrical (E_2 - J/cm) and spherical (E_3 - J) cases of symmetry are equal 0.7 J/cm^2, 0.3 J/cm and 1.9 J for FOE and 8.4 J/cm^2, 37.3 J/cm and 3354 J (about ~ 0.8 g TNT) for FAE. Let's note the extremely fast increase of all these values near to limits. In Fig.8 correlation of calculated and experimental data are presented for H_2-air mixtures (m_3 - TNT

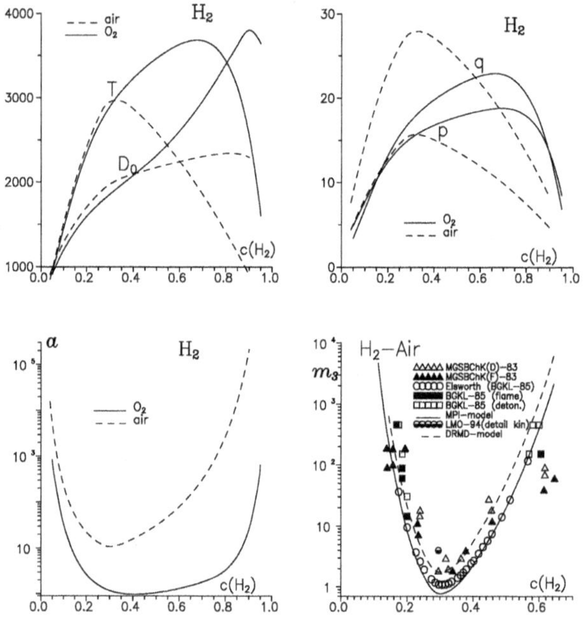

Figs.5-8.

charge in gram): MGSBChK - [36], Elsworth - from [52], BGKL - [52], ABS - [53], LM - accounts [54] with use «generalized» kinetic of an induction period, LMO - accounts [55] on detailed kinetic model. For stoichiometric hydrogen-air mixtures the experimental value of minimal charge weight m_* for initiation of self-sustaining detonation is about 1.1 gram [12], the calculation predicts about 0.8 gram TNT.

Hydrogen peroxide. Pure hydrogen peroxide (HP) as monofuel is characterized by

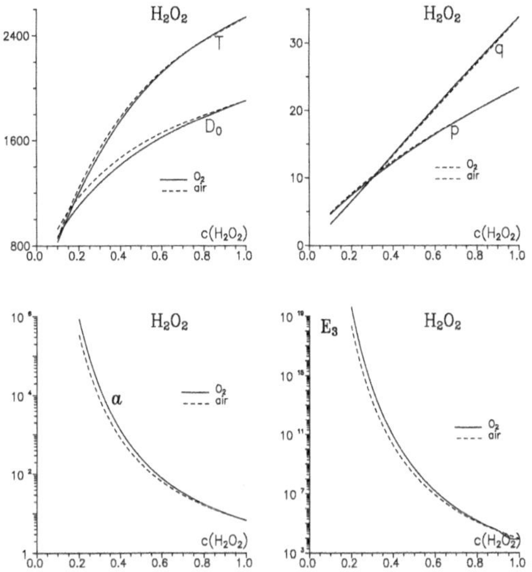

Figs. 9-12

higher detonation hazard in comparison with mixtures of HP-oxygen or HP-air (Figs.9-12).

Example 1: for gaseous HP at $P_0 = 10^5$ Pa and $T_0 = 298$ K the detonation velocity equals 1904 m/sec, product temperature - 2539 K, pressure ratio - 23.4, chemical energy-release on unit mass Q = 720 cal/g, dimensionless energy-release $q = Q/c_0^2 = 33.8$, cell size a = 6.7 mm, critical initiation energies are equal $E_1 = 19.4$ J/cm^2, $E_2 = 64$ J/cm and $E_3 = 8500$ J for plane, cylindrical and spherical cases. One can see in Figs.9-12 that DW parameters are closed for FAE and FOE.

Example 2: at $P_0 = 10^5$ Pa and $T_0 = 423$ K (temperature of phase transition of liquid HP to vapor) the detonation velocity equals 1910 m/sec, product temperature - 2591 K, pressure product ratio - 16.8, chemical energy-release on unit mass Q = 705 cal/g, dimensionless energy-release q = 23.8, cell size a = 4.1 mm, critical initiation energies are equal $E_1 = 9.9$ J/cm^2, $E_2 = 18$ J/cm and $E_3 = 1500$ J for plane, cylindrical and spherical cases.

DW velocity, temperature, pressure ratio, energy-release... are decreased too at decreasing of initial pressure or hydrogen peroxide concentration, while cell size, critical initiation energies, critical diameter... are increased.

If HP in initial mixture assumes in vapor form then the calculated results are the next: detonation velocity equals 1533 m/sec, product temperature - 1685 K, pressure ratio - 14.9, chemical energy-release on unit mass Q = 765 cal/g, dimensionless energy-release q = 36, cell size a = 1620 mm, critical initiation energies are equal $E_1 = 2650$ J/cm^2, $E_2 = 2\ 10^6$ J/cm and $E_3 = 5.6\ 10^{10}$ J for plane, cylindrical and spherical symmetry. In this case the part of chemical energy-release expends on phase change of liquid HP to gaseous. The energy values demonstrate that liquid HP may be classified as hard-detonation substance, but gaseous HP is more dangerous (and may be compared with many fuel-air mixtures).

Dependence of critical diameters for diffraction, self-sustaining propagation in free, or limiting gaseous charge, are similar to cell size line qualitatively. Analogous situation observes among E_3, and E_1 or E_2 -lines.

Ozone. Pure ozone as monofuel is characterized by higher detonation hazard in

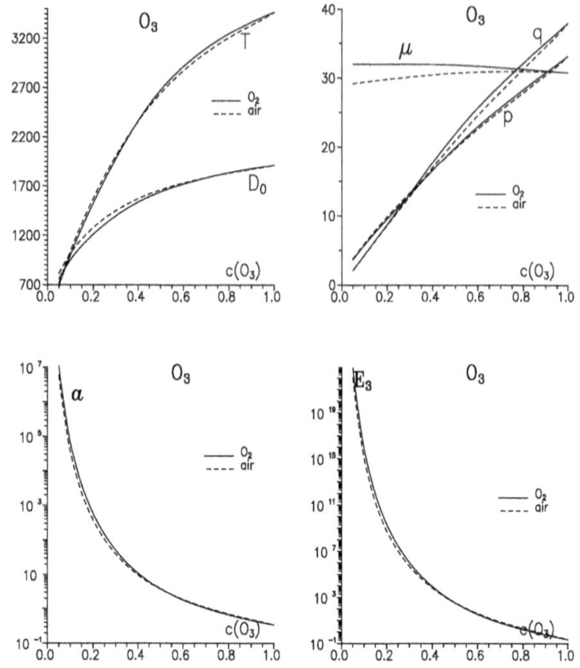

Figs.13-16

comparison with ozone-oxygen or ozone-air mixtures. Example: (at $P_0 = 10^5$ Pa and $T_0 = 298$ K) detonation velocity D_0 equals 1915 m/sec, product temperature T - 3469 K, pressure ratio p - 33.1, chemical energy-release on unit mass Q = 587 cal/g, dimensionless energy-release q = = 37.9, cell size *a* = 0.34 mm, critical initiation energies are equal $E_1 = 0.52$ J/cm^2, $E_2 =$ 0.082 J/cm and $E_3 = 0.22$ J for plane, cylindrical and spherical cases (Figs. 13-16).

DW velocity, temperature, pressure ratio, energy-release... are decreased too at decreasing of initial pressure or ozone concentration, while cell size, critical initiation energies, critical diameter... are increased. Ozone-oxygen and ozone-air mixtures have approximately similar dependence of the main DW parameters on fuel concentration.

The calculated results correspond adequately to available (not numerous) experimental data (for example, [56].)

Hydrazine and its methyl derivatives. Pure hydrazine as monofuel is characterized (at $P_0 = 10^5$ Pa and $T_0 = 298$ K) by the detonation velocity $D_0 = 2482$ m/sec, product temperature T - 1992 K, pressure ratio p - 34.9, chemical energy-release on unit mass Q = 814 cal/g, dimensionless energy-release $q = Q/c_0^2 = 36.7$, cell size *a* = 2.4 mm; critical initiation energies are equal $E_1 = 7.7$ J/cm^2, $E_2 = 13.0$ J/cm and $E_3 = 780$ J for plane, cylindrical and spherical cases. For pure hydrazine *a* and E_i decrease at T_0 increasing while D_0 and T increase. Hydrazine-oxygen and hydrazine-air mixtures demonstrate extreme-shape dependence of the main DW parameters on fuel concentration (see Figs.17-20): *a* and E_3 decrease up to values 0.1 mm and 0.02 J (more hazardous mixture) at molar concentration $c(N_2H_4) = 0.65$ in N_2H_4-O_2 system, and up to *a*=1.05 mm and E_3=22 J at molar concentration $c(N_2H_4) = 0.365$ in N_2H_4-air system.

Parameters of methyl derivatives of hydrazine (methyl hydrazine - mhz; 1,1- and 1,2-dimethylhydrazine - (1-dhz and 2-dhz); 3-methylhydrazine - thz; aerozin) are demonstrated in

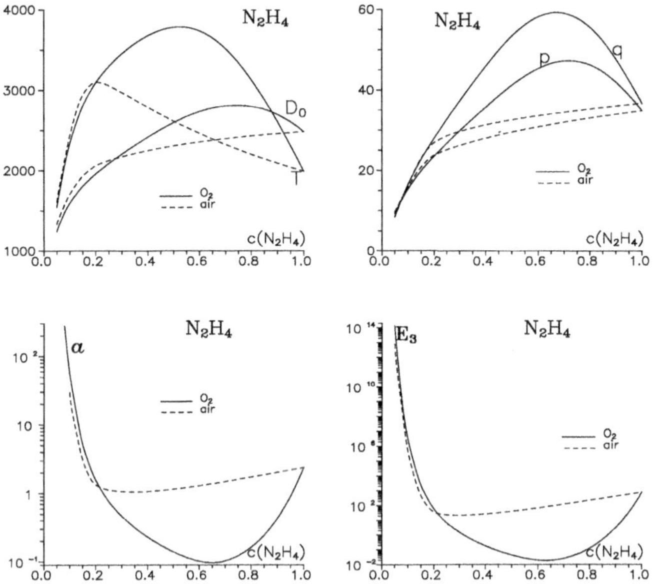
N2H4
4000
3000
2000
1000
D0
T
O2
air
c(N2H4)
0.0 0.2 0.4 0.6 0.8 1.0
N2H4
60
40
20
0
p
q
O2
air
c(N2H4)
0.0 0.2 0.4 0.6 0.8 1.0
N2H4
α
10 2
10
1
10 -1
O2
air
c(N2H4)
0.0 0.2 0.4 0.6 0.8 1.0
N2H4
E3
10 14
10 10
10 6
10 2
10 -2
O2
air
c(N2H4)
0.0 0.2 0.4 0.6 0.8 1.0

Figs.17-20.

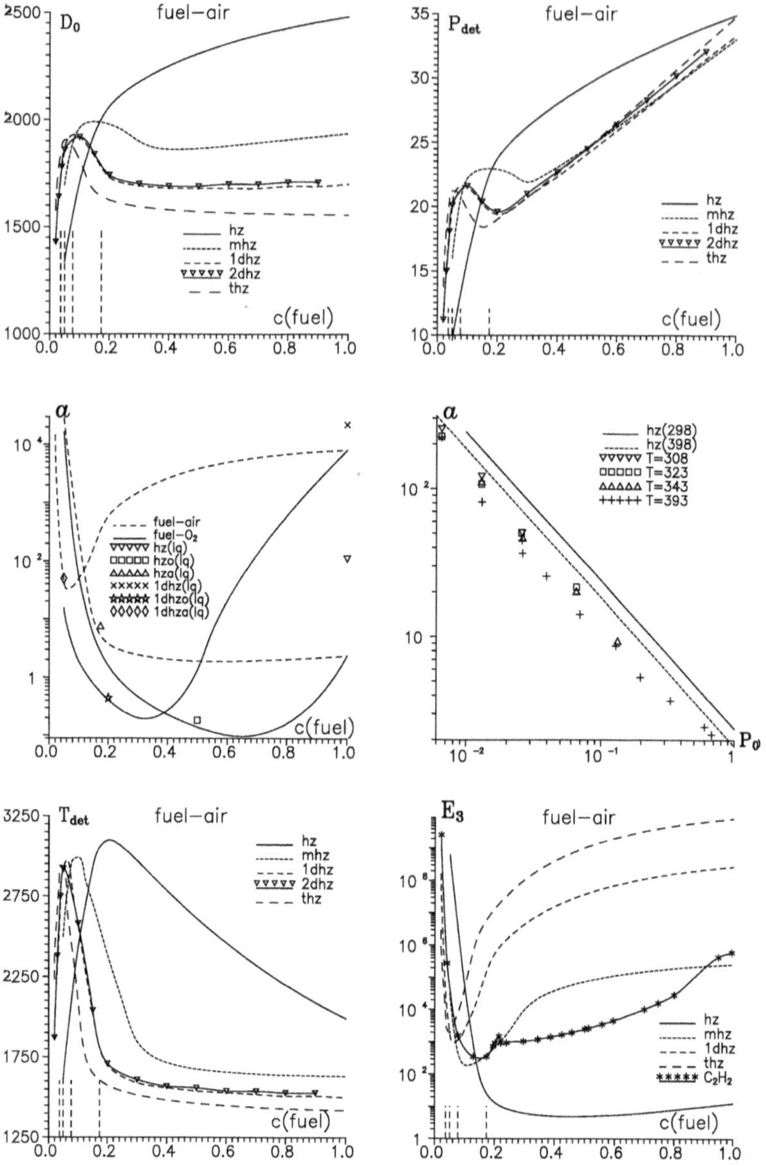

Figs. 21-26

Figs.21-26: detonation velocity D_0 and pressure ratio for detonation products, cell size on molar fuel concentration *c*, and on initial pressure P_0 (at different initial temperature T_0), product temperature T, critical initiation energy for spherical case E_3 (J). Symbols in Fig. 23 correspond to case when fuel in initial mixture is in vapor form. One can see that cell size for vapor monofuels (and critical initiation energy) differs in comparison with gaseous condition, but for stoichiometric mixtures such discrepancy became negligible. In Fig. 24 the experimental data from [57] are used also. In Fig.26 the E-line for acetylene-air mixtures is presented for comparison with hydrazine-air data. One can see that at some concentration range the hazard of hydrazine mixtures is the same as acetylene mixtures. Detailed results on detonation hazards of hydrazine and its methyl derivatives was published in [51].

Methane. For FOE maximal $D_0 \sim 2635$ m/sec is observed in a mixture near $c = 0.5$, for FAE the maximum $D_0 \sim 1833$ m/sec is located with $c = 0.174$ (Fig.27). For stoichiomertic FOE and FAE velocity D_0 are equal 2390 and 1801 m/sec respectively. Maximal temperature of products T ~ 3737 K for FOE and T ~ 2783 K for FAE is reached at near-stoichiometric ratio (Fig.27). At stoichiometry T ~ 3725 K for FOE and T ~ 2780 K for FAE. For stoichiometric mixtures p = 29.3 (FOE) and p = 17.2 (FAE) - Fig.28. The minimal $a \sim 2.2$ mm for FOE and $a \sim 182$ mm for FAE also are close to a-values for stoichiometric mixtures - Fig.29. The minimal calculated energies of a detonation initiation for plane (E_1 - J/cm^2), cylindrical (E_2 - J/cm) and spherical (E_3 - J) cases of symmetry are equal 2.7 J/sm^2, 3.8 J/cm and 100 J for FOE and 270 J/cm^2, 2.3 10^4 J/cm and 7 10^7 J (~ 16.9 kg of TNT) for FAE - Fig.30.

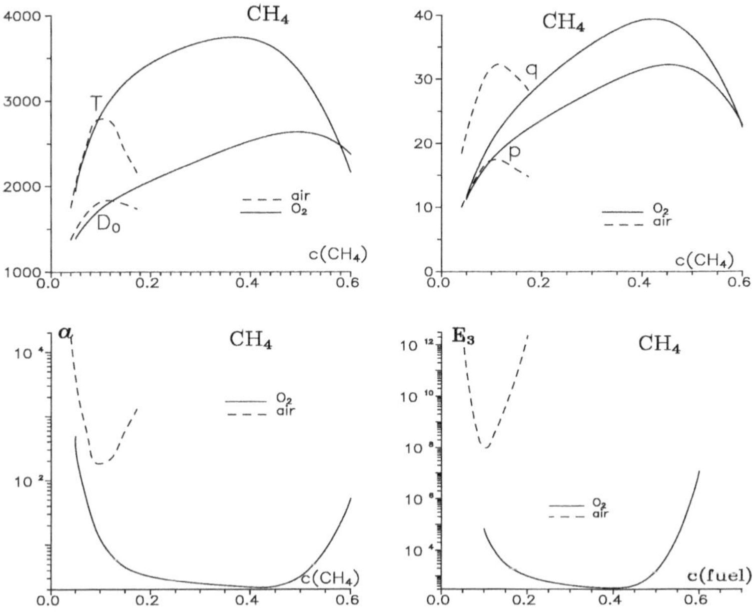

Figs.27-30.

Acetylene. For FOE detonation temperature T is maximal at equimolar composition - 4507 K; at this cell size is 0.08 mm, detonation velocity - 2934 m/sec, detonation pressure ratio - 46.1 (Figs.31-34), critical diameters of detonation reinitiation («exit» diameter) - 1.0 mm... A condense carbon formation takes place at concentrations of acetylene - about 60% in oxygen (and about 20% in air) up to 100 % - starting from points of bifurcation of solid and dash lines on Figs.31-34. In this concentration area the carbon condensation energy plays the important role for stationary propagation of detonation wave, without C-condensation the intensity of DW is sufficiently lower. For clean acetylene (without O_2 or air) D_0=1977 m/sec, T=3130 K, p=20.4, the condense carbon in products equals about 87.5 % from initial quantity... The minimal initiation energy for spherical detonation is observed for FOE at equimolar concentration E3 - 0.0018 J. For FAE the minimal energy equals 350 J, the cell size - 4.7 mm, for stoichiometric acetylene-air mixture the cell size equals 8 mm, the critical diameter - 144 mm, the critical initiation energy - 1600 J. The experimental data on Fig.31 - from [58], on Fig.34 - from [59-60].

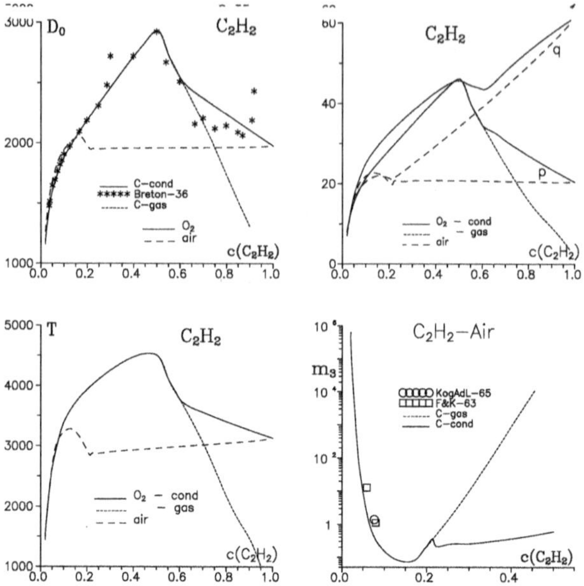

Figs.31-34.

Cyanogen. For FOE detonation temperature T is maximal at equimolar composition -

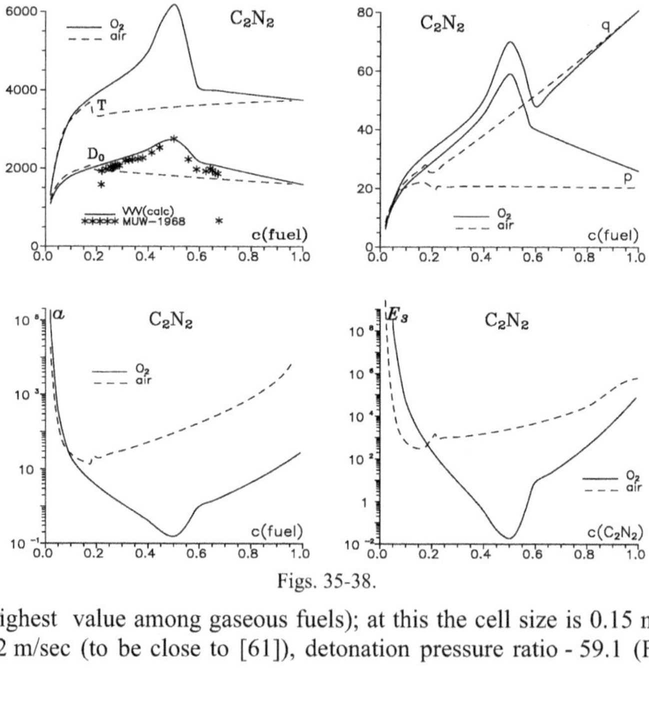

Figs. 35-38.

6178 K (the highest value among gaseous fuels); at this the cell size is 0.15 mm, detonation velocity - 2732 m/sec (to be close to [61]), detonation pressure ratio - 59.1 (Figs.35-38), critical diameters of detonation reinitiation («exit» diameter) - 1.9 mm... A condense carbon formation takes place at concentrations of cyanogen about 60% in oxygen (and about 20% in air) up to 100 %. For clean cyanogen (without O_2 or air) D_0=1591 m/s, T=3746 K, p=26.0, the condense carbon in products equals about 89.4 % from initial quantity... The minimal initiation energy for spherical detonation is observed for FOE at equimolar concentration E_3 - 0.018 J. For FAE the minimal energy equals 12500 J, the cell size - 13.5 mm. The data for nonequimolar cyanogen-oxygen mixtures are similar approximately to hydrogen-oxygen system (for comparison).

Ethylene. For FOE maximal D_0 = 2701 m/sec is observed in mixture with c = 0.45, for FAE the maximum D_0 = 1875 m/sec is located at $c \sim 0.11$. For stoichiomertic FOE and FAE D_0 are equal 2374 and 1822 m/sec and T - 3937 and 2926 K respectively. Maximal temperature of products T = 4002 K for FOE is reached at c = 0.3, the maximum T ~ 2984 K for FAE is reached with c = 0.077. For stoichiometric mixtures p = 33.4, q = 40.7 and a=0.2 mm (FOE), and p = 18.4, q = 19.7 and a=14.1 mm (FAE). The minimal $a \sim 0.16$ mm for FOE

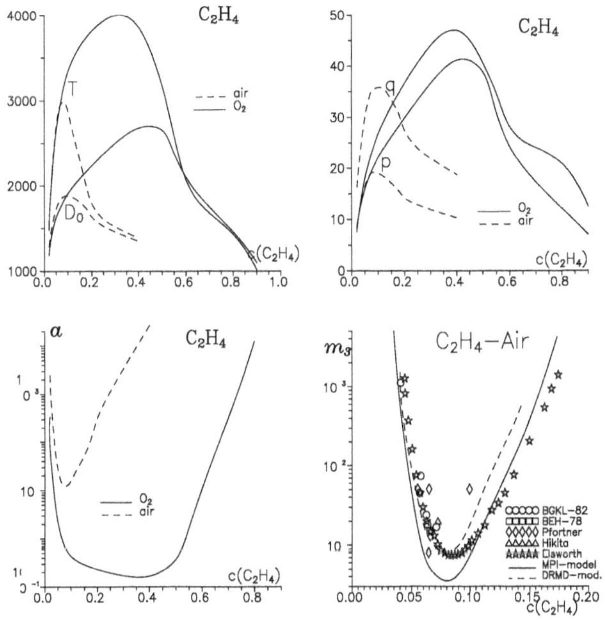

Figs.39-42.

is closed to c=0.35 and $a \sim$ 12.5 mm for FAE - at c=0.077. The critical energies of a detonation initiation for plane (E_1 - J/cm^2), cylindrical (E_2 - J/cm) and spherical (E_3 - J) cases of symmetry for stoichiometric mixtures are equal 0.2 J/cm^2, 0.024 J/cm and 0.04 J for FOE and 17.9 J/cm^2, 117 J/cm and 21650 J for FAE. On Fig.42 correlation of calculated and experimental data are presented for C_2H_4-air mixtures (m_3 - TNT charge in gram): solid and dashed lines correspond to model of multipoints initiation and model of diffraction initiation.

Ammonia. In Fig.39-40 the detonation velocities D_0 (m/s) and temperatures T (K) in DW products are presented for FOE and FAE on ammonia molar concentration c. For FOE maximal $D_0 \sim$ 2545 m/sec is observed in a mixture with c = 0.65, for FAE the maximum $D_0 \sim$ 1870 m/sec is located with c = 0.24. For stoichiomertic FOE and FAE D_0 are equal 2453 and

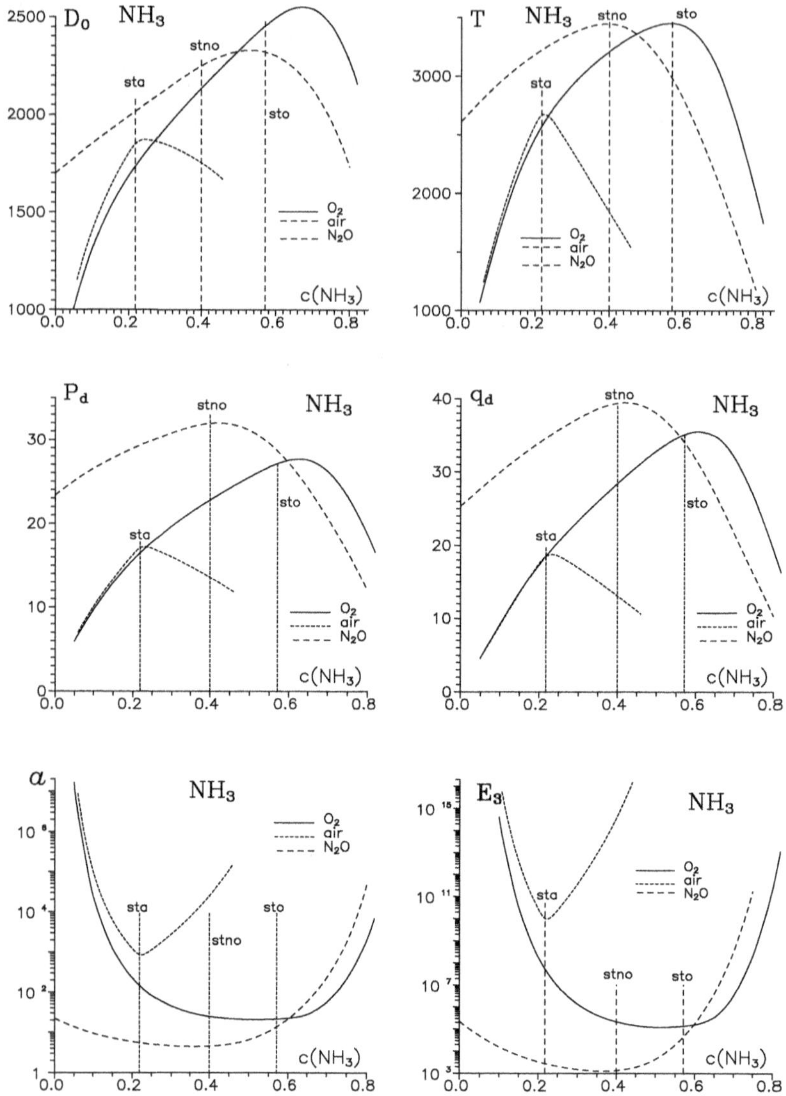

Figs.43-48.

1854 m/sec respectively. Maximal temperature of products T ~ 3450 K for FOE, and T ~ 2670 K for FAE are reached at stoichiometric ratio. Pressure ratio $p = P/P_0$ and dimensionless chemical energy-release q are equal 27.2 and 35.1 for FOE, and 17.1 and 18.6 for FAE at stoichiometry (Q=1197 and 580 cal/g) - Figs.45-46. The dependence of the cell size *a* of a multifront detonation and critical initiation energies for any symmetry from $c(HN_3)$ have characteristic U - form - Figs.47-48. The minimal *a* ~ 22 mm for FOE, and *a* ~ 850 mm for FAE also are close to *a*-values for stoichiometric mixtures. The minimal calculated energies of a detonation initiation for plane (E_1), cylindrical (E_2) and spherical (E_3) cases of symmetry are equal 35 J/cm^2, 440 J/cm and 1.35 10^5 J for FOE, and 1434 J/cm^2, 5.8 10^5 J/cm and 9.8

10^9 J for FAE. For ammonia-nitrous oxide mixtures new dependence are observed in comparison with FOE or FAE.

Nitrous oxide, nitric oxide. Pure nitrous oxide as gaseous monofuel is characterized (at $P_0 = 10^5$ Pa and $T_0 = 298$ K) by the detonation velocity D_0=1699 m/sec, products temperature T=2614 K, pressure ratio $p=P/P_0$=23.4, chemical energy-release on unit mass Q=433 cal/g, dimensionless energy-release q=25.4. Critical initiation energies are equal E_1=40 J/cm^2, E_2=430 J/cm and E_3=1.4 10^5 J for plane, cylindrical and spherical cases. At instantaneous explosion in V_o=const T=2360 K, p=11.9, Q=444 cal/g.

Pure nitric oxide as gaseous monofuel is characterized (at $P_0 = 10^5$ Pa and $T_0 = 298$ K) by the detonation velocity D_0=1913 m/sec, products temperature T=3350 K, pressure ratio $p=P/P_0$=20.83, chemical energy-release on unit mass Q=564 cal/g, dimensionless energy-

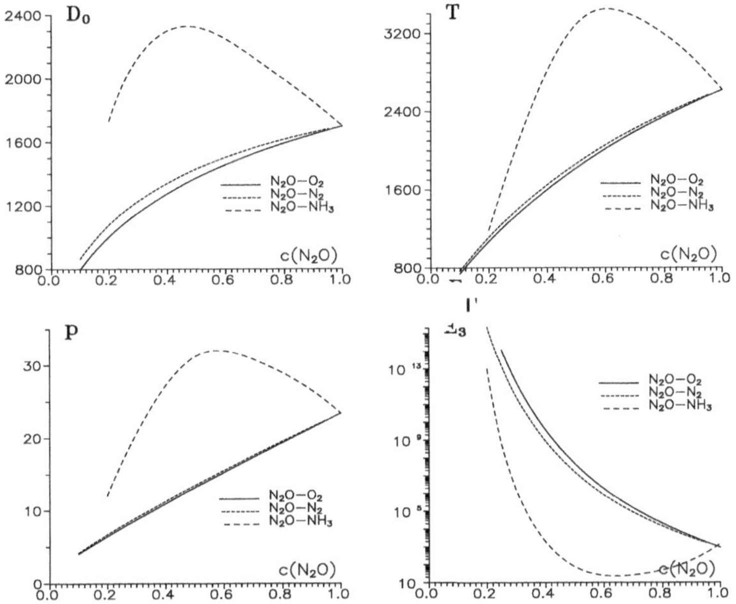

Figs.49-52

release q=20.3. Critical initiation energies are equal E_1=8 J/cm^2, E_2=18.6 J/cm and E_3=1310 J for plane, cylindrical and spherical cases. At instantaneous explosion in V_o=const T=3112 K, p=10.6, Q=604 cal/g. Nitric oxide is more hazardous as nitrous oxide.

Nitrous oxide - oxygen and nitrous oxide - nitrogen mixtures (and nitric oxide - oxygen, and nitric oxide - air also) demonstrate similar dependence of the main parameters on molar N_2O concentration: velocity, pressure, temperature... are decreased and critical initiation energies are increased at decreasing of N_2Oconcentration.

Nitrous oxide - ammonia mixtures demonstrate another character: extreme-shape dependence of the main DW parameters on molar fuel concentration $c(N_2O)$ is typical for this system (see Figs. 49-52, where D_0 m/s, T K, pressure ratio p, critical initiation energy for spherical symmetry E_3 J are demonstrated). Its seen that energy E_3 decreases up to value about 1.3 10^3 J (more hazardous nitrous oxide - ammonia mixture, approximately as ozone-oxygen) at molar concentration $c(N_2O) = 0.65$ in N_2O-NH_3 system. Last result indicates that in hospitals, where nitrous oxide, oxygen and liquid ammonia are used in reanimating room, the situation may be quite dangerous from fire and explosion hazards of view, if additional high voltage defibrillation technique is used for heart stimulation.

Hydrocarbon gaseous hydrate. Gaseous hydrates are new class of changeable chemical substances with complex space-orientation structure (likely fulleren), when the molecules of hydrocarbon fuel-«quest» are surrounded by water molecules. Hydrocarbon hydrates pay a great attention as a new energy sourse. On today, the hydrocarbon reserves in hydrate form may be estimated as 2 10^{16} m^3, which exceeds the total reserves of all other fuels in any form. In this paper some results are presented for acetylene and methane hydrates -

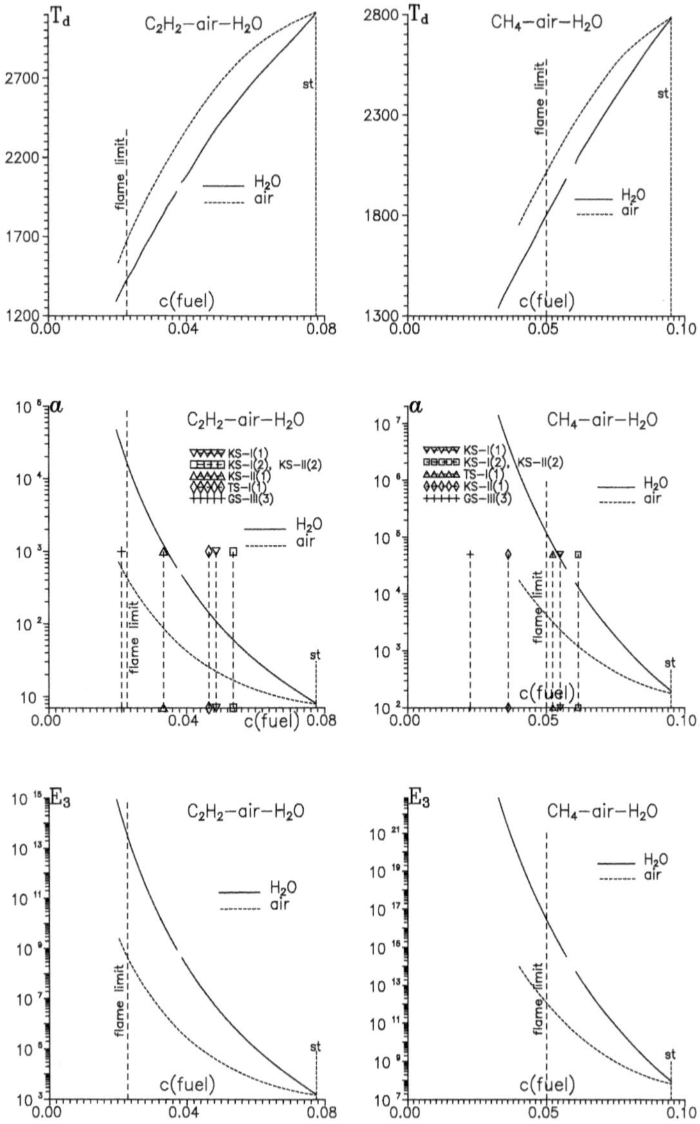

Figs. 53-58.

Figs.53-58. The next structures are known for hydrates: KS-I with h=23/3 (1) or h=23/4 (2); KS-II with h=17 (1) or h=17/3 (2); GS-III with h=34 (3), and TS-I with h=43/5 (1), where h is some coefficient, which characterizes the number of water molecules on unit fuel-«quest» molecule in individual hydrate structure (for example, [62]).

In Figs.53-54 the detonation velocity D_0 (m/sec) and products temperature T (K) are demonstrated: the dotted curve corresponds fuel-air mixtures, solid curve corresponds to stoichiometric fuel-air mixture diluted by water vapor (at its variation). A long vertical dotted line corresponds to lower concentration flame limit: c=0.025 for acetylene-air mixture, and c=0.05 - for methane-air; short vertical dotted line corresponds to stoichiometric molar concentration of fuel.

In Figs.55-56 the cell size dependence can be observed for analogous cases as in Figs.53-54, and additional vertical dotted lines are presented, which illustrates the location of different hydrate structures on concentration axis. It is seen, that some structures are located in inner area of the flame limit region, and so hydrates with such structures may be not only to fire, but also to detonate in principle. Of course, the charge size and the initiator energy must be sufficient at such experiments. It is seen also, that some hydrate structures are located outside the flame limit region, and so such hydrates may be classified as non-combustible. But it must be mentioned especially that at the destruction of hydrate structures the molecules of fuel-«guest» became free, and hydrate system transforms to usual gaseous fuel-air system.

In Figs.57-58 the critical initiation energy for spherical case is demonstrated. One can see, that fuel-air system is more dangerous then hydrate system. Acetylene hydrate may be classified as quite hazards system, which close to typical gaseous systems. For example: acetylene hydrate of KS-I(2) structure may be initiated by E_3~150 g TNT

Alkanes C_nH_{2n+2}. Alkenes C_nH_{2n}. Stoichiometric mixtures with air.

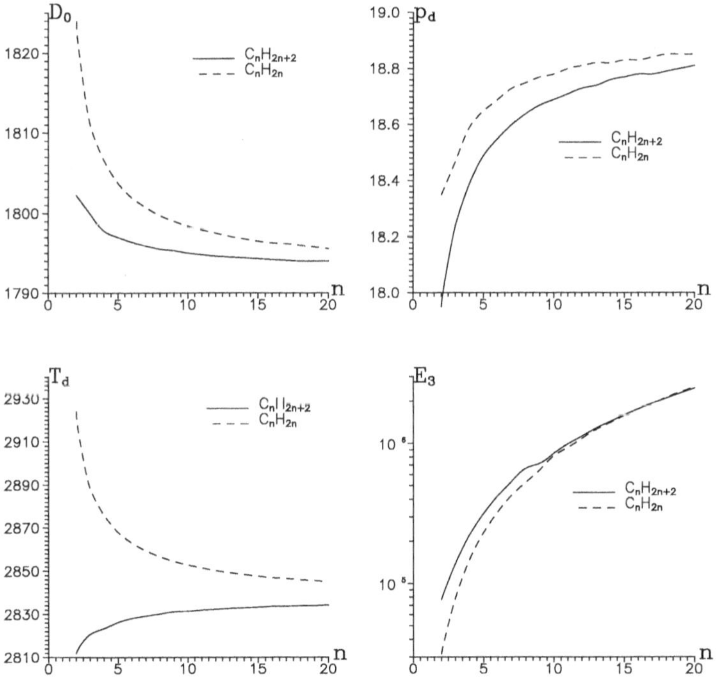

Figs.59-62.

On Figs. 59-62 the detonation velocity D_0 (m/sec), pressure P_d (atm) and temperature T_d (K) of products, critical energy E_3 (J) of spherical initiation are presented for stoichiometric fuel-air mixture for saturated C_nH_{2n+2} and unsaturated C_nH_{2n} hydrocarbon fuels on n-value.

On Fig.63 the total graph of critical initiation energies for different FAE and FOE on molar fuel concentration is presented, which illustrated the comparable relations among different mixtures from hazard point of view.

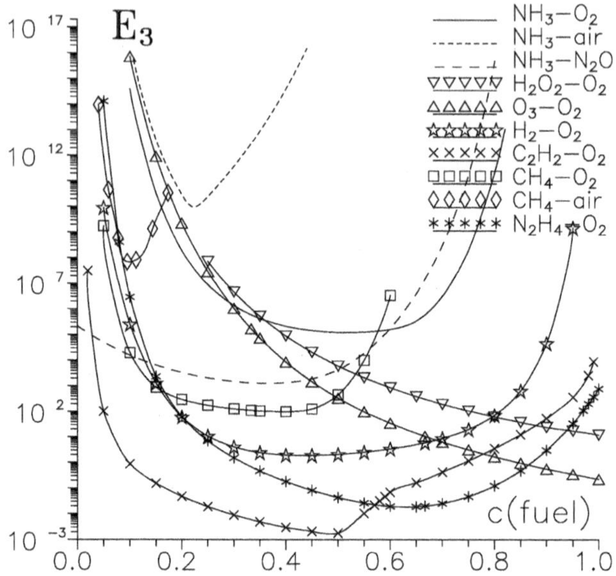

Fig.63.

Conclusions.

The analysis of various models and their comparison with experiment for different in structure and chemical activity fuels in mixes with oxygen and air in a wide range of various parameters (pressure, concentration of fuel, adding degree of inert gases) have shown that the available experimental data are described most adequately by the model of «multi-point» initiation.

The data of detonation hazards of typical combustion systems are presented.

The comparison of calculated and experimental data demonstrates, that the "SAFETY" Program predicts quite well the basic parameters of multifront DW, which are necessary for estimations of detonation and explosion hazard of combustible systems.

References.

1. **Lee J.H.** *Initiation of gaseous detonation*//Ann. Rev. Phys. Chem. -1977. -28. -p.75-104.
2. **Vasil'ev A.A.** *A spatial initiation of multifront detonation (in Russian)*// Comb. Expl. Shock Waves.-1989.-25,1.-p.113-119.

3. **Sedov L.I.** *The methods of similarity and dimensions in mechanics (in Russian)//*M., Nayka. -1987.
4. **Korobejnikov V.P.** *Some problems of the point explosion theory in gases (in Russian)//*M., Nayka. -1973.
5. **Vasil'ev A.A.** *Near-critical regimes of gaseous detonation (in Russian)//* Novosibirsk. -1995.
6. **Lee J.H., Matsui H.** *A comparison of the critical energies for direct initiation of spherical detonations in acetylene-oxygen mixtures//*Combust. Flame. -1977. -28. -p.61-66.
7. **Lee J.H., Ramamurthi K.** *On the concept of the critical size of a detonation kernel//*Comb. Flame. -1976. -27. -p.331-340.
8. **Korobeinikov V.P., Levin V.A., Markov V.V., Chernyi G.G.** *Propagation of blast waves in a combustible gas//*Astr. Acta. -1972. -17,4-5. -p.529-537.
9. **Levin V.A., Markov V.V.** *The initiation of detonation wave at concentrating energy admission (in Russian)//* Comb. Expl. Shock Waves.-1975.-11,4.-p.623-633.
10. **Levin V.A., Osinkin S.F., Markov V.V.** *Direct initiation of detonation in a hydrogen-air mixture//*in "Combustion, Detonation, Shock Waves", v.2. Proceedings of the Zeldovitch Memorial. Russian Section of the Combustion Institute, M. -1994. -p.363-365.
11. **He L., Clavin P.** *On the direct initiation of gaseous detonations by an energy source//*J. Fluid Mech. -1994. -277. -p.227-248.
12. **Zeldovich Ya.B., Kogarko S.M., Simonov N.N.** *The experimental investigations of spherical gaseous detonation (in Russian)//* J. Tech. Phys.-1956. -26,8. -p.1744-1768.
13. **Edwards D.H., Hooper G., Morgan I.M.** *An experimental investigation of the direct initiation of spherical detonation//*Acta Astr. -1975. -3. -p.117-130.
14. **Sichel M.** *A simple analysis of the blast initiation of detonations//*Acta Astr. - 1977. -4,3-4. -p.409-424.
15. **Bach G.G., Knystautas R., Lee J.H.** *Initiation criteria for diverging gaseous detonation//*13-th Symp. (International) on Combust. -1970. -p.1097-1110.
16. **Ul'yanitsky V.Yu.** *The model of direct initiation of gaseous detonation with unstable effect. The point initiation (in Russian)//* Comb. Expl. Shock Waves.-1980.-16,3.-p.101-113.
17. **Bohon Yu.A., Shulenin Yu.V.** *The minimal initiation energy for spherical detonation dor some hydrogen mixtures (in Russian)//*Report of USSR Academy of Sciences. -1979. -245,3. -p.623-626.
18. **Shulenin Yu.V., Bohon Yu.A.** *The minimal initiation energy of gaseous mixtures in unlimited area (in Russian)//*Report of USSR Academy of Sciences. -1981. -257,3. -p.680-683.
19. **Lee J.H., Knystautas R., Guirao C.** *The link between cell size, critical tube diameter, initiation energy and detonability limits//*in "Fuel-Air Explosions" ed. by Lee and Guirao. University of Waterloo press. -1982. -p.157-189.
20. **Benedick W.B., Guirao C.M., Knystautas R., Lee J.H.** *Critical charge for direct initiation of detonation in gaseous fuel-air mixtures//*in "Dynamics of Explosion" ed. by Bowen, Leyer and Soloukhin; v.106 of "Progress in Astronautics and Aeronautics". -N.Y. -1986. -p.181-202.
21. **Zhdan S.A., Mitrofanov V.V.** *A simple model for calculation of initiation energy for geterogeneous and gaseous detonation (in Russian)//* Comb. Expl. Shock Waves.-1985.-21,6.-p.98-103.
22. **Westbrook C.K., Urtiew P.A.** *Chemical kinetic prediction of critical parameters in gaseous detonations//*19-th Symp. (International) on Combust. -1982. -p.615-623.
23. **Troshin K.Ya.** *The initiation energy for divergent detonation waves (in Russian)//* Report of USSR Academy of Sciences. -1979. -247,4. -p.887-889.
24. **Borisov A.A., Zamansky V.M., Lisyansky V.V., Skachkov G.I., Troshin K.Ya.** *The estimation of critical initiation energy for detonable gaseous system by the ignition delays (in Russian)//*Chem. Phys.-1986. -5,12. -p.1683-1689.
25. **Vasil'ev A.A.** *The estimation of initiation energy for cylindrical detonation (in Russian)//* Comb. Expl. Shock Waves.-1978.-14,3.-p.154-155.
26. **Vasiljev A.A., Nikolaev Ju.A.** *Closed theoretical model of a detonation cell//*Acta Astr. -1978. -v.5. -p.983-996.
27. **Vasil'ev A.A., Nikolaev Yu.A., Ulyanitsky V.Yu.** *The critical initiation energy for multifront detonation (in Russian)//* Comb. Expl. Shock Waves.-1979.-15,6.-p.94-104.
28. **Vasil'ev A.A., Grigor'ev V.V.** *The critical conditions for detonation propagation at sharp expanded channel (in Russian)//* Comb. Expl. Shock Waves.-1980.-16,5.-p.117-125.
29. **Vasil'ev A.A.** *A new diffraction estimation of critical initiation energy//*Proceeding of the Colloquim on Gas, Vapor, Hybrid and Fuel-Air Explosions. Safety Consulting Engineers, Inc., Schaumburg, Illinois USA, 1998. P.470-481.
30. **Kailasanath K., Oran E.S., Boris J.P., Young T.R.** *Determination of detonation cell size and the role of transverse waves in two-dimensional detonations*, Comb. Flame. 1985. **61**. pp.199-209.
31. **Gelfand B.E., Frolov S.M., Nettleton M.A.** *Gaseous detonations - a selective review*, Prog. Energy Combust. Science. 1991. **17**. pp.327-371.

32. **Knystautas R., Guirao C., Lee J.H., Sulmistras A.** *Measurement of cell size in hydrocarbon-air mixtures and predictions of critical tube diameter, critical initiation energy and detonation limits*//in "Dynamics of Shock Waves, Explosions and Detonations" ed. by Bowen, Manson, Oppenheim and Soloukhin; v.94 of "Progress in Astronautics and Aeronautics". N.-Y. -1983. -p.23-37.
33. **Matsui H., Lee J.H.** *On the measure of the relative detonation hazards of gaseous fuel-oxygen and air mixtures*//17-th Symp. (International) on Combust., -1978. -p.1269-1280.
34. **Vasiljev A.A.** *Initiation of gaseous detonation by a high speed body*//Shock Waves. -1994. -3,4. -p.321-326.
35. **Vasil'ev A.A., Kulakov B.I., Mitrofanov V.V., Silvestrov V.V., Titov V.M.** *The initiation of gaseous mixtures by high-velocity bullet (in Russian)*// Report of Russian Academy of Sciences. -1994. -338,2. -p.188-190.
36. **Makeev V.I., Gostintsev Ju.A., Strogonov V.V., Bochon Ju.A., Chernushkin Ju.N., Kulikov V.N.** *Burning and detonation of hydrogen-air mixes in free volumes (in Russian*)// Phys. Comb. Expl. -1983. -19, 5. -p.16-18.
37. **Bull D.C., Elsworth J.E., Shuff P.J., Metcalfe E.** *Detonation cell structures in fuel-air mixtures*// Combust. Flame. -1982. -45,1. -p.7-22.
38. **Knystautas R., Lee J.H., Guirao C.M.** *The critical tube diameter for detonation failure in hydrocarbon-air mixtures*//Combust. Flame. -1982. -48. -p.63-83.
39. **Kogarko S.M., Adyshkin V.V., Ljamin A.G.** *Investigations of spherical detonation in gaseous mixtures(in Russian*)//Phys. Comb. Expl. -1965. -2.-p.22-34.
40. **Freiwald H., Koch H.W.** *Spherical detonation on acetylene-oxygen-nitrogen mixtures as a function of nature and strength of initiation*//9-th Symp. (International) on Combust. -1963. -p. 275-281.
41. **Bull D.C., Elsworth J.E., Hooper G., Quinn C.P.** *A study of spherical detonation in mixtures of methane and oxygen diluted by nitrogen*//J. Phys. D: Appl. Phys. -1976. -9. -p.1191-
42. **Moen I.O., Murray S.B., Bjerketvedt D., Rinnan A., Knystautas R., Lee J.H.** *Diffraction of detonation from tubes into a large fuel-air explosive cloud*//19-th Symp. (International) on Combust. -1982. -p.635-644.
43. **Bull D.C., Elsworth I.E., Hooper G.** *Initiation of spherical detonation in hydrocarbon-air mixtures*//Acta Astr. -1978. -5. p.997-1008.
44. **Nicholls J.A., Sichel M., Gabrijel Z., Oza R.D., Vandermolen R.** *Detonability of unconfined natural gas-air clouds*//17-th Symp. (International) on Combust. -1978. -p.1223-1234.
45. **Atkinson R., Bull D.C., Shuff R.I.** *Initiation of spherical detonation in hydrogen-air*//Comb. Flame. -1980. -39.-p.287-300.
46. **Ohyagi S., Yoshihashi T., Harigaya Y.** *Direct initiation of planar detonation waves in methane/oxygen/nitrogen mixtures*//in "Dynamics of Shock Waves, Explosions and Detonations" ed. by Bowen, Manson, Oppenheim and Soloukhin; v.94 of "Progress in Astronautics and Aeronautics". N.-Y. -1983. -p.3-22.
47. **Vasil'ev A.A., Valishev A.I., Vasil'ev V.A., Panfilova L.V., Topchian M.E.** *Detonation waves parameters at increased pressure and temperatures*//Chem. Phys. Reports, 1997, vol. 16 (9), p. 1659-1666.Vasil'ev A.A., Vasil'ev V.A. *The steam influence on hydrogen-oxygen and hydrogen-air detonation*// Conference Proceeding 16-ICDERS. University of Mining and Metallurgy AGH. Cracow, Poland, 1997. P.152-155.
48. **Vasil'ev A.A., Vasil'ev V.A.** *The steam influence on hydrogen-oxygen and hydrogen-air detonation*// Conference Proceeding 16-ICDERS. University of Mining and Metallurgy AGH. Cracow, Poland, 1997. P.152-155.
49. **Vasil'ev A.A., Valishev A.I., Vasil'ev V.A.** *The detonation safety of gaseous fuels. Acetylene and cyanogen*// Conference Proceeding 16-ICDERS. University of Mining and Metallurgy AGH. Cracow, Poland, 1997. P.594.
50. **Vasil'ev A.A., Valishev A.I., Vasil'ev V.A., Panfilova L.V., Topchian M.E.** *Detonation hazards of hydrogen mixtures*//Proceeding of the Colloquim on Gas, Vapor, Hybrid and Fuel-Air Explosions. Safety Consulting Engineers, Inc., Schaumburg, Illinois USA, 1998. P.391-413.
51. **Vasil'ev A.A., Valishev A.I., Vasil'ev V.A., Panfilova L.V**. *Parameters of combustion and detonation of hydrazine and its methyl derivatives*// Combustion, Explosion, and Shock Waves. 2000, vol. 36(3)
52. **Benedick W.B., Guirao C.M., Knystautas R., Lee J.H.** *Critical charge for direct initiation of detonation in gaseous fuel-air mixtures* // in "Dynamics of Explosion" ed. By Bowen, Leyer and Soloukhin; v.106 of "Progress in Astronautics and Aeronautics". -N.Y. -1986. -p.181-202.
53. **Atkinson R., Bull D.C., Shuff R.I.** *Initiation of spherical detonation in hydrogen-air* //Comb. Flame. -1980. -39. -p.287- 300.
54. **Levin V.A., Markov V.V.** *Occurrence of a detonation with concentrated energy input* (in Russian)// Physics of Burning and Explosion. -1975. -11,4. -p.623-633.

55. **Levin V.A., Osinkin S.F., Markov V.V.** *Direct initiation of detonation in a hydrogen-air mixture //* in "Combustion, Detonation, Shock Waves", v.2. Proceedings of the Zeldovitch Memorial. Russian Section of the Combustion Institute, M. -1994. -p.363-365.

56. **Getzinger R.W., Bowen J.R., Oppenheim A.K., Boudart M**. *Steady detonations in gaseous ozone//*10-th Symp. (Int.) on Comb., The Combustion Institute, 1965, pp.779-784).

57. **M.D.Pedley, C.V.Bishop, F.J.Benz, C.A.Bennett, R.D.McClenagan, D.L.Fenton, R.Knystautas, J.H.Lee, O.Peraldi, G.Dupre, J.E.Shepherd**. *Hydrazine vapor detonations//*in «Dynamics of explosions» (Eds. A.L.Kuhl, J.R.Bowen, J.-C.Leyer, A.Borisov). V.114 of «Progress in Astronautics and Aeronautics», AIAA, Washington, 1988, p.45-63.

58. **Breton J**. *Recherches sur la Detonation des Melanges Gazeux//*Theses Faculte des Sciences, Univ. Nancy. 1936. (and Lewis B., Von Elbe G. *Combustion, Flames and Explosions of Gases//*Academic Press Inc., New York - London. -1961.).

59. **Kogarko S.M., Adyshkin V.V., Ljamin A.G.** *Investigations of spherical detonation in gaseous mixtures*(in Russian)//Phys. Comb. Expl. -1965. -2.-p.22-34.

60. **Freiwald H., Koch H.W.** *Spherical detonation on acetylene-oxygen-nitrogen mixtures as a function of nature and strength of initiation//*9-th Symp. (International) on Combust. -1963. -p. 275-281.

61. **Munday G., Ubbelohde A.R., Wood I.F.** *Marginal detonation in cyanogen-oxygen mixtures//*Proc. Roy. Soc.: A. -1968. -306,№1485. -p.179-184.

62. **T.V.Rodionova, D.V.Soldatov, Yu.A.Dyadin** *Gaseous hydrates in Earth ecosystem* (in Russian)//Chemistry in interests of stable development. 1998, **6**,1. P.51-74

COOKOFF TEST MODELS AND RESULTS

J. C. Góis, F. Chaves

P. Simões*, L. Pedroso*, I. Plaskin, R. Mendes, J. Ribeiro, A. Portugal*, J. Campos

Laboratory of Energetics and Detonics
*Mechanical Eng. Department and Chemical Eng. Department**
Faculty of Sciences and Technology of University of Coimbra, Pólo II
3030 Pinhal de Marrocos, Coimbra, PORTUGAL

ABSTRACT

This paper described the laboratory tests performed to generate input data and the development and improvement of a computer model to predict fast and slow cookoff test results. The kinetics parameters of a PBX, based in RDX, was obtained and their results were used as input data in a computer model to predict ignition temperature and time-to-explosion. The results obtained with the 1-step kinetic model, based on the empirical model of Sestak-Berggreen and the 3-steps McGuire-Tarver kinetic model were compared and the discussion is conducted to improved design practices for insensitive munitions.

1. INTRODUCTION

The need to meet insensitive munitions (IM) cookoff requirements has brought many researchers to investigate the thermal behaviour of energetic material (EM). The knowledge of the responses of EM to those threats and the effects of the responses on the environment has particular importance to design practices for IM.

Small-scale tests are often employed to characterise the thermal behaviour of EM. The selection of cookoff tests needs to be based on the mechanisms that support thermal decomposition of EM. Experimental efforts should be concentrated on the understanding of the phenomenon and the generation of data to simulation and large-scale tests.

The main objective of this work is to develop and improve tests methods for predicting the response of a PBX, based in RDX, to the perceived threats of fast and slow cookoff. As important data are lacking different laboratory tests were performed to characterise the energetic components and PBX's materials.

This worked was performed by LEDAP under EUCLID RTP 14.2 contract.

P. Castelo Branco et al. (eds.),
Defense Industries: Science and Technology Related to Security, 65–81.

DSC and TG thermal analysis were carried out to determine the specific heat and the kinetic parameters. Cookoff modelling was focused on the thermal runaway of the reaction. A thermal/chemical model is implemented in ABAQUS 2D code to predict time-to-explosion, ignition temperature and ignition point.

2. PBX CHARACTERISATION

2.1. PBX composition

A PBX based in a mass fraction of 85% of RDX was selected for this study. The filler is composed by a bimodal distribution of RDX particles (70% - d_{50}=71,1 μm and 30% - d_{50}=18.8 μm). In order to evaluate the effect of the composition of the binder on the thermal behaviour of the PBX three different PBX's based in hydroxyl-terminated polybutadiene (HTPB) (PolyBd R45HT, supplied by Atochem) isophorone diisocianate (IPDI) (Ref 59182, supplied by Fluka), as curing agent, and Bis (2, etylhexil) Sebacate (DOS, supplied by Merck Farma), as plasticiser, were manufactured.

Table 1 shows the composition and density of each PBX.

Table 1. PBX's composition and density.

PBX	Density (kg.m^{-3})	TMD (%)
RDX/HTPB (85/15)	1564	99.89
RDX/HTPB/DOS (85/11.5/3.5)	1565	99.90
RDX/HTPB/DOS/IPDI (85/10.46/3.49/1.05)	1569	99.88

2.2. Thermal conductivity of PBX

The thermal conductivity measurements of PBX's were performed with an experimental heat conduction unit (HCU). This HUC has capabilities to perform measurements at ambient temperatures. Thermal conductivity is calculated considering convection and radiation negligible. Table 2 shows the obtained results.

Table 2. Thermal conductivity of PBX's, based in RDX.

Energetic material	Thermal conductivity (W.m^{-1}.K^{-1})
RDX	0.26 @ 293 K
RDX/HTPB (85/15)	0.46 @ 298 K
RDX/HTPB/DOS/IPDI (85/10.46/3.49/1.05)	0.26 @ 297 K

2.3. Specific heat of PBX

The differential scanning calorimeter (DSC) was used to perform the measurement of specific heat of PBX. Measurement results of the heat capacity are presented within a temperature range. Two different methods (method A and method B) were applied for the measurement of the specific heat.

Method A

Method A uses the results of a standard reference material under the same protocol to determine the specific heat of the PBX[1]. A first DSC running must be performed with the empty alumina crucible. A second running at the same heating rate is done with a reference material, which has known specific heat. A last running also at the same heating rate is performed with the PBX sample. The specific heat is calculated from the following equation:

$$\left(\frac{\Delta T}{c_p\, m}\right)_{sample} = \left(\frac{\Delta T}{c_p\, m}\right)_{reference} \qquad (1)$$

Where ΔT represents the temperature range between the sample or reference and the empty alumina crucible and m corresponds to the mass of the sample or the reference.

Method B

Method B is based on the principle that the heat flux absorbed by the sample is proportional with its specific heat[2], which is described by the following equation:

$$\frac{dH}{dt} = m\, c_p\, \frac{dT}{dt} \qquad (2)$$

Where dH/dt represents the heat flux, m is the mass of the sample and dT/dt the heating rate value.

Test methodology

Samples with a mass range between 20 mg to 30 mg were submitted to a constant heating rate until a temperature lower than the onset temperature of chemical reaction or phase transition. Heating rates of 2.5, 5, 7.5 and 10 $K.min^{-1}$ were used to perform measurements. Table 3 shows the specific heat of the different materials.

Table 3. Specific heat of PBX components and mixtures.

Material	c_p ($J.kg^{-1}.K^{-1}$) Method A	c_p ($J.kg^{-1}.K^{-1}$) Method B	dT/dt ($K.min^{-1}$)	Range of temperature (K)
HTPB	1903 ± 65[1]	2138 ± 55	10	343 ~ 413
RDX	966 ± 2	1032 ± 15	5	343 ~ 407
	1148 ± 10	1248 ± 10	10	343 ~ 407
RDX/HTPB (85/15)	1246 ± 86	1445 ± 122	2.5	351 ~ 405
	1337 ± 39	1440 ± 41	7.5	343 ~ 414

The specific heat results obtained with method A are near 10% lower that its obtained with method B and are in good approach with its presented in the bibliography[3].

2.4. Kinetic parameters of PBX's and its compounds

Simultaneous thermal analysis (DSC and TG) were employed to investigate the thermal decomposition in non-isothermal conditions of PBX's mixtures.

The measurements were carried out using a *Rheometric Scientific*®STA 1500 equipment. The samples (sizes ranges between 1.6 mg to 2.6 mg) were carefully loaded into open alumina crucibles and a dry nitrogen purge flows of 50 $ml.min^{-1}$ at 0.1 MPa absolute pressure was used in all measurements.

Non-isothermal kinetic analysis was used to estimate Arrhenius parameters of PBX's and components. The results are discussed on a comparative basis taking into account the dependencies of the activation energy upon the degree of conversion.

Thermal decomposition of the individual components

Each component was firstly studied alone providing a comparative basis for comparative purposes in the viz., of regarding the subsequent study on their mixtures. Figure 1 shows the DSC curves of the different individual components of PBX's materials.

[1] Standard deviation.

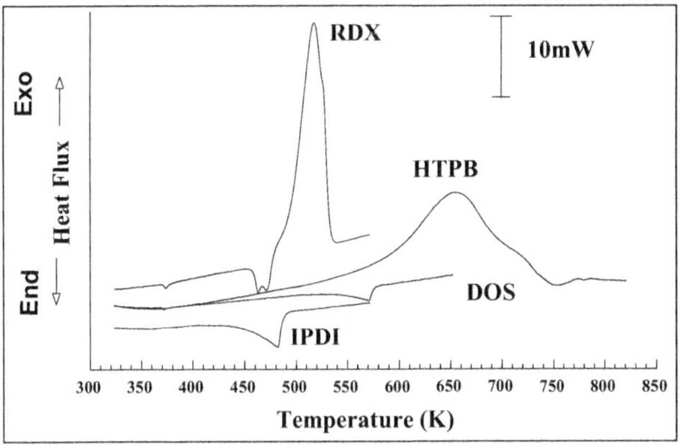

Figure 1. Typical DSC curves of individual compounds for an heating rate of 10 K.min^{-1}.

The obtained thermogram for RDX is in agreement with previous studies[4]. The endothermic peak in Figure 1 corresponding to the melting process starts at 497±1 K (HR of 10 K.min^{-1}). From the TG curves was observed that the mass loss occurs in the temperature range 496 K to 524 K. The TG curves obtained for HTPB shows the presence of two stages. IPDI and DOS DSC curves show an endothermic peak ascribable to the evaporation.

Thermal decomposition of PBX's based on RDX

Figures 2 and 3 present typical thermoanalytical curves for the tested PBX's. Differences in the rates of the global exothermic decomposition process are observed when the composition of the binder change. The mass loss increases as HTPB, HTPB/DOS and HTPB/DOS/IPDI are added to RDX. In the same way the exothermic peaks appear more and more abrupt and the correspondent peak temperatures shifted to lowers temperatures.

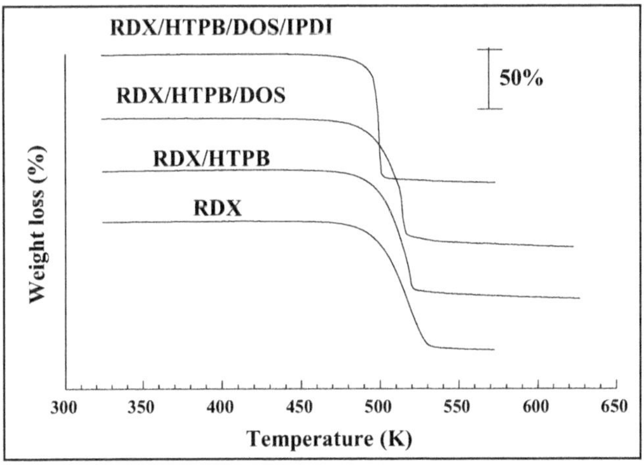

Figure 2. Typical TG curves for PBX based on RDX at an heating rate of 10 K.min^{-1}.

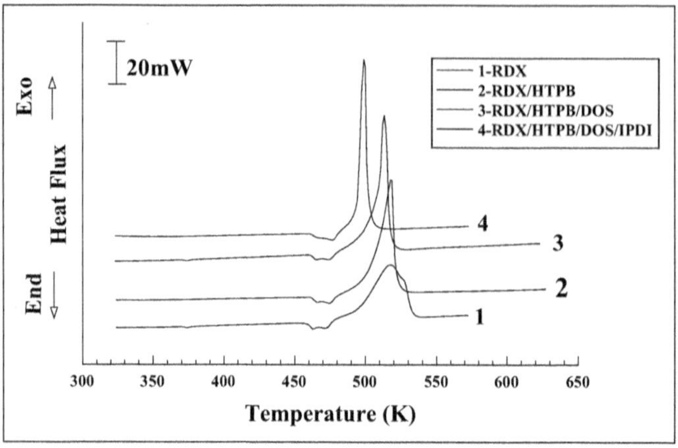

Figure 3. Typical DSC curves for PBX based on RDX at a heating rate of 10 K.min^{-1}.

No relevant departures from RDX alone are encountered in the temperature at which the mass loss begins. Table 4 shows the extrapolated onset temperature (T_{on}) and peak temperature (T_p).

Table 4. Extrapolated onset temperature and peak temperature for PBX, based on RDX, and RDX.

Material Studied	β (K.min^{-1})	T_p [2] (K)	T_{on} range (K)
RDX	5	507.63	488.43 – 488.56
	10	517.69	496.64 – 497.95
	15	524.64	499.87 – 500.38
RDX/HTPB	5	506.87	487.57 – 487.88
	10	516.79	495.18 – 497.02
	15	521.73	502.87 – 504.02
RDX/HTPB/DOS	5	506.20	487.76 – 490.56
	10	512.78	491.44 – 492.47
	15	517.83	505.08 – 505.19
RDX/HTPB/DOS/IPDI	5	495.90	480.05 – 489.78
	10	500.03	491.50 – 496.37
	15	500.80	495.75 – 496.56

The kinetic analysis of the RDX based mixtures were based on the TG curves. TG curves were normalised to establish a degree of conversion, α, ranging from 0 to 1. Friedman method[5] was applied to estimate the activation energy at different levels of α.

The theory supporting the kinetic analysis is based in the Arrhenius expression,

$$\frac{d\alpha}{dt} = k(T)f(\alpha) \quad (3)$$

[2] Average of peak temperature for each heating rate.

Where, $k(T) = A\exp\left(-\frac{E}{RT}\right)$,

α is the degree of conversion, E the activation energy, A the pre-exponential factor, T the absolute temperature, R the gas constant, t the time and $f(\alpha)$ the conversion model function[6].

Applying logarithms to Eq. (3) we get,

$$\ln\left(\frac{d\alpha}{dt}\right) = \ln\left[A\, f(\alpha)\right] - \frac{E}{RT} \tag{4}$$

Eq.(4) permits to estimate the value of E without any assumption other than the rate of the process being described by the Arrhenius law.

With the purpose to obtain a simple kinetic model function, $f(\alpha)$ able to describe the process, the kinetic equations were transformed to a generalized kinetic equation at infinite temperature[7,8]. Due of its flexibility and reduced number of kinetic exponents the empirical function $h(\alpha)$ of Sestak-Berggren model, SB (M,N)[9,10] was selected to reproduced $f(\alpha)$.

$$h(\alpha) = \alpha^M (1-\alpha)^N \tag{5}$$

The kinetic exponents M and N in the SB (M,N) model are evaluated according to the relationship[9]

$$\ln\left(\frac{d\alpha}{d\theta}\right) = \ln A + N \ln\left[\alpha^P (1-\alpha)\right] \tag{6}$$

Where

$$P = \frac{M}{N} = \frac{\alpha_M}{1-\alpha_M} \tag{7}$$

A plot of ln $(d\alpha/d\theta)$ against ln $[\alpha^P (1-\alpha)]$ gives us a slope corresponding to the kinetic exponent N and the value of A is obtained from the axis intersection value.

The invariance of E with α is a prerequisite considering the applicability of Eq.(4). Figure 4 shows the evaluation of E as a function of α for the mixtures RDX/HTPB and RDX/HTPB/DOS. The observed pattern for the RDX/HTPB/DOS reactive system is a clear indication that the process is too much complex to be tractable within the framework of kinetic analysis, based on the available thermoanalytical data. Worst results (not shown in Fig. 4) were obtained for the RDX/HTPB/DOS/IPDI mixture.

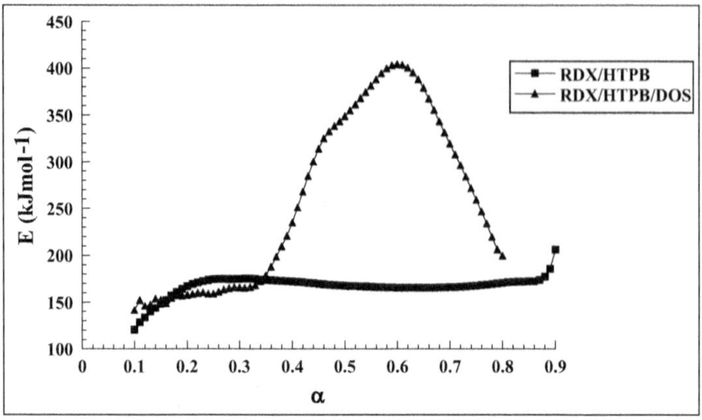

Figure 4. Activation energy as a function of the degree of conversion calculated from the Friedman method for PBX's based on RDX.

Concerning to RDX/HTPB the thermal decomposition suggests that it would be reasonable to proceed with the kinetic analysis following Friedman method. Tab. 5 shows the mean value of E calculated from those values of E satisfying the criterion of a deviation from the mean value within the interval of ±10%.

The kinetic model function h(α) able to describe the thermal decomposition of RDX/HPTB get M=0.9 and N=0.8.

Table 5. Kinetic parameters of the materials studied.

Material	α range	Temperature range[3] (K)	E $(kJmol^{-1})$	A[4] (s^{-1})
RDX	0.14 – 0.84	495.90 – 523.84	200.25 ± 3[5]	4.51×10^{18}
DOS	0.08 – 0.27 0.47 – 0.90	505.73 – 536.48 552.16 – 572.52	60.56 ± 1 86.70 ± 1	—
IPDI	0.05 – 0.90	408.61 – 473.82	59.98 ± 1	3.42×10^{4}
HTPB– 1st Stage	0.33 – 0.62	649.04 – 676.70	77.62 ± 6	—
HTPB– 2nd Stage	0.04 – 0.90	708.53 – 758.72	262.56 ± 7	—
RDX/HTPB	0.17 – 0.89	492.54 – 512.69	170.02 ± 3	5.72×10^{15}
RDX/HTPB/DOS	0.11 – 0.34	490.23 – 503.95	158.95 ± 5	—
RDX/HTPB/DOS/IPDI	0.2 – 1[6]	471.40 – 486.78	117.17 ± 15	—

[3] Average temperature at different heating rates β for the same degree of conversion α.
[4] The values here presented are an average of all measurements performed.
[5] Standard deviation.
[6] This range of α corresponds to 10 percent of total mass loss during decomposition process.

3. COOKOFF MODELLING OF PBX CYLINDER CHARGE

3.1. Computer model implementation

Modelling cookoff involves the coupling of thermal, chemical and mechanical computer codes. LEDAP has selected an extra-small test vehicle with 1.57 cm^3 volume to perform the prediction of the thermal trial response of energetic material.

A cookoff model was implemented in ABAQUS 2D code[11] to predict the temperature distribution inside the charge and confinement, the ignition point and the time-to-explosion. The size of the charge was fixed in agreement with our future plan to test experimentally the response of a small cylinder test vehicle with the same size.

Cylindrical symmetry geometry is applied for the mesh of test vehicle. The model mesh is shown in Fig. 5. Some parts of the test vehicle were omitted and simplified to build the grid model. The different colour in the center of Fig 5 represents the explosive charge.

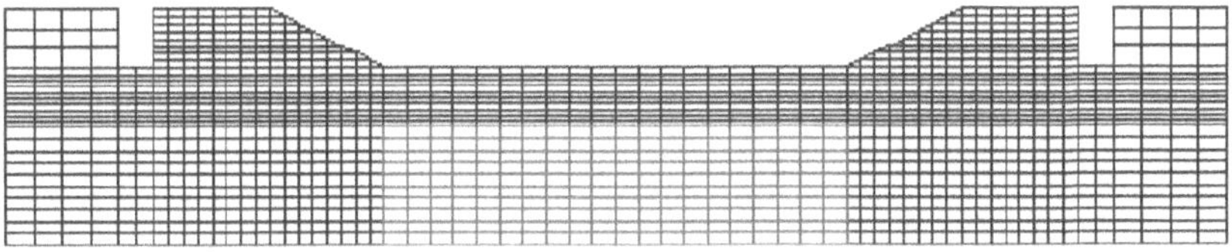

Figure 5. Grid model and mesh of test vehicle.

The properties of steel casing of test vehicle (42CrMo4) and PBX selected for this simulation are shown in Tab. 6.

Table 6. Material properties of steel casing (42CrMo4) and PBX (RDX/HTPB).

	Thermal conductivity ($W.m^{-1}.K^{-1}$)	Density ($kg.m^{-3}$)	Specific heat ($J.kg^{-1}.K^{-1}$)
Steel casing	42.7	7900	460.9
RDX/HTPB (85/15)	0.32 @ 298 K	1564	1291.5 @ 378 K

The single step kinetic model, based on the empirical model of Sestak-Berggreen[8,9], $h(\alpha)$, refereed above and the 3-steps McGuire-Tarver kinetic model[4] were used as options of kinetic model used in the simulation of thermal decomposition of PBX.

The equation used to predicting thermal decomposition is eq. 8., the Frank-Kamenetskii differential equation.

$$- \lambda \nabla^2 T + \rho c_p (dT/dt) = S \qquad (8)$$

Where S is the chemical heat source during the process reaction, λ the thermal conductivity, ρ the density and c_p the specific heat.

The McGuire-Tarver kinetic model for RDX is a 3-steps reaction scheme. Mass is converted from one chemical species to another via a system of simulated chemical reactions. In these reduced chemical reaction systems the intermediate species and their linking reactions do not correspond to any real product, but instead represent the combined effect of the numerous real chemical reactions which occur when real explosives decompose. McGuire-Tarver model reaction scheme is described below:

$$A \rightarrow B \rightarrow 2C \rightarrow D$$

Where,

$A \equiv$ RDX

$B \equiv H_2C = N\text{-}NO_2$

$C \equiv CH_2O + N_2O$ (and/or $HCN + HNO_2$)

$D \equiv$ Final products.

In the 3-steps reaction scheme the species C and the final products are gaseous[4]. In this model each of the reduced chemical reactions follows the Arrhenius equation:

$$\frac{d\alpha}{dt} = f(\alpha).A \exp\left(-\frac{E}{RT}\right) \qquad (9)$$

For each of the steps of reaction scheme exist E and A values. These values characterise the chemical reaction and govern the reaction rate of the reaction. The input kinetic parameters and the reaction heat for each of the 3-steps kinetic model used in the simulation were taken directly from the McGuire-Tarver paper without any modifications[4]. Tab. 7 shows the input kinetic parameters for the simulation of thermal decomposition of PBX, based in RDX.

Table 7. RDX input kinetic parameters[4].

Reaction step	Kinetic parameter	RDX
1	E (kJ.mol^{-1})	197.2
	A (s^{-1})	5.76x10^{19}
	Q (J.kg^{-1})	- 0.418x10^{6}
2	E (kJ.mol^{-1})	184.6
	A (s^{-1})	4.74x10^{17}
	Q (J.kg^{-1})	1.254x10^{6}
3	E (kJ.mol^{-1})	142.8
	A (s^{-1})	1.59x10^{15}
	Q (J.kg^{-1})	5.02x10^{6}

In the 3-steps kinetic model of McGuire-Tarver, S is the summation of 3-steps of the chemical heat source during the process reaction.

$$S = \sum_{1}^{3} \rho\, Q_i\, A_i \exp(-\mathrm{E}_i / R\, \mathrm{T}) \qquad (10)$$

For the 1-step kinetic model the term for the reaction heat source is described by the following equation:

$$S = \rho\, Q\, \alpha^{M} (1-\alpha)^{N} A \exp(-\mathrm{E}/R\, \mathrm{T}) \qquad (11)$$

With M=0.9 and N=0.8.

The parameters Q, E and A depend of the composition of the explosive. For the single step kinetic equation those parameters were adjusted with the applicability of extended Friedman analysis and Sestak-Berggren method.

Table 8 shows the input kinetic parameters for the simulation of thermal decomposition of RDX and RDX/HTPB.

Table 8. Input kinetic parameters of 1-step kinetic model based on the empirical model of Sestak-Berggreen.

Energetic Material	Activation energy E (kJ.mol^{-1})	Pre-exponential factor A (s^{-1})	Heat of decomposition Q (J.kg^{-1})
RDX	200.3	4.51x10^{18}	1.555x10^{6}
RDX/HTPB (85/15)	170.3	5.72x10^{15}	1.54x10^{6}

In the simulation of heating over the external surface of the cylinder test vehicle was adopted a prescribed temperature rise over the central region of the tube and the half length of the slope surfaces at each side of this central region. The temperature over the external diameter of the tube is exactly the same everywhere of the region mentioned above, and is corresponding to the node temperature.

The heat loss from the test vehicle near the plugs is due to convection and radiation to the surrounding environment. The heat loss due to convection is proportional to the temperature difference between the surrounding environment and the external surface of the tube and plugs, expressed by the following equation:

$$Q = - h\,(T_s - T_\infty) \qquad (12)$$

The heat loss due to radiation is calculated using the Stefan-Boltzmann law:

$$Q = \varepsilon\sigma(T_s^4 - T_\infty^4) \qquad (13)$$

Parameters used in the simulation are shown in Table 9.

Table 9. Heating rate and heat loss parameters and coefficients used in the simulation of thermal de composition of PBX.

Parameters and coefficients	Value
Slow temperature rise	3.3 °C.h^{-1}
Fast temperature rise	3.3 °C.min^{-1}
Convection heat coefficient	7.5 W.m^{-2}.K^{-1}
Emissivity	0.8
Stefan-Boltzaman constant	5.67x10^{-8} W. m^{-2}.K^{4}

3.2. Thermal decomposition results and discussion

The simulation of cookoff with ABAQUS 2D code has allowed to identifying the evaluation of temperature and mass loss of PBX as a function of time for both kinetic models.

Figure 6 shows sequential images of temperature distribution inside steel casing and the PBX obtained using the McGuire-Tarver kinetic model for a prescribed slow temperature rise (3.3 °C.h^{-1}). Temperature distributions at the plan correspondent to the middle of the cylinder tube and at the axis have no significant changes.

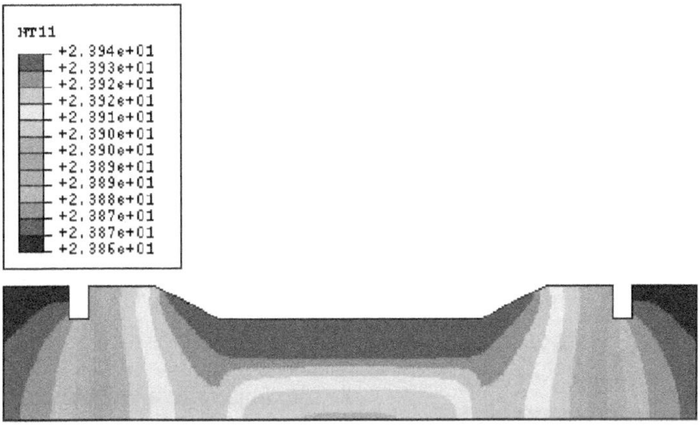
NT11
+2.394e+01
+2.393e+01
+2.392e+01
+2.392e+01
+2.391e+01
+2.390e+01
+2.390e+01
+2.389e+01
+2.389e+01
+2.388e+01
+2.387e+01
+2.387e+01
+2.386e+01

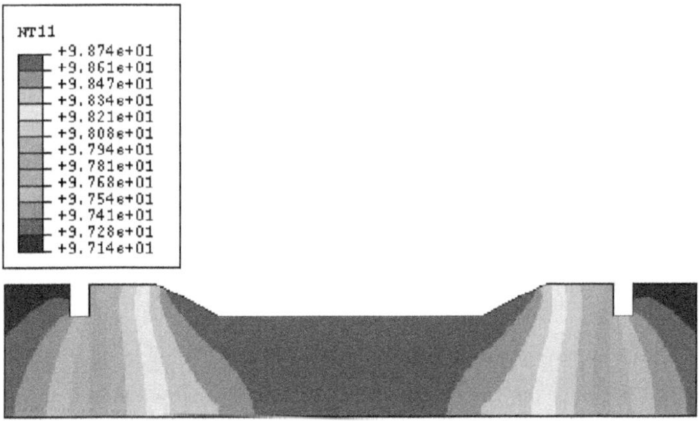
NT11
+9.874e+01
+9.861e+01
+9.847e+01
+9.834e+01
+9.821e+01
+9.808e+01
+9.794e+01
+9.781e+01
+9.768e+01
+9.754e+01
+9.741e+01
+9.728e+01
+9.714e+01

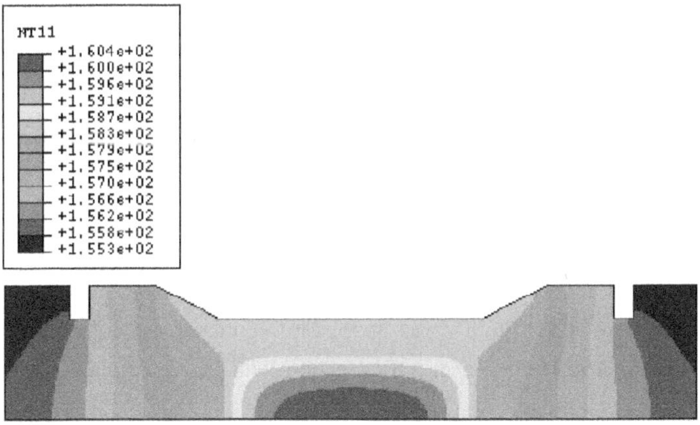
NT11
+1.604e+02
+1.600e+02
+1.596e+02
+1.591e+02
+1.587e+02
+1.583e+02
+1.579e+02
+1.575e+02
+1.570e+02
+1.566e+02
+1.562e+02
+1.558e+02
+1.553e+02

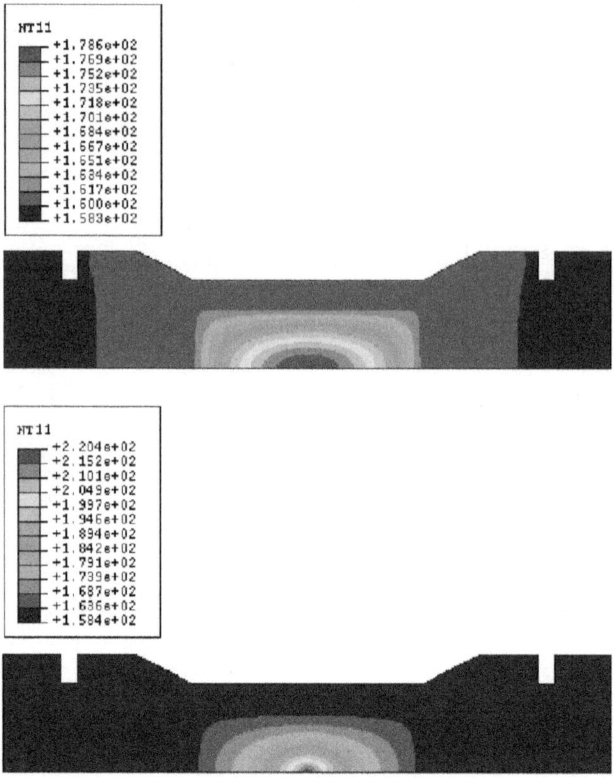

Figure 6. Sequential image of temperature distribution inside test vehicle under slow cookoff, when McGuire-Tarver kinetic model is applied.

The distribution of temperature in different locals of the test vehicle (steel casing and charge) is almost coincident due of the small size of test vehicle. For slow cookoff the ignition point is reached near the center of the charge.

Considering the single step kinetic model, described above, the process of temperature rising and ignition point are closed to its obtained with the 3-setps kinetic model. The differences associated with the kinetic model used appears on the values of ignition temperature and time-to-explosion. Differences were also observed for the shape of mass loss curve. Comparing the mass loss curve trajectory for both kinetic models we verify a smooth and long mass loss process for the 3-setp kinetic model in contrast with the abrupt decay for the single step kinetic model.

A temperature rise of 3.3 °C.min^{-1} was applied to simulate fast cookoff. Figure 7 shows sequential images of temperature distribution inside steel casing and PBX obtained using the McGuire-Tarver kinetic model.

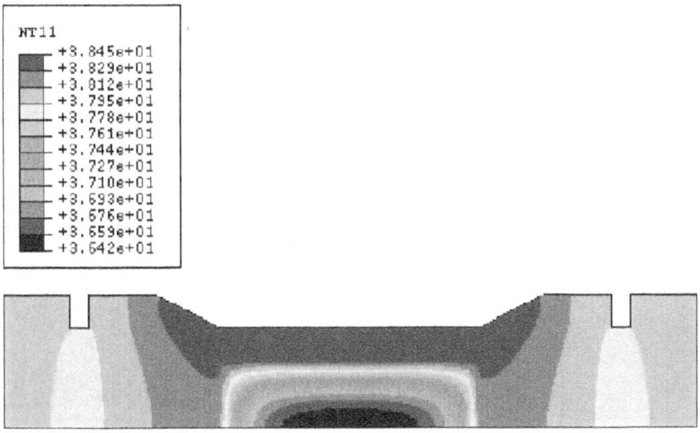
NT11
+3.845e+01
+3.829e+01
+3.812e+01
+3.795e+01
+3.778e+01
+3.761e+01
+3.744e+01
+3.727e+01
+3.710e+01
+3.693e+01
+3.676e+01
+3.659e+01
+3.642e+01

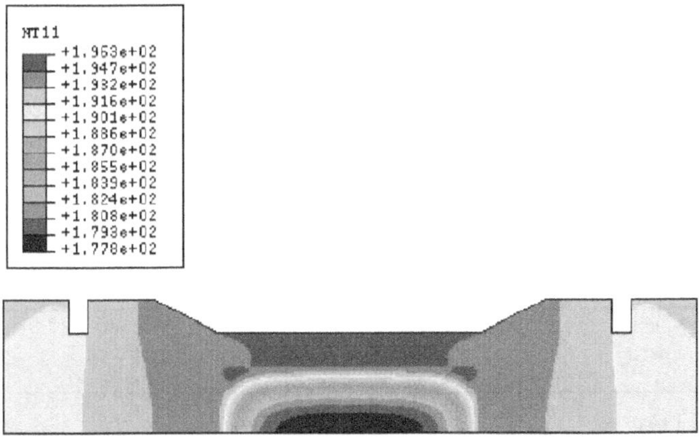
NT11
+1.963e+02
+1.947e+02
+1.932e+02
+1.916e+02
+1.901e+02
+1.886e+02
+1.870e+02
+1.855e+02
+1.839e+02
+1.824e+02
+1.808e+02
+1.793e+02
+1.778e+02

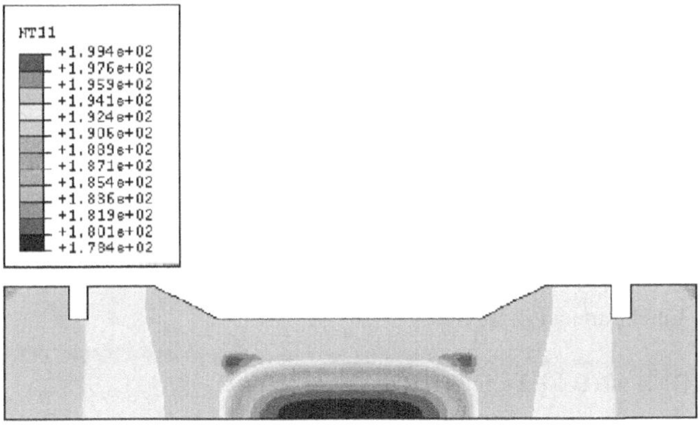
NT11
+1.994e+02
+1.976e+02
+1.959e+02
+1.941e+02
+1.924e+02
+1.906e+02
+1.889e+02
+1.871e+02
+1.854e+02
+1.836e+02
+1.819e+02
+1.801e+02
+1.784e+02

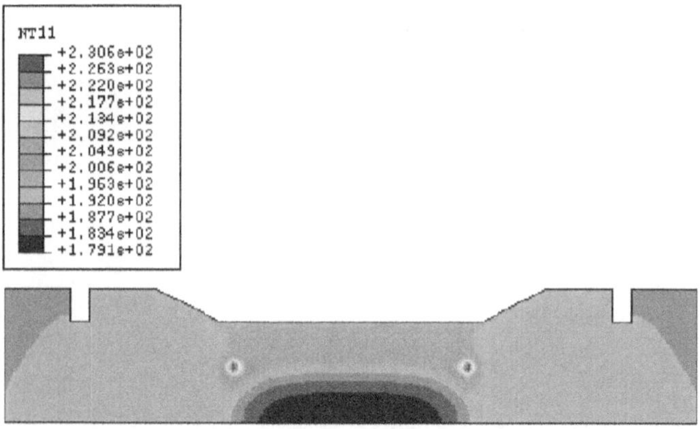

Figure 7. Sequential image of temperature distribution inside test vehicle under fast cookoff, when McGuire-Tarver kinetic model is applied.

When the 3-steps McGuire-Tarver kinetic model is applied the ignition point starts close to the radial limit of the charge and near the plugs. When the single step kinetic model is considered the ignition point starts at the center of the charge, like is observed in slow cookoff. Tab. 10 summarizes the results of simulation for fast and slow cookoff, for two different kinetic models.

Table 10. Predicted ignition temperature and time-to-explosion of PBX under slow and fast cookoff.

Heating Regime	Kinetic model	Ignition Point (mm)	Time of initiation (min)	Temperature of initiation (°C)	Time-to-explosion (min)	Temperature of ignition (°C)
SCC	3-steps	0, 0	2415.9	132.7	2939.5	164.2
	1-step	0, 0	3221.2	177.2	3445.9	193.5
FCC	3-step	±9, 4.5	47.5	156.2	58.7	193.7
	1-step	0, 0	67.0	222.7	67.3	226.2

For both kinetic models, the mass loss of PBX in the simulation due to thermal decomposition is higher for slow cookoff than for fast cookoff. The delay time between the starting of the initiation of the decomposition, characterized by mass loss, and the time-to-explosion is higher when the 3-step kinetic model is applied.

4. CONCLUSIONS AND FUTUR WORK

The characterization of a PBX was performed to generating input data for the implementation in a computer model based in ABAQUS 2D code. A single step kinetic model of decomposition, based on the empirical model of Sestak-Berggreen, was developed for using in the computer model simulation of cookoff of a small test vehicle. This 1-setp kinetic model was compared with the 3-steps kinetic McGuire-Tarver model in terms of ignition temperature and time-to-explosion.

In slow cookoff simulation, both kinetic models give the same ignition point at the center of the charge. Different local of ignition point appears when both kinetic models are applied to simulate fast cookoff. The comparison of two kinetic models on the ignition temperature and time-to-explosion shows higher values when the 1-step kinetic model is used.

Trials of slow and fast cookoff with the small test vehicle will be performed to confirm the results of simulation. The small size of test vehicle not allow to controlling the temperature at different points inside the charge. Thus information will be only providing at the center and axis of the charge.

A gas pressure model will be performed to implementation in the computer model to predicting the increasing of the internal pressure induced by the thermal decomposition.

5. REFERENCES

1. Spears R. – Thermal analysis of materials, Marsel Dekker, Inc., 1994.
2. Cuimei Y, Ziru L, Ganghe W, Chengyun W et al. - "*Determination of heat capacity of explosives and related materials by DSC*", Proc. of the 17th IPS, Beijing Institute of Technology Press, vol. 1, 1991, pp. 515-521
3. Victor, A – "Exploring cookoff mysteries", *Cook-off and XDT Mechanisms*", Vol I (Revised) by P. Kernen, A. Sanderson and B. Stokes. Presentations of the Join NIMIC/KTA 4-20 Workshop at NAWC, China Lake, USA on 12-15 March, 1996, pp. 217-239
4. McGuire, R.R., Tarver, C.M., "Chemical decomposition models for the explosion of confined HMX, TATB, RDX and TNT explosives", Proc. 7th symposium on Detonation, Annapolis, 1981, p. 56.
5. Friedman H.L., J. Polym. Science. C6, 1964, p. 183.
6. Kissinger H.E., Anal. Chem. 29 1957, p. 1702.
7. Ozawa T., J. Thermal Analysis 31, 1986, p. 547.
8. Ozawa T., Thermochimica Acta 203, 1992, p. 159.
9. Koga N. and Málek J., Thermochimica Acta 282/283, 1996, p. 69.
10. Málek J., Thermochimica Acta 200, 1992, p. 257.
11. ABAQUS Manual, version 5.8, Hibitt, Karlsson&Sorensen, Inc, (1998).

MICRO SLAPPER INITIATORS

R. Mendes, J. Campos, I. Plaksin and J. Ribeiro

Lab. of Energetics and Detonics, Mechanical Eng. Dept., Fac. of Sciences and Technology, University of Coimbra; 3030 Coimbra, Portugal

ABSTRACT

The well known Exploding Foil Initiator (EFI), also called slapper detonator, is formed by a small capacitor that is used to burst a very thin copper bridge to accelerate a dielectric flyer plates to high velocities enough to initiate the secondary explosive. The presented particular case uses an EFI formed by a capacitor up to 0.2μF charged until 3kV, and the copper bridges with 0.3x0.3mm and 0.4x0.3mm with 0.005 mm of thickness, allows to accelerate Kapton flyer plates with 25 μm of thickness until 5 km/s. The process of Shock to Detonation Transition (SDT) in explosive samples with 5mm of diameter by 10mm of height was characterized by an optical method based on 64 optical fibbers ribbon (250μm of diameter each fibber) connected to a fast electronic streak camera. The obtained results, a (x,t) diagram, with (1 ns) resolution, show continuously the shock to detonation transition regime and allowed the evaluation of the detonation velocity and the detonation wave front curvature. In that regime DW propagation presents oscillations in velocity around a mean detonation velocity value. The results also show the effects of flyer plate velocity, the initial density on the regime of the SDT and the DW irregularities.

P. Castelo Branco et al. (eds.),
Defense Industries: Science and Technology Related to Security, 83–93.

MICRO SLAPPER INITIATORS

R. Mendes, J. Campos, I. Plaksin and J. Ribeiro

LEDAP - LAB. ENERGÉTICA E DETÓNICA
FCT - University of Coimbra

- **Advanced Research Workshop**
Defence Industries: Science and Technology Related Security - Impact of Conventional Munition on Environment and Population

- ***29-31 October 2001, Porto, Portugal***

INITIATOR FUNCTION

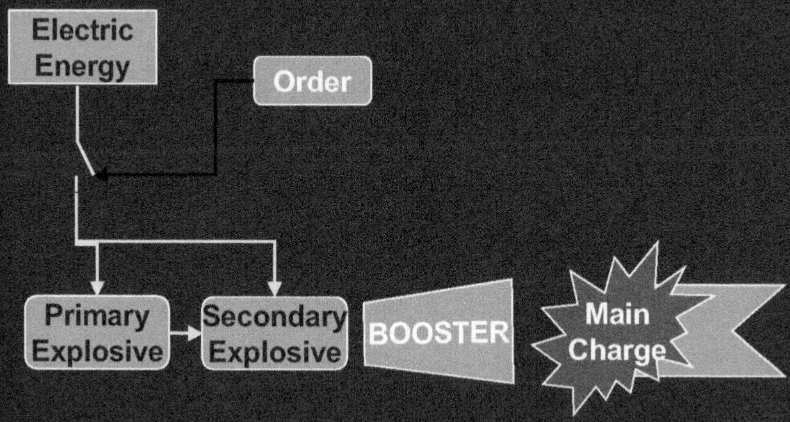

REQUIREMENTS FOR A DETONATOR

- Correct functioning
- Small in size
- Safe
- Fast response, short delay time
- No restrictions after environmental testing
- Long life
- Low price

ELECTRIC INITIATORS

- HOT WIRE - HW
  - Heating the metallic wire by electric current
  - In contact with high sensitive explosive
- ELECTRIC BRIDGE WIRE – EBW
  - Electric explosion of the bridge wire by electric current
  - Itiniation of the secondary explosive by shock wave

ELECTRIC INITIATORS

EXPLODING FOIL INITIATOR- EFI

- Electric explosion of the mettalic bridge, using high electric current
- Dielectric flyer plate acceleration
- Initiation of the high isensitive explosive by flyer plate impact

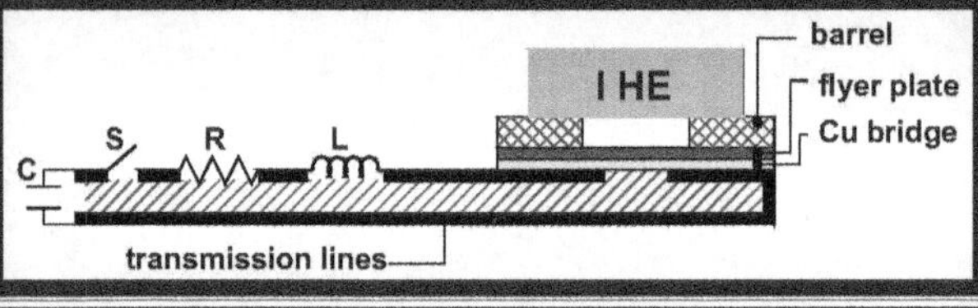

COMPARISON

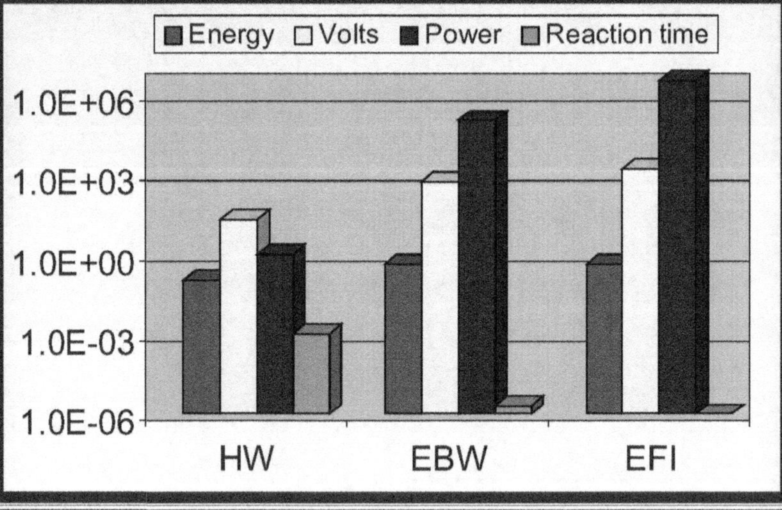

CRITICAL COMPONENTS

- CAPACITORS
- HIGH VOLTAGE SWITCH
- EFI
- EXPLOSIVE MATERIAL

CAPACITORS

- High energy density
  - Capacitance
  - Operating Voltage
  - Small dimensions and weight
- Low inductance
- Cost

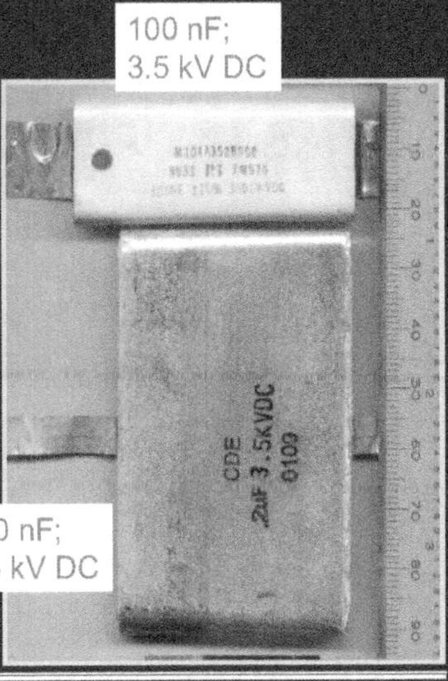

HIGH VOLTAGE SWITH

- SPARK GAP
  - Gas Filled
  - Vacuum

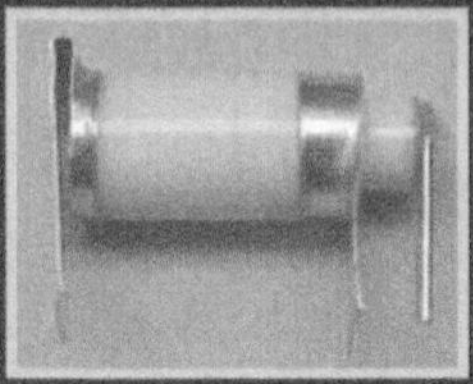

- SOLIDE STATE COMPONENTS
  - MOSFET - Metal oxide sc field effect transistor
  - N-MTC - MOS controlled thyristor

EFI

- COPPER BRIDGE
  - Thickness (μm)
  - Legth, Width (100's μm)

- OPERATING VOLTAGE
  - 2 - 3 kV
- KAPTON BASE
- KAPTON FLYER

EXPLOSIVE MATERIAL

- HNS-IV
  - Most popular
  - Accepted by MIL-STD-1316D
  - High cost

CURRENT WAVEFORM

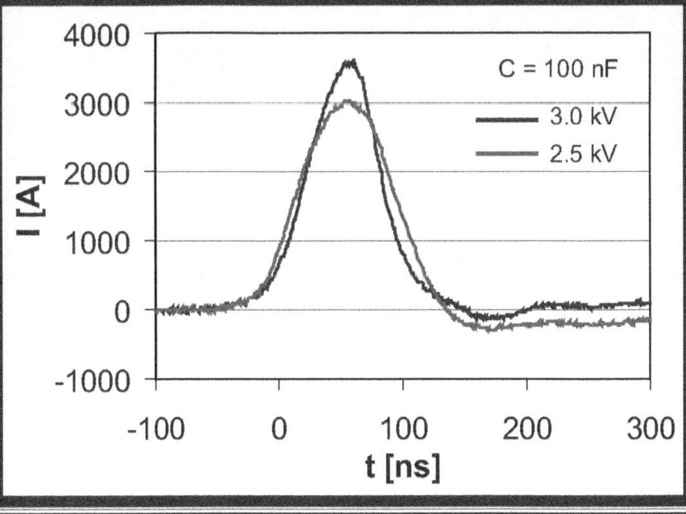

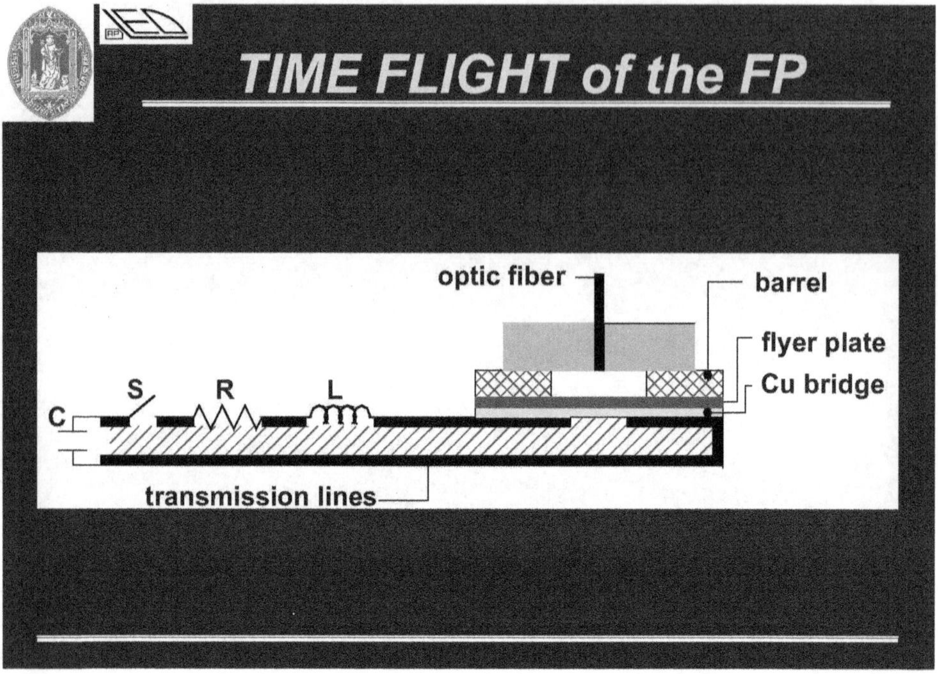
TIME FLIGHT of the FP
optic fiber
barrel
flyer plate
Cu bridge
S
R
L
C
transmission lines

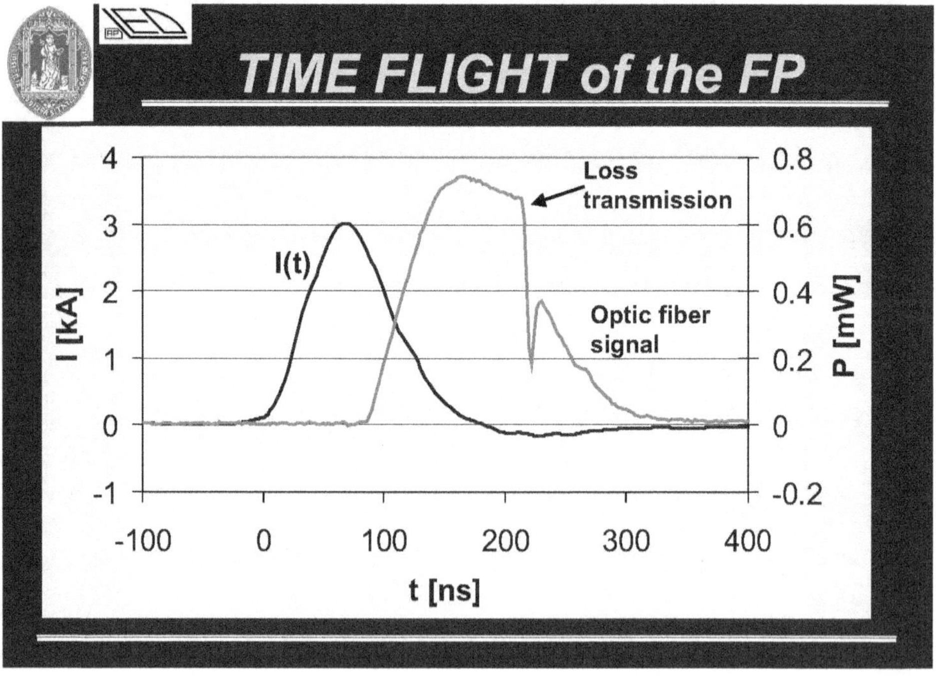
TIME FLIGHT of the FP
Loss transmission
I(t)
Optic fiber signal
I [kA]
P [mW]
t [ns]
4
3
2
1
0
-1
0.8
0.6
0.4
0.2
0
-0.2
-100
0
100
200
300
400

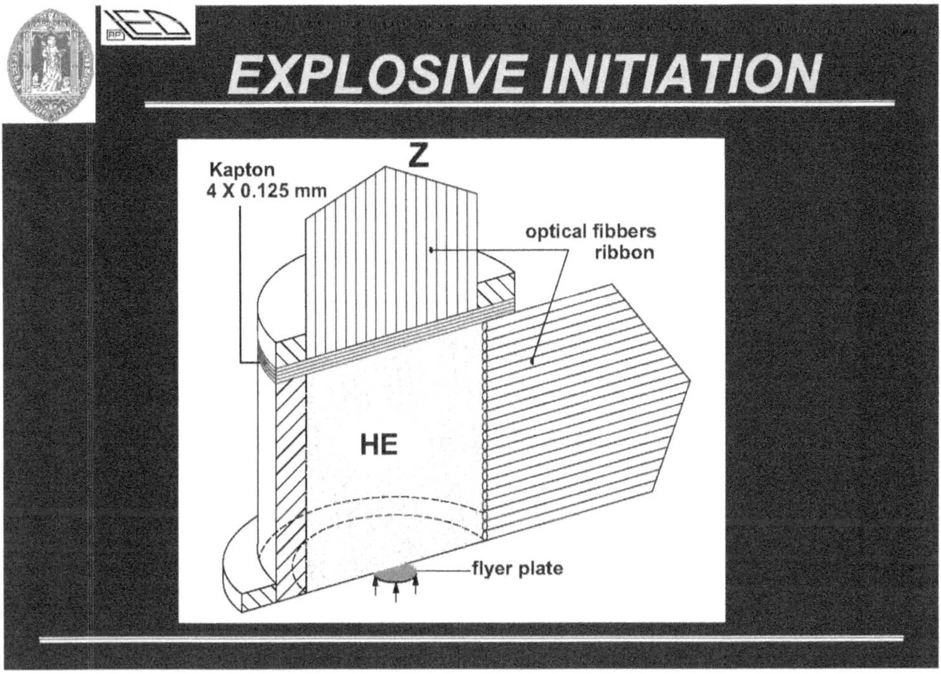
EXPLOSIVE INITIATION
Z
Kapton
4 X 0.125 mm
optical fibbers
ribbon
HE
flyer plate

EXPLOSIVE INITIATION
DW FRONT
CURVATURE
SW ATTENUATION
INERT MONITOR
TIME
(Z,t) DIAGRAM

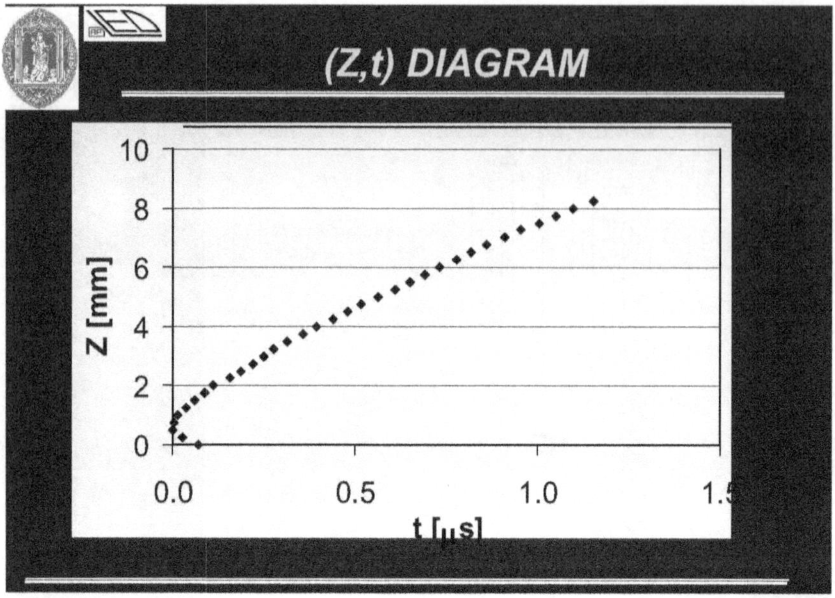

DISADVANTAGES

- More complicated than normal detonatores
- Reduced number of EFI's that can be fired by the same energy system
- High Electric Power
  - Low inductance
  - Short distance (energy system - EFI)

ADVANTAGES

- **SECURITY**
  - No content Primary Explosive
  - Secundary HE is not in contact with electric system
  - Isensitive to ESD
  - Isensitive to RF
- **HIGH REPEATIBILITY**
  - From shot to shot

ADVANTAGES

- **HIGH REPRODUTIBILITY**
  - Simultaneous Detonations
- **HIGH TEMPERATURE**
- **Can be used in a explosive train (MIL - STD - 1316D)**

Insensitive and „Green“Munition

ENVIRONMENTAL PROBLEMS RELATED TO EXPLOSIVES AND AMMUNITION – THE PRESENT SITUATION IN ROMANIA

Constatin Neagu

Abstract

The article deals with the present situation regarding the environmental problems related to the industrialization of explosives and ammunition in Romania.

After presenting the general legal framework concerning the safety problems within the Romanian explosives and ammunition industry, the article focuses on the present technologies in Romania. Thus, there are presented the responsibilities of the National Authority in charge with the ratification of the production technologies of the explosive materials and ammunition in Romania and also its strategy to comply with the NATO standards and the national ones concerning the safety of the production technologies of ammunition and explosive materials, as a part of the general strategy meant to achieve the NATO integration.

At the end, there are briefly presented the concerns of the Romanian defence industry regarding the implementation of the new types of explosives and ammunition with attenuated risk.

P. Castelo Branco et al. (eds.),
Defense Industries: Science and Technology Related to Security, 97–111.

THE IMPACT ON THE ENVIRONMENT
OF THE AMMUNITION MANUFACTURING PROCESS

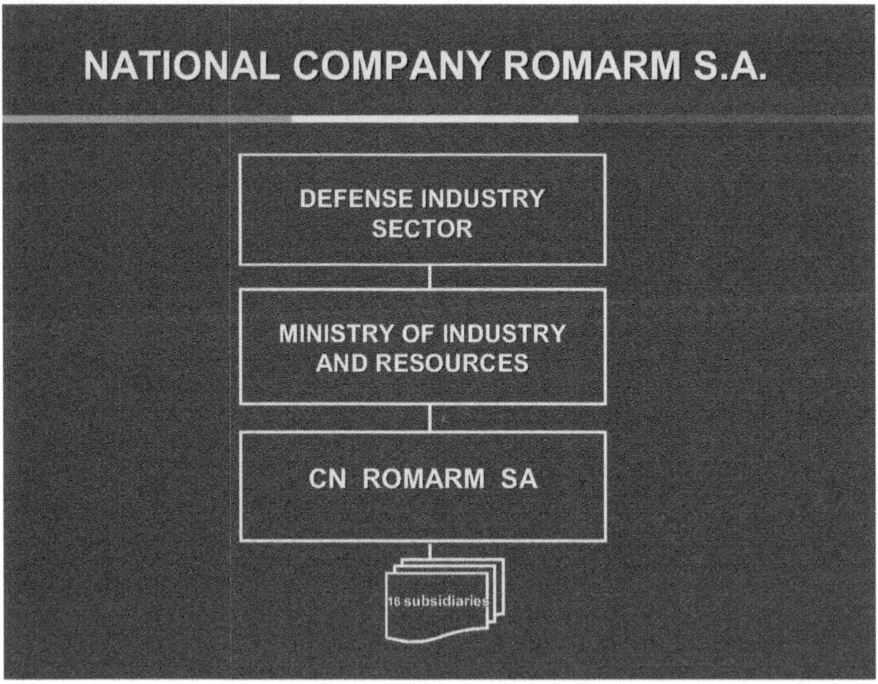
NATIONAL COMPANY ROMARM S.A.
DEFENSE INDUSTRY SECTOR
MINISTRY OF INDUSTRY AND RESOURCES
CN ROMARM SA
16 subsidiaries

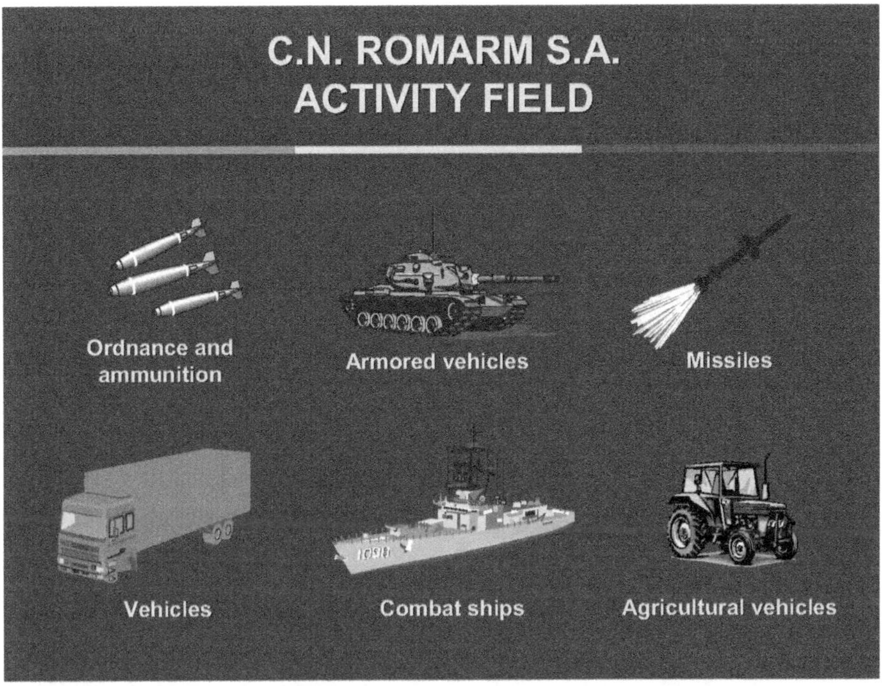
C.N. ROMARM S.A.
ACTIVITY FIELD
Ordnance and ammunition
Armored vehicles
Missiles
Vehicles
Combat ships
Agricultural vehicles

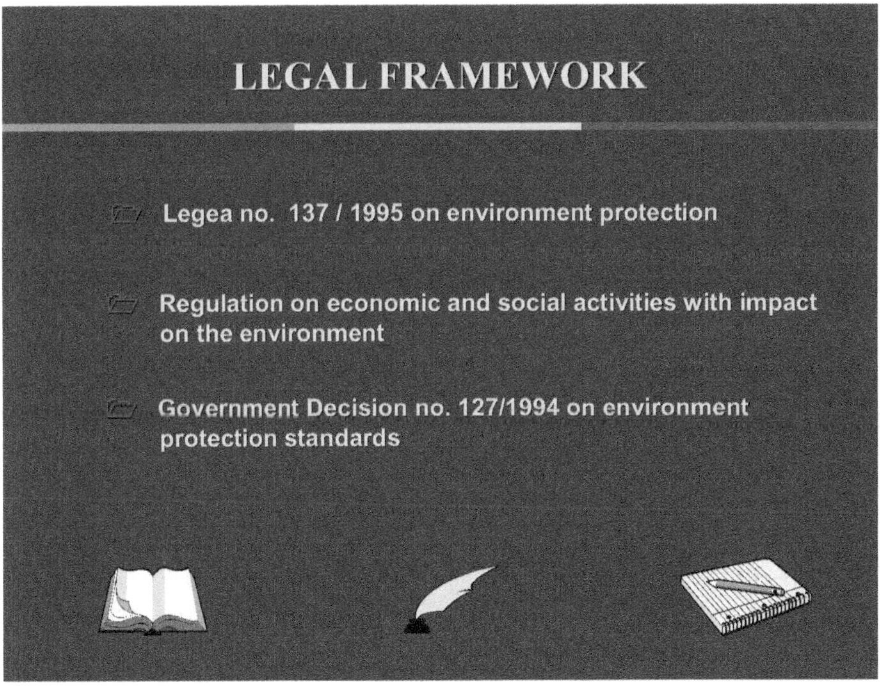
LEGAL FRAMEWORK
Legea no. 137 / 1995 on environment protection
Regulation on economic and social activities with impact on the environment
Government Decision no. 127/1994 on environment protection standards

LEGAL FRAMEWORK
(continuation)

- Government Decision no. 462/1993 on technical conditions for atmosphere protection
- Regulations on the earth contamination with various polluters
- Government Decision no. 138/1994 on sanctioning of the offenders in the field of water
- Standards and regulations on environment protection

TYPES OF POLLUTION

POLLUTERS

- waste resulted in the explosives manufacturing processes
- waste resulted in the processes of explosives incorporation into ammunition
- waste resulted during the industrial explosives use
- waste resulted in the explosives cleaning and removal processes

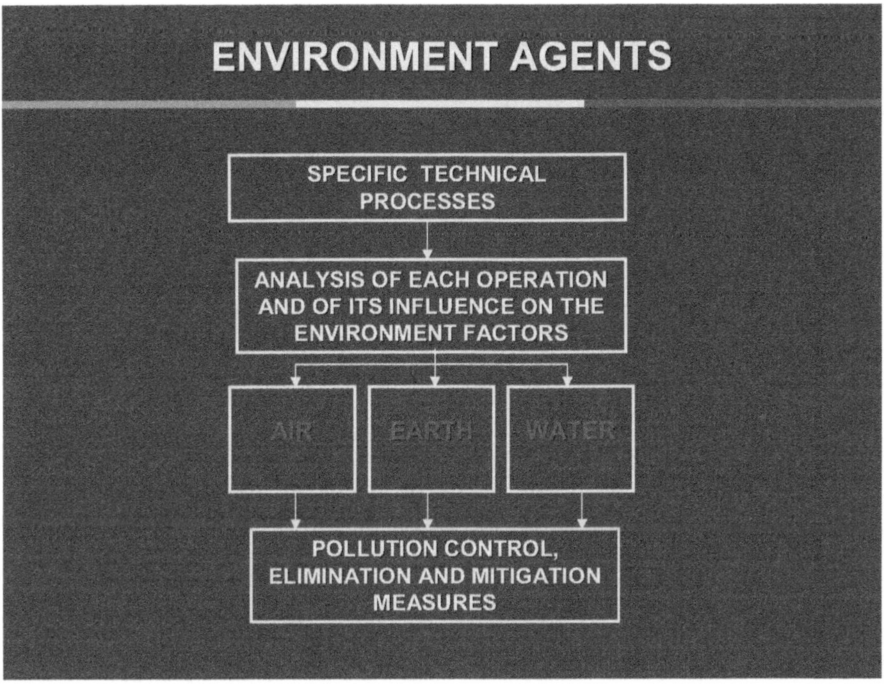
ENVIRONMENT AGENTS
SPECIFIC TECHNICAL PROCESSES
ANALYSIS OF EACH OPERATION AND OF ITS INFLUENCE ON THE ENVIRONMENT FACTORS
AIR
EARTH
WATER
POLLUTION CONTROL, ELIMINATION AND MITIGATION MEASURES

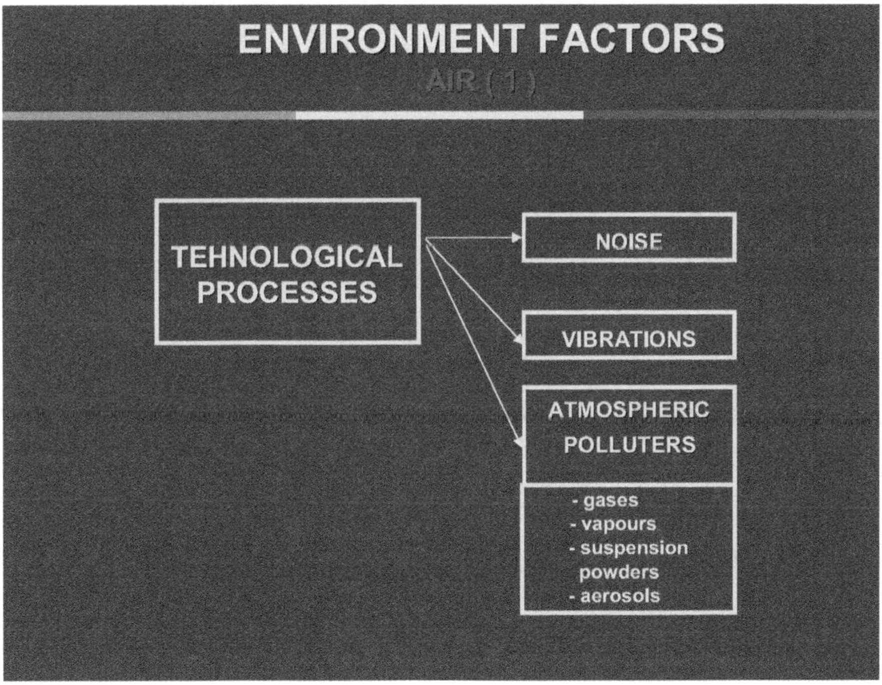
ENVIRONMENT FACTORS
AIR (1)
TEHNOLOGICAL PROCESSES
NOISE
VIBRATIONS
ATMOSPHERIC POLLUTERS
- gases
- vapours
- suspension powders
- aerosols

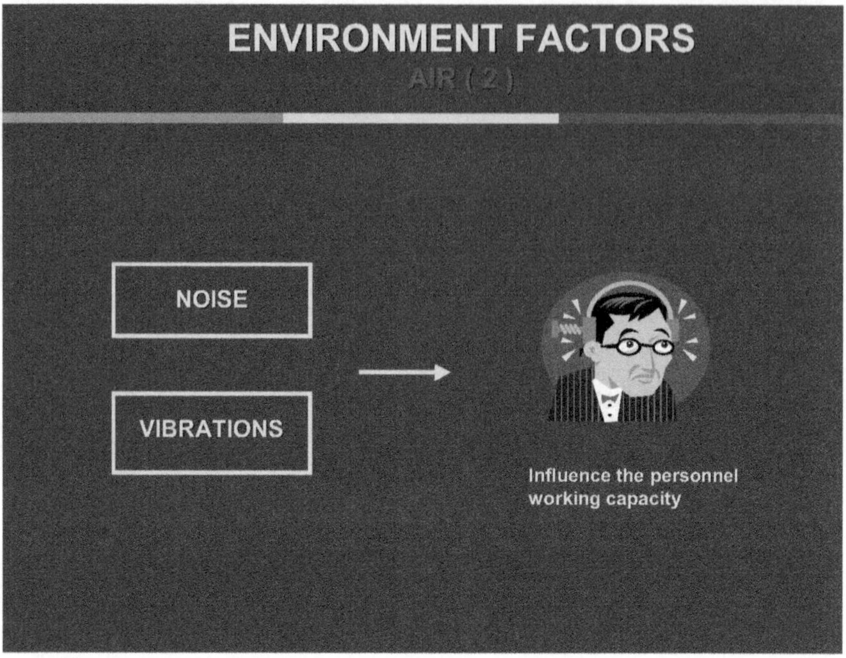
ENVIRONMENT FACTORS
AIR (2)
NOISE
VIBRATIONS
Influence the personnel working capacity

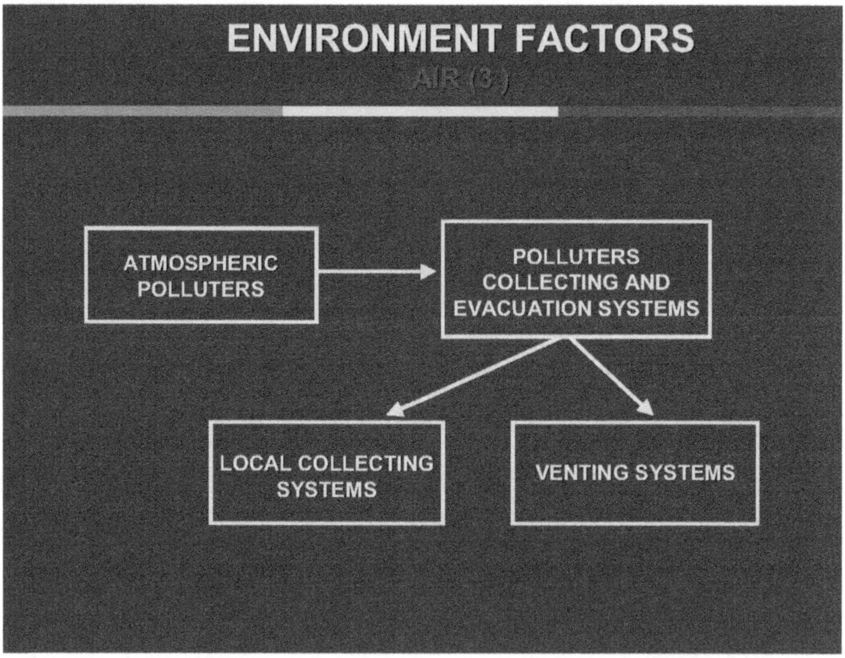
ENVIRONMENT FACTORS
AIR (3)
ATMOSPHERIC POLLUTERS
POLLUTERS COLLECTING AND EVACUATION SYSTEMS
LOCAL COLLECTING SYSTEMS
VENTING SYSTEMS

ENVIRONMENT FACTORS

AIR (4)

EMISSION SAMPLES GATHERING

- excesses
  - blasting powders
  - carbon oxide (metal builbings, thermal treatment processes)
  - solvent mixture emitted as a result of the painting processes
  - gases resulted in the thermal treatment processes
- according to regulations
  - sedimentary powders (0,62-1,82 g/m /lmonth)
  - CO from gases burned in plants (0,1%)
  - CO resulted in the burning of fuel oil (14%)

ENVIRONMENT FACTORS

EARTH (1)

POLLUTERS WHICH CAN MODIFY THE CHARACTERISTICS OF THE ANALYZED EARTH

- HEAVY METALS (Cr, Ni, Zn, Cd, Cu)
- AEROSOLS (acids, hydro-acids)
- ALKALINE CYANIDES

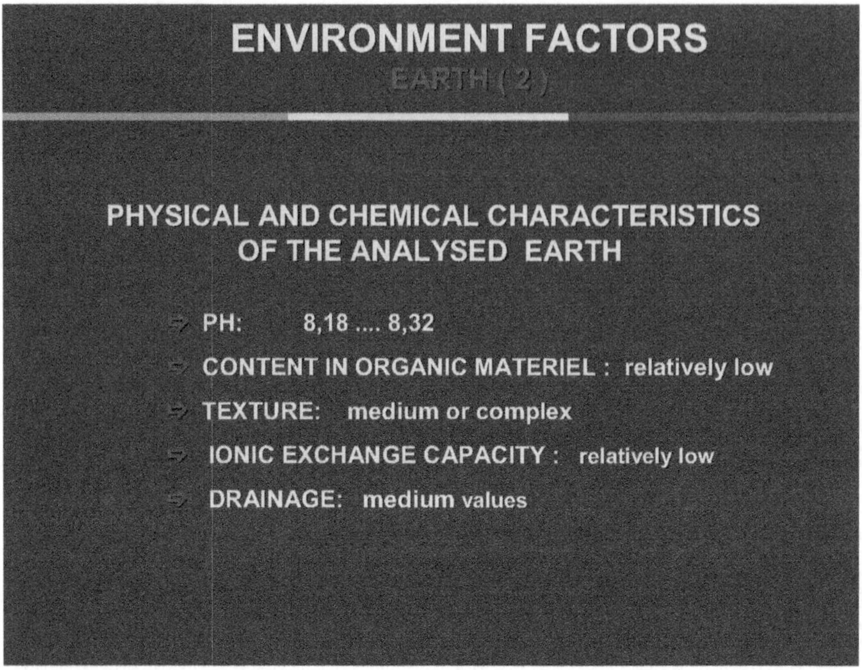
ENVIRONMENT FACTORS
EARTH (2)
PHYSICAL AND CHEMICAL CHARACTERISTICS OF THE ANALYSED EARTH
PH: 8,18 8,32
CONTENT IN ORGANIC MATERIEL : relatively low
TEXTURE: medium or complex
IONIC EXCHANGE CAPACITY : relatively low
DRAINAGE: medium values

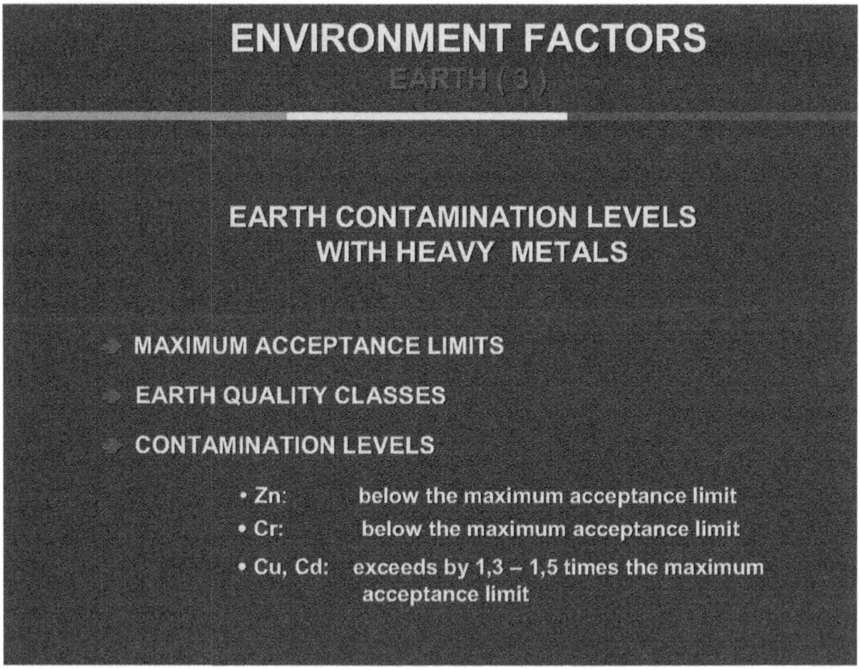
ENVIRONMENT FACTORS
EARTH (3)
EARTH CONTAMINATION LEVELS WITH HEAVY METALS
MAXIMUM ACCEPTANCE LIMITS
EARTH QUALITY CLASSES
CONTAMINATION LEVELS
• Zn: below the maximum acceptance limit
• Cr: below the maximum acceptance limit
• Cu, Cd: exceeds by 1,3 – 1,5 times the maximum acceptance limit

ENVIRONMENT FACTORS

WATER (1)

- TYPES OF INDUSTRIAL WATERS
  - FAECALOID-GENERAL-USE
  - PRE-PURIFIED
  - CONVENTIONALLY CLEAN AND PLUVIAL
- INDUSTRIAL WATERS PURIFICATION PHASES
  - CHEMICAL
  - MECHANICAL
  - MUD EQUIPMENT

ENVIRONMENT FACTORS

WATER (2)

CHEMICAL PURIFICATION

- ACIDS AND BASES NEUTRALIZATION
- Cr^{+6} REDUCTION FROM CHROMIC ACID
- CYANIDES OXIDATION
- HEAVY METAL IONS PRECIPITATION

ENVIRONMENT FACTORS

WATER (3)

- ALKALINE AND ACID NEUTRALIZATION
  - sol. 40 % NaOH
  - sol. 40 % H_2SO_4
- CHROMIC WATER NEUTRALIZATION
  - DIRECT PRECIPITATION
  - REDUCTION OF Cr^{+6} TO Cr^{+3} and
  - ITS PRECIPITATION

ENVIRONMENT FACTORS

WATER (4)

- CYANIDE WATERS NEUTRALIZATION
  - OXIDATION
  - TRANSFORMATION INTO CYANIDE ACID AND BURNING
  - TRANSFORMATION INTO LESS TOXIC COMPOUNDS
  - SEPARATION FROM IONIC EXCHANGERS
  - POLIMERIZATION WITH FORMALDEHYDE
  - BIOLOGICAL TREATMENT WITH ALGAE OR FUNGUS
  - IRRADIATION
- PRECIPITATION OF HEAVY METAL IONS
  - AFTER REDUCTION
  - PH CORRECTION

PRODUCTS RESULTED IN THE PURIFICATION PROCESS

MUDS (1)

DEFINITION

- *physical* - Complex colloidal systems with heterogeneous composition,
  consisting in:
  - ⇒ **colloidal particles (d < 1 μm)**
  - ⇒ **dispersed particles (d: 1 μ m-100 μ m)**
  - ⇒ **aggregates**
  - ⇒ **gelatinous suspension, containing very much water**
- *technological* - Final water purification phase, incorporating:
  - ⇒ **products resulted in metabolic activities**
  - ⇒ **raw material**
  - ⇒ **intermediary products**
  - ⇒ **industrial final products**

PRODUCTS RESULTED IN THE PURIFICATION PROCESS

MUDS (2)

PHYSICAL AND CHEMICAL CHARACTERISTICS

- GENERAL INDICES
  - **humidity**
  - **specific weight**
  - **pH**
  - **mineral-volatile ratio**
  - **specific filtering resistance**
  - **compressibility**
  - **heavy metal content**
  - **caloric power**
- SPECIFIC INDICES
  - **fertilizers**
  - **detergents**
  - **metals**
  - **oils**
  - **greases**

PRODUCTS RESULTED IN THE PURIFICATION PROCESS
MUDS (3)

PROCESSING METHODS

- THICKENING
- FERMENTATION
- CONDITIONING
- DEHYDRATATION AND/OR DRYING
- INCINERATION

PRODUCTS RESULTED IN THE PURIFICATION PROCESS
MUDS (4)

USE

- TEHNOLOGY — reuse of useful products (metals) in industrial manufacturing processes
- ENERGY — secondary and regenerable energy resourses
- VEGETAL AND ANIMAL PRODUCTION — fertilizers, fodder etc.

PRODUCTS RESULTED IN THE PURIFICATION PROCESS

MUDS (5)

PILOT STATION

- PURPOSE

Processing residual waters and precipitates containing mercury and ethanol in order to recover and reuse them in the manufacturing process

- PRINCIPLE

Elimination of mercury fulminate in residual waters by boiling these with acetylenic sludge

PRODUCTS RESULTED IN THE PURIFICATION PROCESS

MUDS (6)

ELIMINATION OF MERCURY FULMINATE

- $$CaC_2 + 2H_2O \longrightarrow \underbrace{C_2H_2 + Ca(OH)_2 + C}_{\text{acetylenic sludge}}$$

- $$Ca\,(OH)_2 + Hg(C{\equiv}HO)_2 \xrightarrow[\text{boiling}]{} HgO\downarrow + CaH\text{-}CH + CO_2 + H_2O$$

- $$CaH\text{-}CH + 3H_2O \xrightarrow[\text{pH>12}]{} CaCO_3\downarrow + N_2H_3$$

PRODUCTS RESULTED IN THE PURIFICATION PROCESS

MUDS (7)

MERCURY RECOVERY

- PRINCIPLE

Dissolution of the precipitates containing mercury (HgS, HgO) and/or metallic mercury in sodium sulphide, followed by the electrolysis of the diluted polysulphide solutions

- APPLICATION

• Mercury sulphide storages
• Solutions containing fulminate

- ENVIRONMENT PROTECTION

Residual waters are free of mercury ions

CONCLUSIONS

- Development of a permanent control program in co-operation with the Environment Protection Agency
- Development or adequate equipment of the control laboratories specialized in environmental agents
- Setting up an integrated guidance concerning the identification of all categories of waste resulted in the manufacturing process and their monitoring, storage, neutralization and reuse

CONCLUSIONS
(continuation)

- Technological processes optimization (welding, blasting, solvent-free painting)

- Acquisition of environmental protection specific equipment

    - sand-blasting powders retaintion equipment
    - solvenț retaintion and elimination equipment (acetates, toluene ...)
    - gaseous flow treatment equipment
    - water pre-purification stations modernized with automated equipment for chemical reagents dosing

- Permanent control of sewerage systems and their maintaining in service

ENVIRONMENTAL IMPACT OF MUNITIONS ON MILITARY PROVING GROUNDS

Drs N.H.A. van Ham, Dr A.C. van der Steen

TNO Prins Maurits Laboratory
P.O.Box 45 2280 AA Rijswijk
Netherlands

This presentation is based on the development of a new methodology for the assessment of environmental risks on large areas like military proving grounds, due to the long-term use of munitions.
Based on historical investigations a sampling plan was made. The area was hereto divided in three classes:

- non polluted,
- slightly polluted
- heavily polluted

The results obtained so far indicate that the methodology leads to acceptable results.

P. Castelo Branco et al. (eds.),
Defense Industries: Science and Technology Related to Security, 113–122.

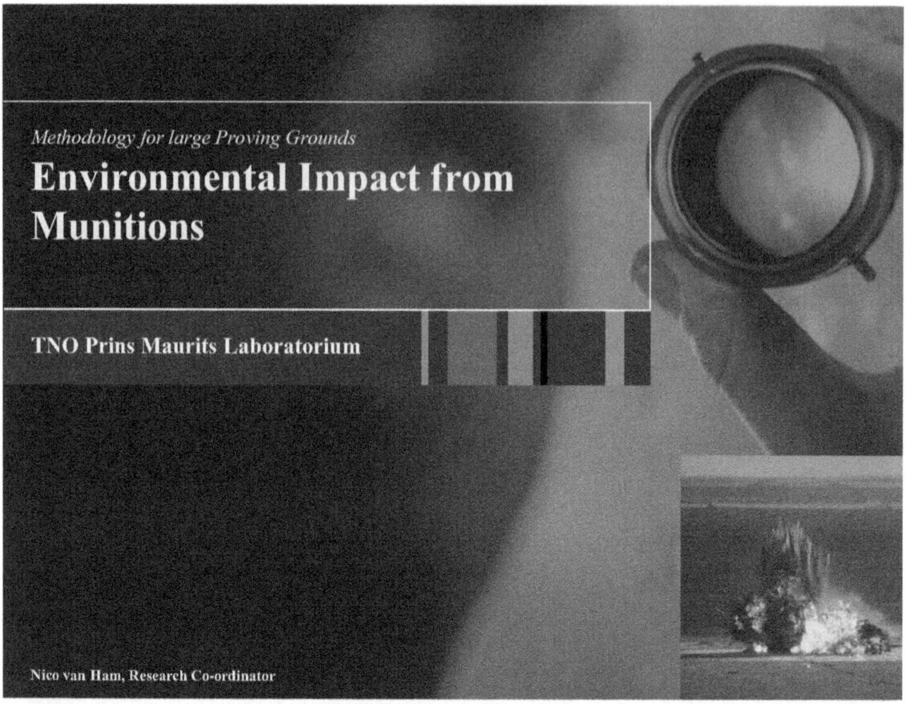
Methodology for large Proving Grounds
Environmental Impact from Munitions
TNO Prins Maurits Laboratorium
Nico van Ham, Research Co-ordinator

t
Environmental Impact from Munitions
02 May 2001
2

t
Environmental Impact from Munitions
02 May 2001
3

t
Environmental Impact from Munitions
02 May 2001
4

t Environmental Impact from Munitions 02 May 2001 5

Methodology large areas

- Historical Investigation
- Classification of Area
- Measuring Plan
- Sampling at Surface
- Sampling below Surface
- Sampling of Ground Water
- Ecological Damage

t Environmental Impact from Munitions 02 May 2001 6

Historical Investigation

- Munitions used
- Numbers and types
- Position of Weapons
- Location of Impact

t Environmental Impact from Munitions 02 May 2001 7

Components from Munitions

- Unreacted Energetic Materials
- Reacted Energetic Materials
- Heavy Metals

t Environmental Impact from Munitions 02 May 2001 8

Weight of Metals In 1000 kg Munition

Component	Weight kg
Steel	617
Lead	8.4
Copper	0.5
Aluminium	25

t Environmental Impact from Munitions 02 May 2001 9

Reaction Products Energetic Materials

- CO
- VOC
- PAK
- Dioxine

t Environmental Impact from Munitions 02 May 2001 10

Classification of Area

- Non polluted
- Slightly polluted
- Heavily polluted

Measuring Plan

- 4-8 samples from non polluted areas
- 4-30 samples from slightly polluted areas
- 4-30 samples from heavily polluted areas
- 4 reference samples from identical soil outside the military proving ground

t
Environmental Impact from Munitions
02 May 2001
13

t
Environmental Impact from Munitions
02 May 2001
14

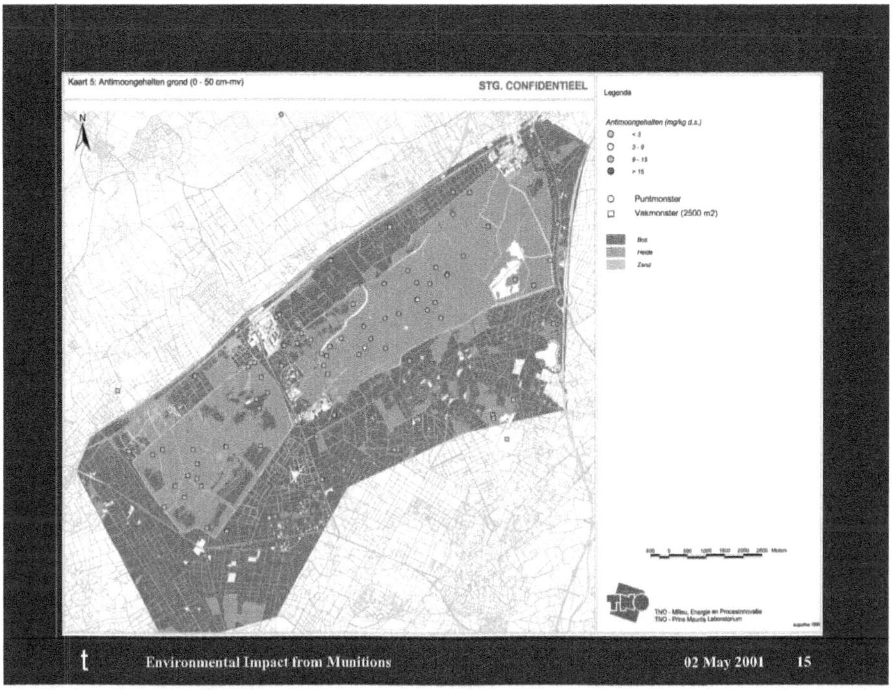
Kaart 5: Antimoongehalten grond (0 - 50 cm-mv)
STG. CONFIDENTIEEL
Legenda
Antimoongehalten (mg/kg d.s.)
< 3
3 - 9
9 - 15
> 15
Puntmonster
Vakmonster (2500 m2)
Bos
Heide
Zand
TNO - Milieu, Energie en Procesinnovatie
TNO - Prins Maurits Laboratorium
Environmental Impact from Munitions
02 May 2001
15

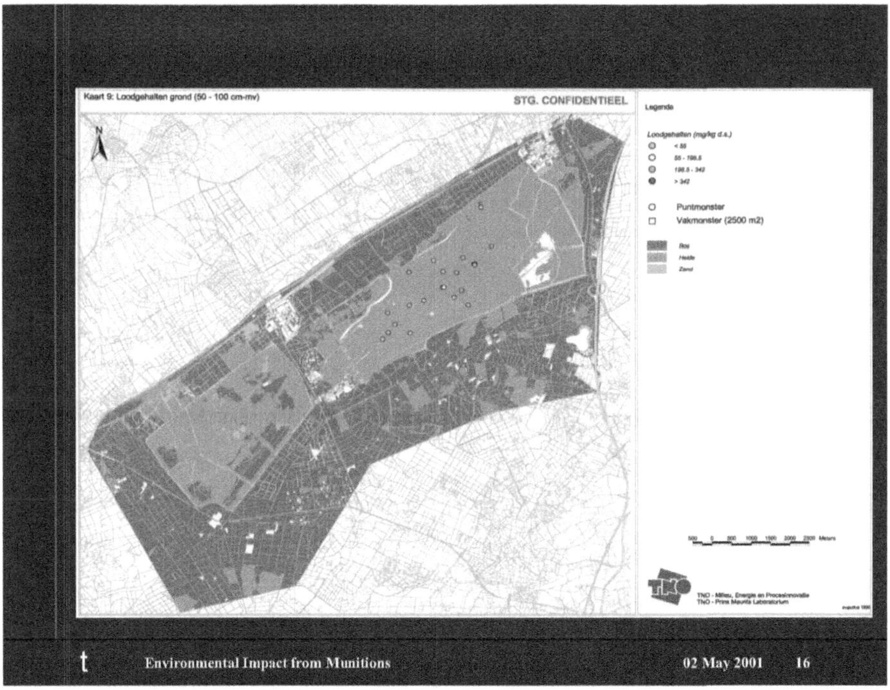
Kaart 9: Loodgehalten grond (50 - 100 cm-mv)
STG. CONFIDENTIEEL
Legenda
Loodgehalten (mg/kg d.s.)
< 55
55 - 198.5
198.5 - 342
> 342
Puntmonster
Vakmonster (2500 m2)
Bos
Heide
Zand
TNO - Milieu, Energie en Procesinnovatie
TNO - Prins Maurits Laboratorium
Environmental Impact from Munitions
02 May 2001
16

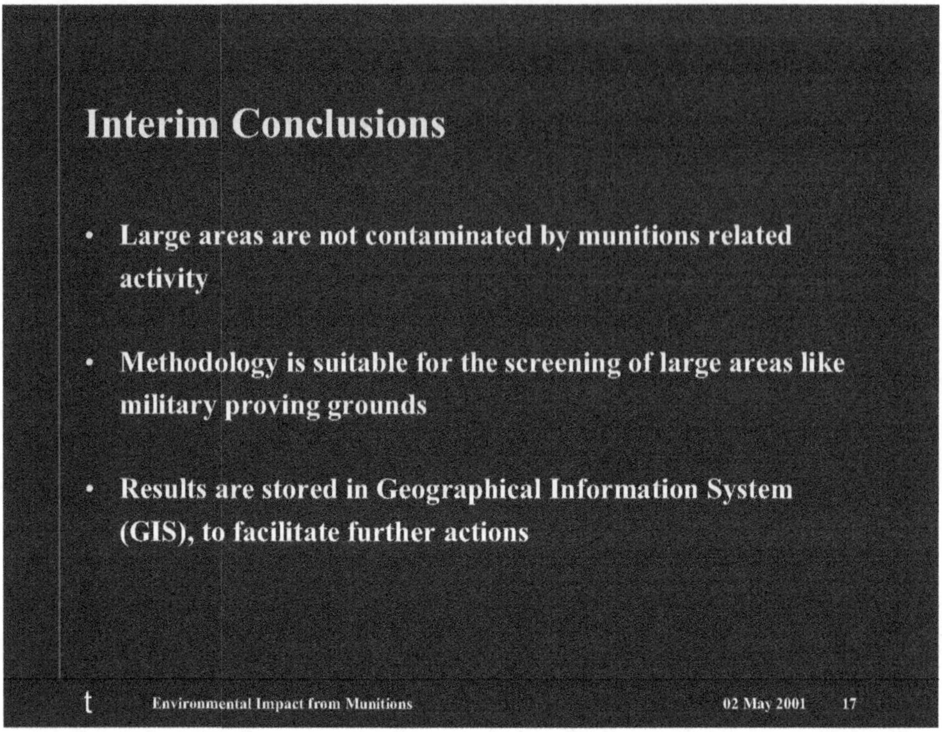
Interim Conclusions
• Large areas are not contaminated by munitions related activity
• Methodology is suitable for the screening of large areas like military proving grounds
• Results are stored in Geographical Information System (GIS), to facilitate further actions
t
Environmental Impact from Munitions
02 May 2001
17

Safety Effects caused by Insensitive Munitions

Richard Wild

Diehl Munitionssysteme GmbH & Co. KG
Werk Maasberg

Abstract

Military history knows many catastrophic losses caused by munition.
Worldwide efforts are made to reduce the vulnerability of munitions and to develop „Insensitive Munitions“ (IM).
IM fulfill their performance requirements, while minimizing the violence of their response to unplanned stimuli, such as heat, shock and impact.
IM objectives, tests and guidelines are established in the MIL-STD 2105 and in the STANAG 4439.

IM exceeds the actual safety levels and offers more safety

- in production	(example: initiation threshold),
- in storage	(example: Sympathetic Reaction),
- in transportation	(example: Liquid Fuel Fire) and
- in use	(example: Bullet Impact).

The safety of munition rapidly increased the last 10 years.
While traditional munition in a fire causes mass detonation, Insensitive Munitions even stimulated by a detonation will only burn:
IM saves lifes and materials.

Introduction

Military history knows many catastrophic losses caused by munition.
Serious accidents with munitions on weapon platforms, in stores, or on transport carriers made evident the vulnerability of conventional munition and caused global efforts to reduce the vulnerability, developing so called „Insensitive Munitions“.

In most cases not the munition itself is the reason for the accidents but fires, which spread to the munition and lead to mass detonations.

The safety policy of the US-Navy, derived from accidents like these on ships, found worldwide interest and in the meantime an IM policy is also formulated in STANAG 4439.

What means Insensitive Munitions (IM) ?

Definitions of IM are included in MIL-STD 2105 and STANAG 4439:

P. Castelo Branco et al. (eds.),
Defense Industries: Science and Technology Related to Security, 123–126.

Insensitive Munitions reliably fulfill their performance, readiness and operational requirements on demand, and minimise the probability of inadvertent initiation and severity of subsequent collateral damage to weapon platforms, logistic systems and personnel when subjected to unplanned stimuli.

In short: IM fulfill the performance requirements at considerably lower vulnerability.

When is Munition insensitive ?

STANAG 4439 lists 6 different threats (stimuli), munition will face during its life cycle:

-Open fire and fire in an adjacent room
-Impact by bullet and fragment attack
-Shaped charge jet attack and
-Sympathetic Detonation.

For every threat a munition test procedure is specified.
Munition is considered as insensitive, when

- the response of fire and impact stimuli is a type V (burning) reaction, (STANAG: for Slow Heating type IV is allowed)

- the response of Sympathetic Reaction is not type I, (STANAG: is not type I and type II),

- the response of a Shaped Charge Jet Impact: no type I reaction (only specified in the MIL-STD).

Insensitive Munitions Technology

IM requirements usually cannot be fulfilled with yesterdays welltried methods, such as traditional TNT or wax bonded explosives.

Basis for IM nowadays are less sensitive PBXs and since some years we use a new generation of pressable PBXs developed by Diehl-MS.
To take advantage of the whole insensitivity potential of these explosives, the manufacturing processes are adjusted to the lower sensitivity.
In case of need the insensitivity of the munition is then guaranteed by a specific IM design of the warhead and the complete munition, together with the packing.

Only the add up of all these steps results in an „Insensitive Munition".

Safety effects caused by Insensitive Munitions

All the IM tests, specified in the standards, simulate threats which are most likely to occur during military operations.

Slow and Fast Heating, as well as Sympathetic Detonation can be considered equally applicable for production, storage and transportation of ammunition during peacetime.

Insensitive Munitions offer more safety in production

The higher safety levels of Insensitive Munitions with PBX affect already the production of warheads.
In the different steps of the processes, explosive powders and unconfined pressed or casted charges have to be handled. Here e.g. improved impact and friction sensitivities, as well as higher initiation thresholds help to achieve better process safety.

Moulding powders which were for sale 10 years ago, with 95% HMX and wax or plastic binders had an impact sensitivity of 10J or less. Better coated materials, used today have impact sensitivities of 30J and higher.

The following example of the shockwave sensitivity of pressed PBX-charges with 96% HMX, shows the rapid ascent in the development of less shock-sensitive energetic materials:
more than 5 years ago: 20 kbar initiation pressure, 3 years ago: 28 kbar initiation pressure, last year: 40 kbar initiation pressure.

Insensitive Munitions offer more safety in storage

The actual advantages Insensitive Munitions have, of course are more safety in storage and transportation.
The Sympathetic Reaction test shows the response of a detonating munition on adjacent warheads as they are packed up in a store or vehicle.

For IM the Sympathetic Reaction test has to make sure, that a warhead detonation will not propagate this detonation to the acceptor warheads.
Test setup of the example:
The donor of a cure cast filled missile warhead is ignited with its own initiation train.
Lilac acceptor without any barrier to the donor.
Yellow acceptor, separated by the original storage packaging (PU-foam between GF-resin).
The witness plates, the warheads were standing on, and the WH parts found after he test, show a full detonation of the donor, a burning reaction of the yellow acceptor and no reaction traces of the lilac warhead.

So for this weapon a detonation of one warhead in a store will not lead to a mass detonation of the stored munition.

Insensitive Munitions offer more safety in transportation

The Liquid Fuel Fire test simulates the fast heating conditions of an open fire engulfing the munition in a flame envelope. For IM no reaction more severe than typeV shall occur.

In the Fuel Fire test of a highly confined warhead (more than 15 mm wall thickness of the steel case), filled with a pressed PBX (92% RDX), the explosive had burnt out, while the metal case is only damaged in the thread.

So for this weapon a fire on a munition truck will not cause reactions more severe than burning of the PBX filling of the transported warheads.

Insensitive Munitions offer more safety in use

Bullet Impact is one of the most probable threats, that may happen during military operations. For IM no reaction more severe than typeV shall occur.

For a cure cast filled missile warhead the response of the impact of a 12,7 mm AP projectile is a hole in the case, where the PBX slowly burnt out .

So for this weapon the impact of an armor-piercing projectile will only lead to a burning of the explosive filling of the warhead.

Safety improvements by Insensitive Munition

While conventional munition in a fire leads to more and more violent reactions up to mass detonation, fire and impact of IM will cause only burning of the munition.
Even a detonation with IM leads from a type III response, or in my example from the detonation direct to burning.

From every possible reaction level Insensitive Munition always leads to a burning reaction.

Conclusions

Accidents with conventional munition in the past caused many catastrophic losses.
The safety of munition has rapidly increased. Traditional explosives like TNT have been almost completely replaced by less sensitive plastic bonded explosives.
IM is less vulnerable to fire, impact or detonation propagation and guarantees more safety in production, storage, transport and use of munition.
IM hightens the survivability of combat platforms and fighting personnel.

Exceeding actual safety levels, IM saves lifes and materials.

Insensitive Munitions and Green Energetics – Ways of Meeting the Need

Dr Adam S Cumming

A requirement has developed for new energetic materials which can provide the necessary performance while meeting the insensitivity and environmental requirements now in force. This poses a significant challenge to the munition designer; one that cannot be easily met using existing technology. New materials and new ways of working can provide an answer to the need, and research in the UK has matched that in progress elsewhere in addressing the issues, though from a UK perspective.

This paper will describe the issues from a UK perspective and illustrate how the available technology has been harnessed to begin to meet the developing needs. It has required a systems approach using technology such as steel strip laminate motors for rockets; combustible cases; as well as new compositions, both using inert rubbery binders and exploring the strengths and limitations of energetic binders. New solids are also being explored; as are clean manufacturing and disposal techniques, and these are described. While there has been some progress, much remains to be done, and priorities need to be assigned. These are discussed.

P. Castelo Branco et al. (eds.),
Defense Industries: Science and Technology Related to Security, 127–140.

Insensitive Munitions and Green Energetics -Ways of Meeting the Need

Dr Adam S Cumming

Outline

- Introduction to UK approach
- Background
- IM Studies
- Environmental Studies
- Future Requirements and Strategy
- Conclusions

Introduction

- The UK has developed an IM policy and is in the process of implementing it
- Environmental issues are becoming more critical
  - Clean-up of sites
  - Demilitarisation
  - Design for Disposal
  - Whole Life Cost issues
- Performance still a major driver
  - What is meant by performance?

Background

- All these issues have formed part of the UK research programme for several years.
- Three main strands:
  - Understanding and Quantitation of Hazard
  - Synthesis of existing materials by clean methods
  - Application of new types of materials to give reduced environmental impact and reduced vulnerability
- All are linked and rely on an integrated approach
- Little work till recently on Disposal

[dstl] 10 September 2004 / 01 May 2004 © Dstl 2001
Dstl is part of the Ministry of Defence

Summary of UK IM Policy

The vulnerability of the munitions in the MOD inventory will be reduced over time to meet the requirements of STANAG 4439 (Policy for Development and Assessment of Insensitive Munitions).

All new munitions requirements are to stipulate compliance with the criteria for IM set out in STANAG 4439.

All legacy munitions are to be kept under review to identify opportunities to achieve IM compliance at MLU, refurbishment or re-provisioning.

In both instances formal dispensations at Two Star level is required for any non-compliance, either in the requirement definition or in the procurement solution.

Routes To IM Systems

- Correct mechanical properties - correct choice of rubbery binder
- Minimisation of solids loading
- Use of less sensitive fillers:
  - NTO, TATB, FOX-7
  - Finer grades of nitramines (avoid large crystal defects giving rise to hotspots > 1 μm)
- Use of energetic binders
- This approach has provided a range of IM explosives, typical of which is CPX413 (HMX 35%, NTO 45 %, 20 % polyNIMMO/K10 PU Binder)
- Careful overall system design

Explosive Performance vs Safety

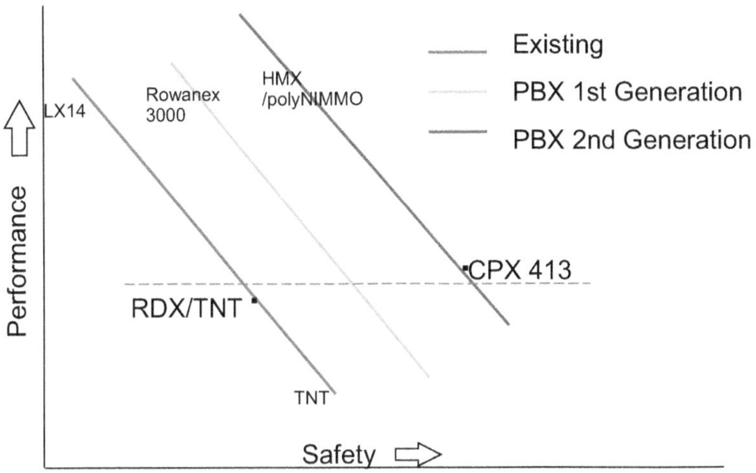

Approaches to IM

- IM requires ***Quantified Risk!***
  - STANAG 4439 and UK policy demands a Threat Hazard Analysis
  - In reality the designer must be able to predict the response before the event
  - A detailed understanding of explosive behaviour is required if prediction is to be possible
  - Modelling assists in defining capability gaps
  - HOWEVER, there is a tendency to study what is easy rather than what is important!

Approaches to IM Research

- UK programme designed to provide predictive tools
  - Sufficiently detailed to meet the need, but adaptable enough to meet future needs - not easy!
  - Results need to be expressed as system benefits, and with cost savings quantified.
  - Expressed in terms understood by Users
  - Needs to be considered at initial design stage
  - A major part of Whole Life Costing

[dstl] 10 September 2004 01 May 2001 ©Dstl 2001

Dstl is part of the Ministry of Defence

UK Approach

- The UK has used a combination of
  - Modelling and small scale tests
  - New materials and processes
  - Large scale tests
  - Collaboration with Allies on materials and procedures
    - Research Collaboration
    - NATO and other groups on procedures and acceptable parameters
- Some examples:

[dstl] 10 September 2004 01 May 2001 ©Dstl 2001

Dstl is part of the Ministry of Defence

IHEs - Effect on weapon and platform vulnerability

Bullet Attack on a simulated warhead
- 3 rounds, .5" Armour Piercing Browning MG

RDX/TNT - Explosion

Yes

Chance of Sympathetic Reaction ?

Platform destruction

CPX413 - minimal response

No

Platform Survival

Sympathetic Detonation Test - CPX413

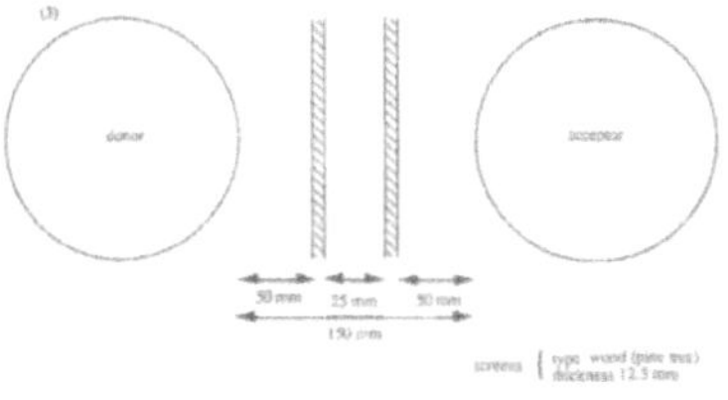

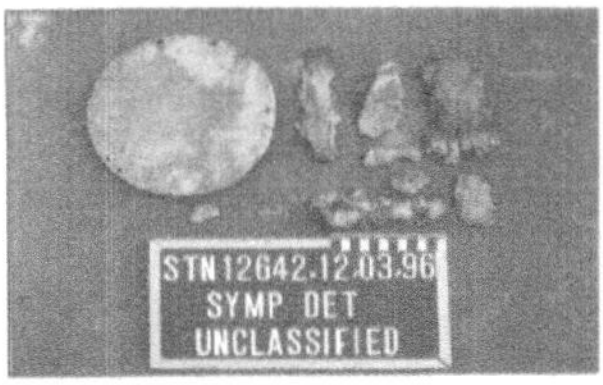

dstl 01 May 2004
© Dstl 2001
Dstl is part of the Ministry of Defence

Energetic Binders

- PolyNIMMO and PolyGlyn have been successfully demonstrated in Explosives, Rocket Propellants and Gun Propellants.

- High energy HMX/polyNIMMO LOVAs
  Show low vulnerability to:-

Bullet Attack
Fragment Attack
Shock (ELSGT)
Thermal Stimuli
Small Scale Shaped Charge Attack

dstl 10 September 2004 01 May 2004 © Dstl 2001

Dstl is part of the Ministry of Defence

Next Generation Rocket Propellants -Ingredients

Fillers	Binders	Plasticisers	Additives
CL20	polyNIMMO	10+	20+
HNF	polyGLYN		
ADN	GAP		
AN	Energetic TPEs		
	Others		

dstl 10 September 2004 01 May 2004 © Dstl 2001

Dstl is part of the Ministry of Defence

Modelling Formulation Interactions

- Modelling of binder / filler interactions to investigate bonding agents to further improve vulnerability.

CL20 Electrostatic Surface

- Plasticiser/ binder interactions helping synthetic chemists purpose design new plasticisers.

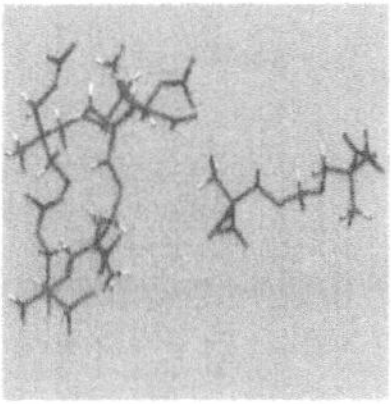

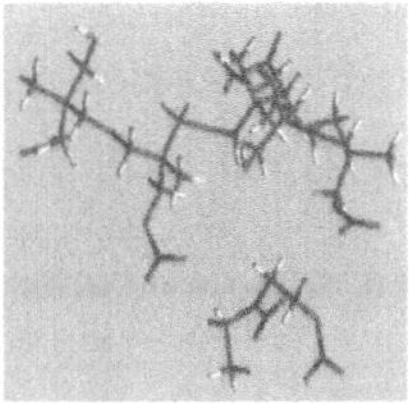

- Predict Tg and viscosity trends

Performance and the Environment

- High quality reinforces traditional performance parameters - example RF 42-08 - now an industrial product - Rowanex 3000.
  - HMX/TPE 95/5 Pressable
  - Lower vulnerability alternative to LX14
  - solvent free processing
  - meltable TPE binder
    - ease of processing
    - ease of recycling
- Synthetic Chemistry team looking at Energetic TPEs for higher performance / lower vulnerability and environmental impact.

Rowanex 3000

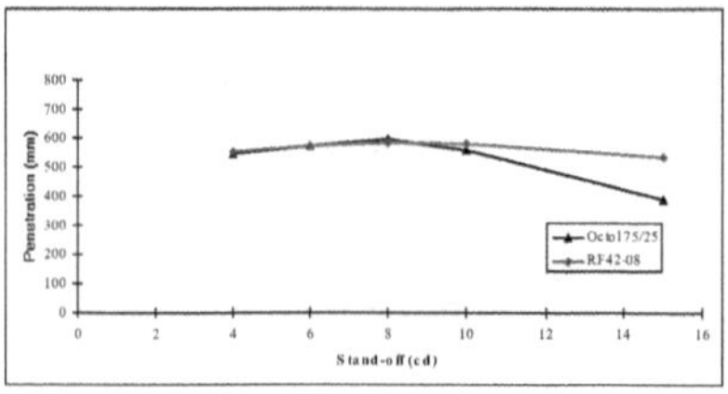

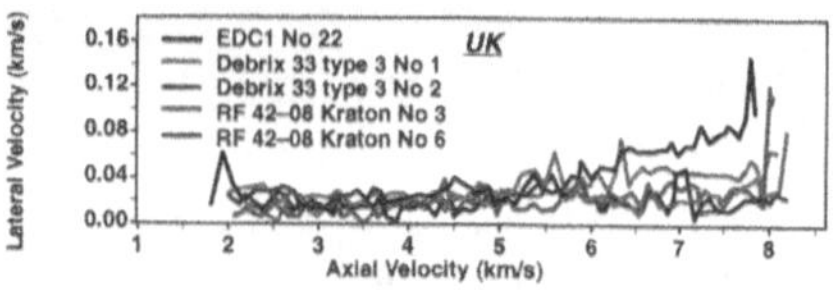

- Higher penetration performance than Octol 75/25 at long stand-offs
- High pressing quality
  - high jet tip velocities
  - low lateral drift velocities
  - good jet particulation

Environmental Studies

- Increasing formal awareness in UK
- HOWEVER
  - Environmental impact properties have been considered in the UK programme for several years
  - Synthesis studies include Dinitrogen Pentoxide applications
  - Formulation studies included reduced impact
  - ‘Design for Disposal’ an advisory requirement
- BUT
  - all the above were secondary to performance

Disposal

- UK has used
  - Sea Dumping - banned by Treaty, but items return!
  - Open Burning/Open Demolition - still in use but increasingly limited, and likely to be more limited
  - Incineration at dedicated incinerator
- Design for Disposal means
  - Preparing for end of life
  - Requires assessment of risks and knowledge of life in service
  - Use of materials and processes appropriate for disposal

Disposal and Contamination

- Proper assessment of risk requires:
  - Knowledge of fate of residues
  - Estimation of what is acceptable contamination
  - Cost effective methods for disposal and clean up
  - Possible methods for recycling and reuse
  - Continual monitoring of technology and risks
  - A model to predict disposal and impact costs.

Environmental Research

- Clean Synthesis
- Formulation and Processing studies
- Assessment of Impact reduction
- Disposability
- Clean up and remediation

Clean Synthesis

- Environmentally Friendly Nitration
  - New, environmentally acceptable methods are needed for the production of energetic materials to ensure continuity of supply as a result of application of the Montreal Protocol
  - Clean synthetic methods will have been assessed for a number of known energetic materials, and advice and recommendations for future production will be available.

Clean Manufacture

- Energetic Thermoplastic Elastomers
  - This project is intended to synthesise and characterise ETPEs, and to provide binders for assessment in formulation studies, including formulations products.
  - These materials not only offer improved performance but also provide easy, environmentally acceptable routes to disposal.

Selected molecules for ETPEs

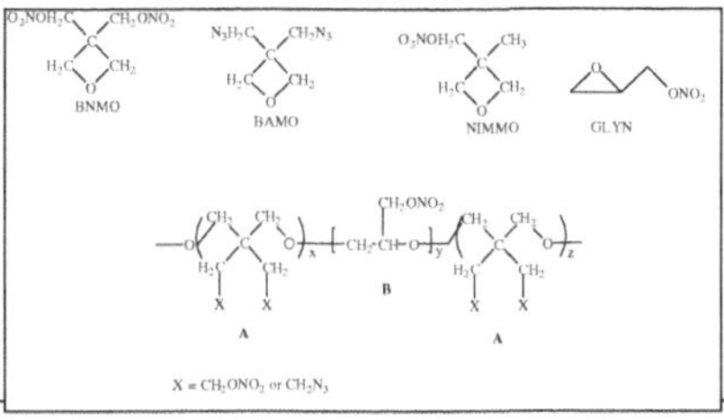

Processing method for ETPE

[dstl] 01 May 2004
' Dstl 2001

Dstlis part of the Ministry of Defence

Environmental Remediation

The work is developing a series of modular bio/chemical and microbiological methods for the remediation of pollution arising from the explosives waste. The aim is to develop a system capable of taking explosives contaminated waste water from a munitions break down facility and process the material so that it is fully compliant with disposal through a conventional sewerage plant.

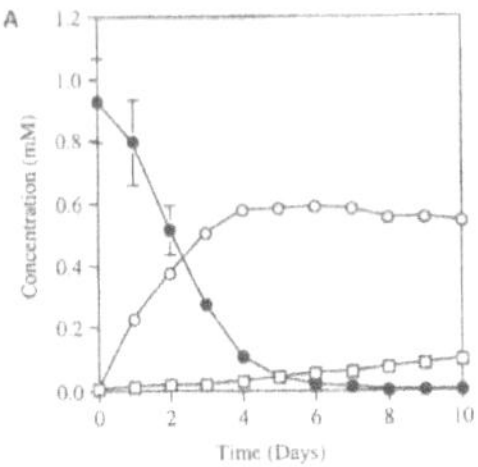

Degradation of GTN by wild-type tobacco seedlings.

Wild type seeds (100 per flask) were surface-sterilised, germinated and grown for 7 days. Seedlings were incubated with 1mM GTN in sterile water. Concentrations of GTN (l) and GDN (m) and GMN(o) were determined by HPLC. Results of measurements shown are the mean and standard deviation of measurements on three independent flasks.

[dstl] 10 September 2004 01 May 2004
©Dstl 2001

Dstl is part of the Ministry of Defence

Environmental Impact

- Rocket Plumes & Nozzles
  - Designed to develop the existing tools to give a means of prediction of plume effects, and to contribute to the design of rocket motor nozzles for greater effectiveness.
  - The studies include examination of novel types of propellant as well as those either in or entering service.

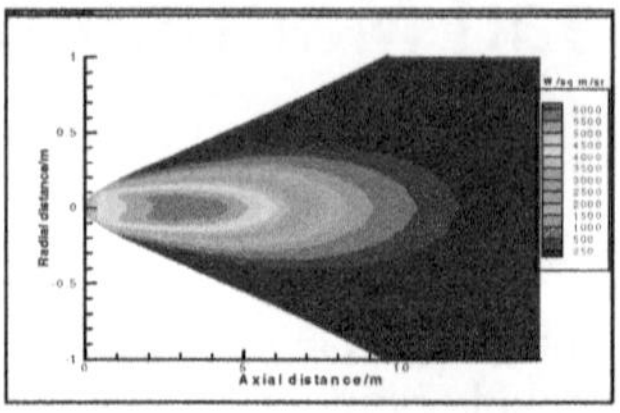

Service Double Base Propellant

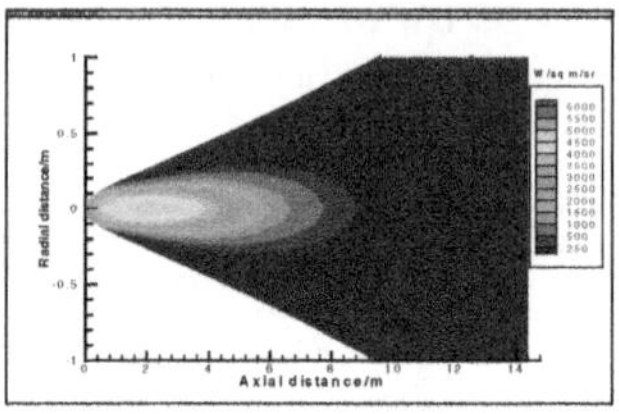

CL20/AN/PolyGlyn - next generation

Conclusions

- The UK has attempted to meet apparently conflicting needs inventively
- There is no need to sacrifice any of performance, vulnerability or environmental impact to satisfy the others
- Does require investment to enable this to happen
  - modelling; experiment; validation are all needed
  - clear guidelines on what is needed/acceptable are also required
  - policy direction required
- A challenge for the future?

Green Energetics at The Naval Surface Warfare Center, Indian-Head Division

Dr. Alfred G. Stern*; Dr. James A. Kenar; Dr. Nirupam J. Trivedi; Dr. William M. Koppes; Mr. Robert E. Farncomb; and Ms. Sibyl Turner
Naval Surface Warfare Center Indian Head Division, Indian Head, MD,

Mr. David C. Bomberger, Mr. Paul Penwell
SRI International, Menlo Park, CA

Mr. G. E. Manser, Mr. S. E. Spas, Dr. B. D. Nahlovsky
GenCorp Aerojet, Sacramento, CA

Advanced Research Workshop of the Scientific Affairs Division of NATO "Science and Technology Related to Security Impact of Conventional Munitions on the Environment and Population" Porto, Portugal, 30 October 2001

Abstract

Biocatalytic synthesis of energetic ingredients and nitration and polymerization reactions in supercritical fluid CO_2 are new technologies that reduce the amount of toxic materials treated, disposed and otherwise released into the environment in energetic materials manufacturing. These new technologies were studied under the SERDP "Green Energetics" program. This report describes the results of peroxidase calalyzed synthesis of energetics, and a series of studies using supercritical fluid as a reaction media in the synthesis of energetic materials.

Biocatalytic processes that rely on naturally occurring enzymes and microbes to effect chemical transformations have become important alternatives to traditional chemical syntheses. Biocatalysis in organic synthesis is important from an environmental perspective since these reactions typically are run under mild reaction conditions, produce high yields and cleaner products, avoid the use of heavy metal catalysts and produce little or no hazardous wastes. Herein, is described an investigation into the peroxidase-catalyzed oxidative coupling of potassium propane-2-nitronate to prepare 2,3-dimethyl-2,3-dinitrobutane, a volatile detection agent required as an additive for marking all plastic explosives.

The use of liquid and supercritical carbon dioxide ($scCO_2$) as a substitute solvent for chemical synthesis and processing is an environmental benign alternative to organic solvents. The use of $scCO_2$ rather than organic solvents offers a number of significant advantages in processing, carbon dioxide has a low impact on the environment, and supercritical fluid processing has good health and safety benefits.

Solubility, homopolymerization and block copolymerization studies were performed on 2-azidomethyl, 2'-methyloxetane (AMMO) and 2,2' bis-azidomethyloxetane (BAMO) in a

P. Castelo Branco et al. (eds.),
Defense Industries: Science and Technology Related to Security, 141–207.

reactor with super critical-CO_2, and the results of the polymerization of cyclic ethers (oxetanes) using supercritical carbon dioxide as a replacement solvent for methylene chloride are reported.

In addition, the nitration reactions of various alcohols were carried out in liquid carbon dioxide using nitrogen pentoxide as nitrating reagent. The nitration of hydroxyl groups of cellulose gave cellulose nitrate. The nitration of cyclodextrin gave the cyclodextrin nitrate in which ~ 22 out of 24 hydroxyl groups were nitrated. The nitration of cyclic alcohols, such as, glycidol and 3-hydroxymethyl-3-methyloxetane (HMMO) also gave the corresponding glycidol nitrate and 3-nitratomethyl-3-methyloxetane (NMMO) in 46 and 67 % yields, respectively. The nitration of a model amine, piperidine, was carried out to obtain moderate yield of 1-nitropiperidine. The results of our findings are reported.

Finally, the hydrogenolysis of TAIW, a precursor to the nitramine explosive CL-20, was investigated in supercritical fluid medium. The low solubility of TAIW in $scCO_2$ adversely affected its efficient conversion to products that could serve as immediate nitratable precursors. Hydrogenolysis with formic acid as a co-solvent in $scCO_2$ did provide the nitratable monoformyl amine, $(Ac)_4C_6N_6(CHO)H$, in high yield.

PEROXIDASE-CATALYZED SYNTHESIS OF 2,3-DIMETHYL-2, 3-DINITROBUTANE (DMDNB): A DETECTION AGENT FOR MOLDABLE EXPLOSIVES

Naval Surface Warfare Center Indian Head Division, Indian Head, MD

INTRODUCTION

Syntheses of energetic ingredients using biocatalysis in aqueous or mixed aqueous organic media substantially reduces source amounts of hazardous waste through the elimination of ozone depleting solvents (ODS), heavy metal catalysts, and toxic chemical reagents used by current synthesis methods. Pollution reductions found in using biocatalysis also reduces the costs associated with the environmental impact of conventional organic reactions.

Biocatalysis used to synthesize energetic ingredients possess several advantages over conventional processes. Enzyme catalyzed reactions typically occur under mild reaction conditions which yields an energy savings. Biocatalyzed reactions generally produce products with high structural selectivity which makes them easier to purify. Also, enzymes are biodegradable and produce little or no hazardous waste as compared to the heavy metal catalysts that are used in performing the same reactions.[1]

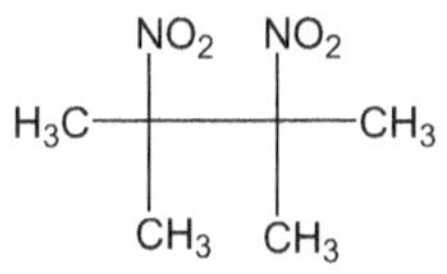

1

DMDNB

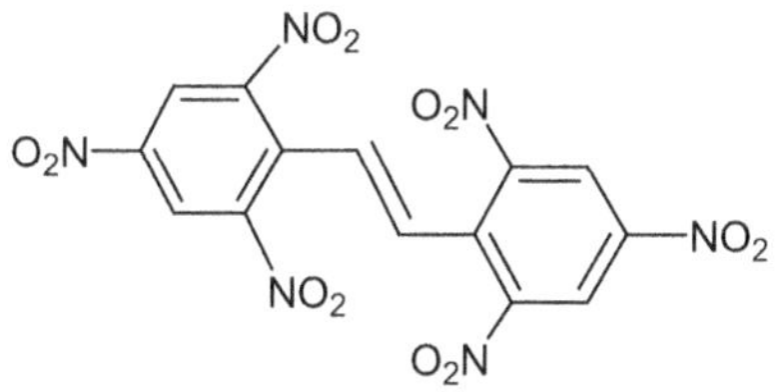

2

HNS

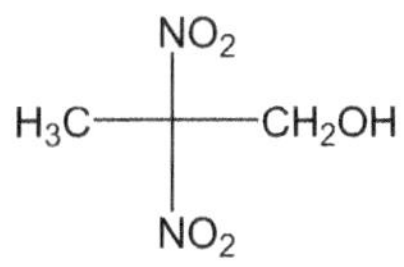

3

DNPOH

An ingredient of importance to Navy/DoD programs, 2,3-dimethyl-2,3-dinitrobutane (DMDNB, 1), a volatile detection agent required by executive order as an additive for all plastic explosives, was targeted for biocatalytic synthesis using

peroxidases. Estimates suggest biocatalytic DMDNB synthesis would reduce the waste streams from approximately ten pounds to less than one pound for every pound of DMDNB produced.

Herein, we describe our experimental results and efforts to optimize the oxidative coupling of 2-nitropropane to produce DMDNB using peroxidase enzymes.

Detection Agents In Explosives (DMDNB)

The recent bombings of the World Trade Center in New York, the Alfred P. Murrah Federal building in Oklahoma and Centennial Park in Atlanta are vivid examples of the ever-increasing threat to the United States towards explosive-related incidents (3,163 total bombings in 1994; a 6.1 % increase from 1993).[2] To combat this terrorist threat, the Antiterrorism and Effective Death Penalty Act of 1996 has been enacted.[3] This law has selected four potential detection agents, 2,3-dimethyl-2,3-dinitrobutane (DMDNB), Ethylene Glycol Dinitrate (EGDN), para-mononitrotoluene (p-MNT) and ortho-mononitrotoluene (o-MNT), for marking explosives.

Because of the short shelf life and carcinogenicity associated with the other three explosives markers, 2,3-dimethyl-2,3-dinitrobutane (DMDNB) was selected by the United States and several other countries as the exclusive detection agent for use in moldable explosives. DMDNB is not carcinogenic, has a 20 year shelf life, and has the best detectability of the detection agents proposed. DMDNB is not used in any other commercial use, therefore, the detection of DMDNB indicates a high probability that explosives are present.

Currently, DMDNB is only available in the U.S. from the Hampshire Chemical Company for \$200/Lb.[4] To date, Hampshire chemical has not revealed the process used to make their DMDNB, but most likely their process relies on coupling two 2-nitropropane molecules via single electron transfer (SET) chemistry.

Conventional DMDNB Syntheses

Several methods to prepare DMDNB using electrochemistry and conventional chemical techniques have been reported.[5] These routes require hazardous solvents and reagents such as dimethyl sulfoxide and carbon tetrachloride or suffer from low DMDNB yields.[5] Of these published routes, the best approach appears to be a process reported in an Imperial Chemical Industries (ICI) patent from 1965, Scheme 1.[6] The reaction of propane-2-nitronate (**5**) with one half equivalent of iodine in the presence of KI (to aid I_2 dissolution) generates halogenated intermediate, 2-iodo-2-nitropropane that subsequently reacts with excess **5** to produce DMDNB.

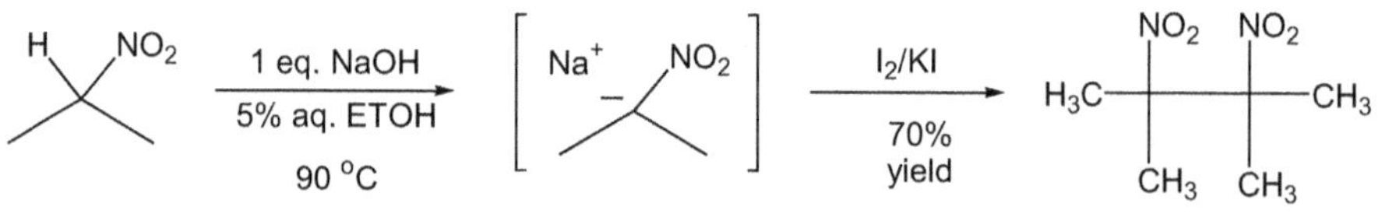

4 **5** DMDNB

Scheme 1. ***Imperial Chemical Industries Patented Route To DMDNB***

The coupling reaction between two propane-2-nitronate molecules using halogens has been studied extensively, and is thought to proceed through a $S_{RN}1$ single electron transfer chemistry mechanism shown in scheme 2.[7] Much controversy surrounds the mechanistic aspects, and the general mechanism shown in equations 1-6 is commonly invoked.

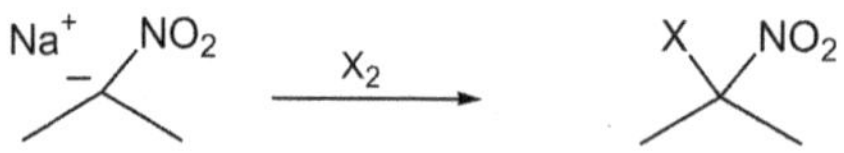

(eq 1)

5 **6**

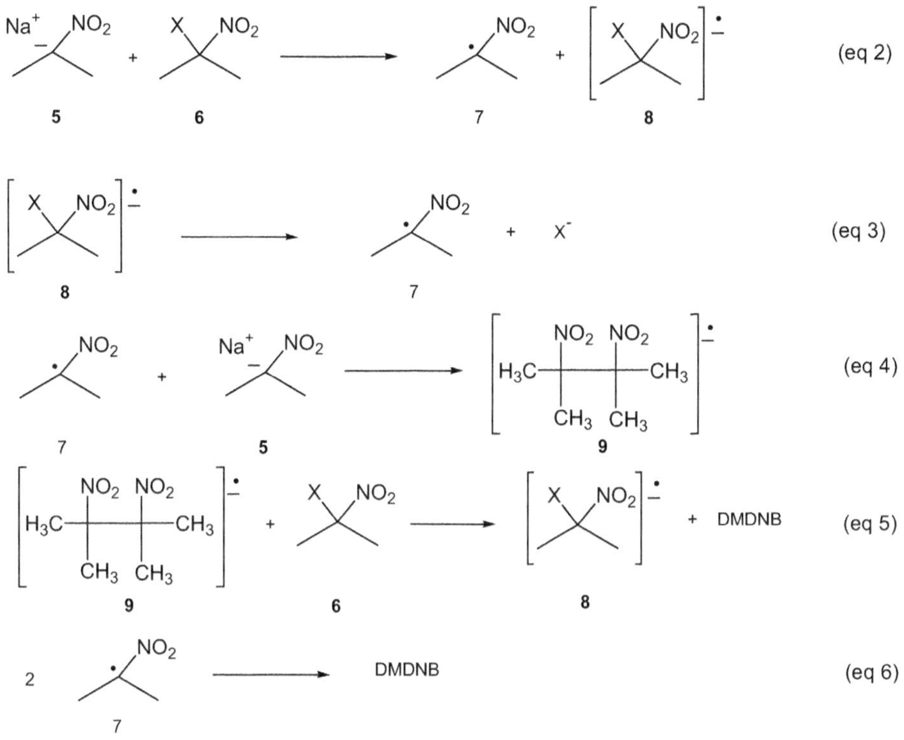

Scheme 2. ***$S_{RN}1$ Reaction Mechanism For DMDNB Formation From Propane-2-Nitronate***

Nitronate anion **5** is partially converted into the halonitro compound **6** by reaction between **5** and halogen (equation 1). Halonitro compounds are known to be good electron acceptors, and readily accept electrons from other molecules. Single electron transfer between anion **5** and **6** generates the 2-nitropropyl radical, **7**, and radical anion, **8** (equation 2). Radical anion **8** can dissociate into halide anion and alkyl radical **7** thereby initiating the chain $S_{RN}1$ mechanism (equation 3). At this point two nitropropyl radicals can couple to terminate the reaction (equation 6). Alternatively, radical **7** can react with anion **5** to give radical anion **9** (equation 4). Radical anion **9** can then react with another molecule of **6** to produce **1** (DMDNB) and regenerate **8** which continues to propagate the reaction (equation 5). Factors such as bond strength, nucleofugal properties, and molecular orbital overlap have been suggested to influence the reaction.[8]

Peroxidase-Catalyzed DMDNB Synthesis

Peroxidases (EC 1.11.1.7) such as chloroperoxidase (CPO), soybean peroxidase (SBP) and horseradish peroxidase (HRP), isolated from plants and microbes, represent commercially available redox enzymes that catalyze the oxidative transformations of organic substrates using a clean oxidant such as H_2O_2.[9] Peroxidases are ideal catalysts to facilitate these transformations because they require no expensive cofactors. Moreover, they are relatively stable and can accept a wide range of substrates. Peroxidases such as HRP and SBP should be capable of oxidatively coupling two 2-nitropropane molecules together to form DMDNB. In fact, literature precedent has demonstrated horseradish peroxidase activated with H_2O_2 utilizes propane-2-nitronate to give acetone under aerobic conditions and DMDNB under anaerobic conditions, Scheme 3.[10a]

Porter and Bright, the researchers who conducted this mechanistic study, showed only the ionized forms of 2-nitropropane, i.e. nitronate and aci-nitro forms, act as substrates for HRP. They also concluded the peroxidase acts as a radical chain initiator, and more than 90% of the oxidation occurs nonenzymatically after initiation.[10a] The aforementioned study did not examine the DMDNB formation using other available peroxidases (SBP or CPO), or conditions needed to optimize DMDNB yields.

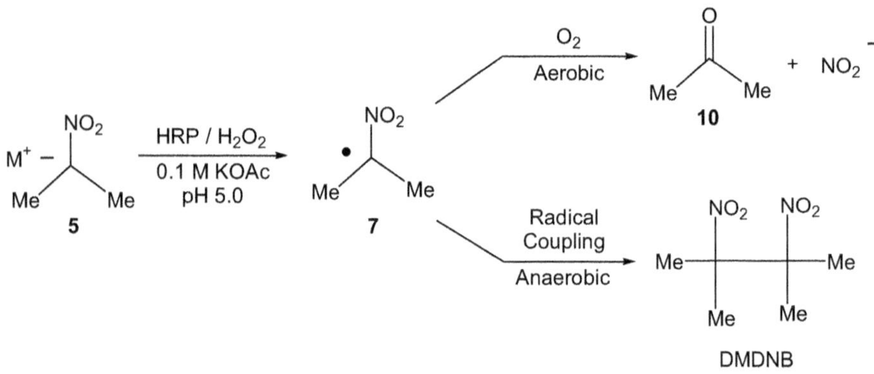

Scheme 3. ***HRP-Catalyzed Reaction Of Propane-2-Nitronate To Give Acetone Or DMDNB***

The peroxidase-catalyzed mechanism proposed to produce DMDNB is shown in scheme 4. The first step in the catalytic cycle is a two-electron oxidation of the resting high-spin Fe^{III} native enzyme by H_2O_2 to give a species commonly called "Compound **I**" and H_2O as a byproduct, scheme 4.[9a] The formal oxidation state of Compound **I** is Fe^{V}, although experimental evidence suggests it is most likely a highly delocalized Fe^{IV} porphyrin radical cation.[11] Compound **I** then proceeds to abstract an electron from propane-2-nitronate to generate the corresponding 2-nitropropyl radical and Compound **II**. Compound **II** subsequently abstracts an electron from another propane-2-nitronate molecule to generate another free radical and regenerate native peroxidase. The ensuing radicals then diffuse into the reaction medium and couple to produce DMDNB.

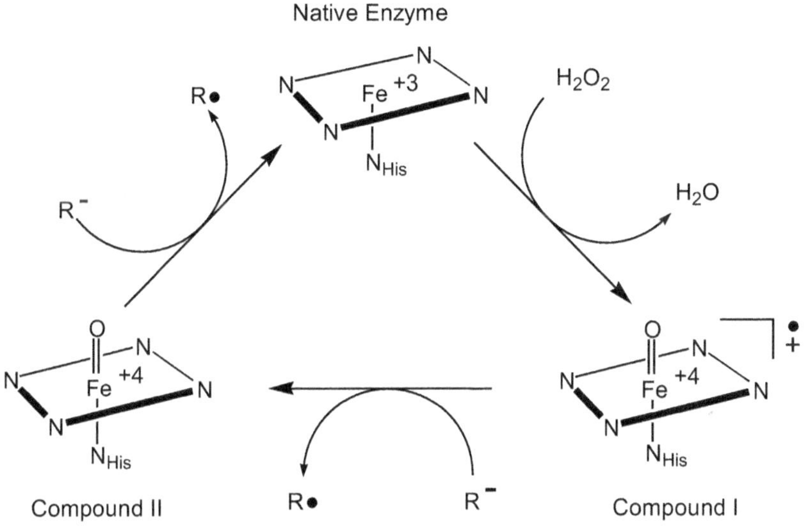

Scheme 4. ***Peroxidase-catalyzed synthesis of DMDNB; R= $(CH_3)_2CNO_2$***

RESULTS AND DISCUSSION

Initial efforts focused on repeating Porter and Bright's peroxidase-catalyzed DMDNB synthesis.[10a] These experiments allowed our technique to be honed and because the DMDNB yields were not previously reported, a general idea of the yield obtainable by this procedure was established. The reaction was repeated, as reported by Porter and Bright, by adding 10-20 mg HRP to a solution made by mixing 0.2 M H_2O_2 (2 mmol) in one portion to 0.1 M propane-2-nitronate in 0.1 M potassium acetate (KOAc) buffer, pH 5.0 (100 mL). The mixture was stirred one hour, the resulting solid was filtered, collected and dried to give a crude greenish white solid in low yields (8-10 %). The isolated product was recrystallized from CH_2Cl_2/hexane and its identity was confirmed to be DMDNB by elemental analysis and by comparison of the product's ^{1}H, ^{13}C NMR, and IR spectra to an authentic DMDNB sample, figures 1 and 2.

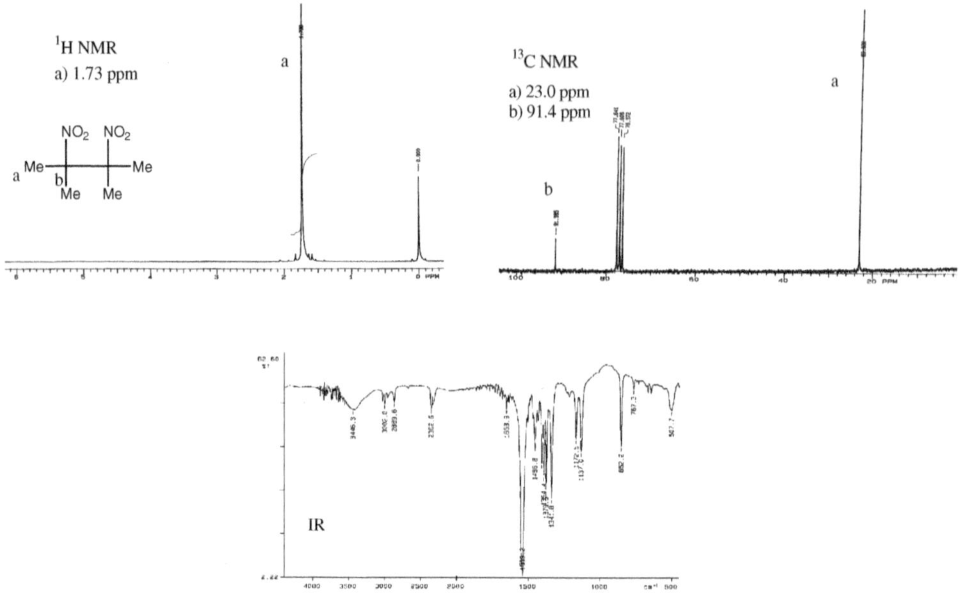

Figure 1. ***^1H, ^{13}C NMR ($CDCl_3$) And IR (KBr) Spectral Data For DMDNB Collected From The HRP-Catalyzed Oxidation Of Propane-2-Nitronate***

Figure 2. ***DMDNB Produced From The Horseradish Peroxidase Catalyzed Propane-2-Nitronate Oxidation***

Examination of the reaction's stoichiometry shows one H_2O_2 molecule generates two 2-nitropropyl radicals that ultimately couple to form one DMDNB molecule. In the initial experiments described, H_2O_2 was used in limiting proportions. When the amount of H_2O_2 was increased to equal the moles of nitronate in the buffered solution, the DMDNB yield was expected to improve significantly. It became apparent after running the reaction several times, an increase in H2O2 did not improve the yields as expected and yields of only 15-20 % were attainable. Because the DMDNB yields obtained from the reaction were not useful for scale-up, investigations into the reaction variables and possible side reactions occurring during the reaction were examined more closely.

Three pathways are available to the propane-2-nitronate intermediate during the HRP-catalyzed reaction. Namely, propane-2-nitronate can protonate to give back 2-nitropropane, it can be converted into acetone in the presence of oxygen, or it can couple to form DMDNB in the absence of oxygen. If propane-2-nitronate reprotonation occurs to a large extent to give 2-nitropropane, the yields of DMDNB would suffer accordingly since 2-nitropropane is known not to be a substrate for HRP.[10a] To elucidate how quickly propane-2-nitronate reprotonates, reprotonation rates were determined at 25 °C spectrophotometrically (230 nm) in potassium acetate buffer (pH 5.0 and 6.0) ranging in concentration from 25-100 mM KOAc. The reprotonation rates followed first order kinetics and the rate constants calculated from these experiments when plotted against KOAc concentration gave a linear relationship as shown in figure 3. The half-life for 100 uM propane-2-nitronate reprotonation in 100 mM KOAc was found to be 157 min at pH 5.0 and 11.5 h at pH 6.0.

Interestingly, comparison of the reprotonation rates between propane-2-nitronate and other nitronate anions shows propane-2-nitronate reprotonates extremely slow ($t_{½}$=157 min) relative to the other nitronate anions, table 1. For example, the reprotonation half-lives for ethane nitronate, propane-1-nitronate, and nitrocyclohexane nitronate in 100 mM KOAc buffer (pH 5.0) are on the order of 1.6, 1.3, and 26.3 min, respectively.[12] It is not apparent why propane-2-nitronate reprotonates so slowly, but hyperconjugation and steric effects of the methyl groups have been postulated as possible reasons.[13] The slow propane-2-nitronate reprotonation rate is fortuitous in our case since the rate at which peroxidase utilizes propane-2-nitronate can be tailored depending upon the peroxidase concentration of the reaction solution. Thus, propane-2-nitronate reprotonation appears to be a minor problem in this reaction.

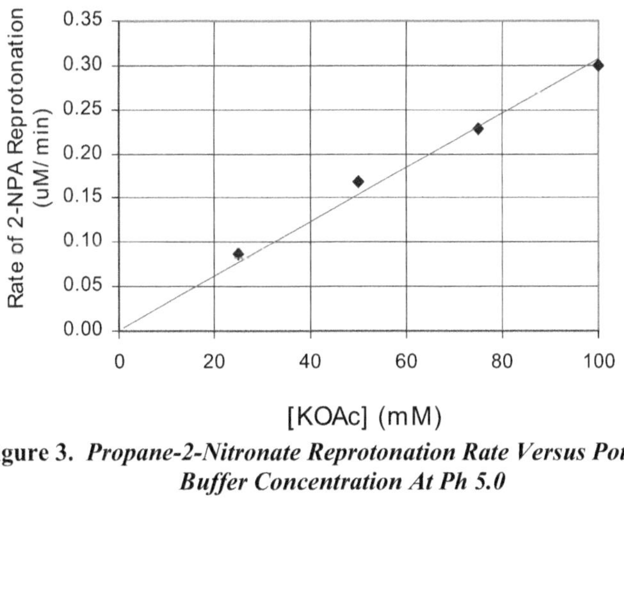

Figure 3. ***Propane-2-Nitronate Reprotonation Rate Versus Potassium Acetate Buffer Concentration At Ph 5.0***

To make relative comparisons between peroxidase activities under different reaction conditions a spectrophotometric method was utilized to analyze peroxidase activities, whereby the rate at which peroxidase consumed propane-2-nitronate was followed at 230 nm. To validate this method, the propane-2-nitronate oxidation rate was examined over a series of peroxidase and nitronate concentrations to determine the linearity of the oxidation rates. Accordingly, when the propane-2-nitronate oxidation rates determined spectrophotometrically were plotted against HRP concentration over 3-30 nM, a linear response was observed over the peroxidase concentrations examined, figure 4A. Likewise, the propane-2-nitronate oxidation rates were also shown to be linearly dependent upon propane-2-nitronate concentration over 50-250 uM, figure 4B. All further assays performed to determine peroxidase activities used peroxidase and nitronate concentrations within the concentration ranges found to give a linear response.

With a general method to assay peroxidase activity using propane-2-nitronate, horseradish and soybean peroxidase's activity was examined over the pH range 4.0-6.0. Accordingly, peroxidase activity was assayed at 25 °C using 100 uM propane-2-nitronate in 100 mM KOAc buffer with 1.0 mM H_2O_2 and 20 nM HRP. Horseradish peroxidase was found to have its greatest activity towards propane-2-nitronate oxidation at pH 5.0 in agreement with literature results,[10a] while soybean peroxidase showed its greatest activity between pH 5.0-5.5.

Table 1. ***Nitronate Anion Reprotonation Rates In 0.1M KOAc Ph 5.0***

Nitroalkane (100 uM)[a]	p_{Kc}[b]	p_{ko}[b]	Half life (min)
Nitromethane	10.2	3.3	-
Nitroethane	8.5	4.4	1.6
1-Nitropropane	9.0	4.6	1.3
2-Nitropropane	7.7	5.1	157
Nitrocyclohexane	10	-	26.3

[a]measured spectrophotometrically at 230 nm in 0.1 M KOAc pH 5.0. [b]p_{Kc} refers to the acidity of the hydrogen on the parent nitroalkane. p_{Ko} refers to the acidity of the hydrogen on the aci-nitro compound.

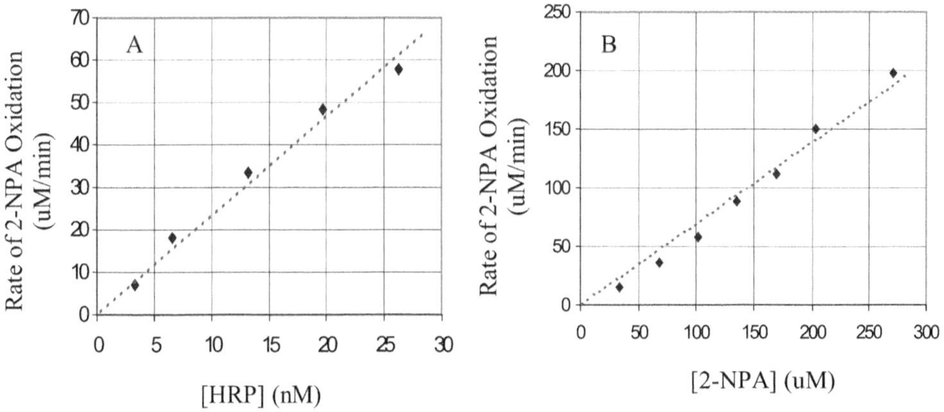

Figure 4. ***A) Plot Of Propane-2-Nitronate Oxidation Versus HRP Concentration; B) Plot Of Propane-2-Nitronate Oxidation Versus Propane-2-Nitronate Concentration***

Because HRP's optimal activity was found to be at pH 5.0, efforts were made at pH 5 to adjust other reaction variables such as propane-2-nitronate and HRP concentrations in an effort to improve the DMDNB yields. Accordingly, the concentrations of propane-2-nitronate (50-220 mM) and HRP (3-60 uM) were varied independently over the ranges shown in parentheses and the change in DMDNB yield recorded. Unfortunately, varying the concentrations of these reagents did not improve the DMDNB yields significantly beyond the 20 % yields previously obtained.

Although peroxidases need H_2O_2 to operate, they are also slowly inactivated by H_2O_2 over time.[14] Because of the known instability of peroxidases towards H_2O_2, the stability of HRP towards H_2O_2 was examined by making a series of solutions, under simulated reaction conditions, in which propane-2-nitronate and HRP concentrations were held constant while H_2O_2 concentration was varied from 0-2000 uM. The prepared solutions were each incubated at 25 °C for approximately 30 minutes, a known aliquot from each H_2O_2 incubated solution was then immediately assayed for HRP activity by spectrophotometrically measuring the rate at which a known amount of propane-2-nitronate was consumed over time. It was found that as the H_2O_2 concentration increased HRP's activity toward propane-2-nitronate decreased, figure 5. These results emphasize HRP's sensitivity towards high H_2O_2 concentrations. To keep the H_2O_2 concentration low and minimize HRP's exposure to H_2O_2 during the reaction, a H_2O_2 solution was slowly added by syringe pump to the reaction solution over time in an effort to extend the utility of the HRP. An H_2O_2 addition rate of 1.5-2.0 mL/h was found to be optimal for the reaction conditions studied. In the future, we hope to address the HRP's instability towards H_2O_2 by using HRP variants engineered to withstand higher H_2O_2 concentrations. These engineered enzymes may be obtained through collaborations established with researchers undertaking these types of modifications.

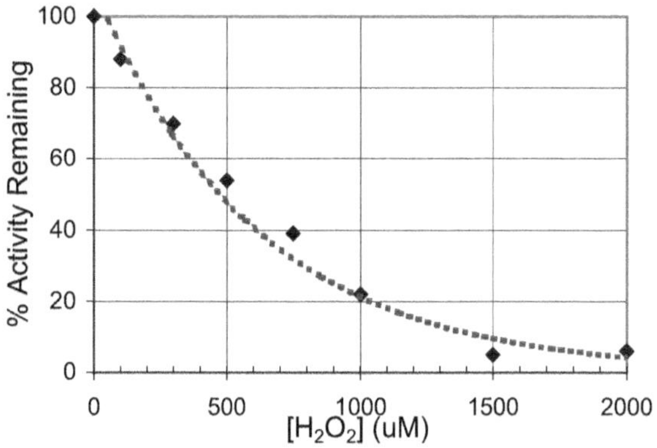

Figure 5. HRP Activity Remaining After A 30-Minute Incubation With Various H_2O_2 Concentrations

Because changing the propane-2-nitronate and HRP concentrations did not improve the DMDNB yield significantly, instrumental techniques were used to examine the reaction products being formed during the reaction. Gas chromatography was initially used as a potential method to follow and quantify the reaction products formed during the reaction course. Unfortunately, residual propane-2-nitronate, present in the reaction aliquots, is readily converted to acetone and 2-nitropropane on the GC's injection port, thereby making accurate quantification of these key reaction byproducts impossible.

Focus then turned to ^1H NMR as a method to follow the reaction course. Initial NMR experiments were performed using D_2O that contained 0.1 M KOAc buffer to maintain the reaction's pH. Reactions were run in a round bottom flask and aliquots were removed at various times during the reaction and analyzed by NMR. Unfortunately, the large signal and the chemical shift of the acetate's methyl group (2.0 ppm) effectively masked the region where the propane-2-nitronate signal (1.91 ppm) appeared making it difficult to accurately follow the reaction's progress.

To circumvent this problem, the reactions were run in D_2O without any buffer. To maintain the desired pH, a pH controller was utilized to deliver acid (HCl) via a syringe pump into the reaction flask. The addition of 8 % HCl (v/v) by this method worked well to maintain the reaction pH at 5.0. Diglyme, ethylene glycol dimethyl ether, was used as an internal standard to quantitate the reaction products observed by NMR. The chemical shifts of diglyme's two singlets (3.37 and 3.60 ppm) were sufficiently downfield to allow easy integration and comparison to all observable reaction product signals.

A typical NMR time course for the peroxidase catalyzed propane-2-nitronate oxidation at pH 5.0 is shown in figure 6. As can be seen at t=0 min, before peroxidase and H_2O_2 are added, the signal due to propane-2-nitronate is the major peak observed accompanied by low levels of 2-nitropropane and lesser amounts of other unidentified peaks. The addition of horseradish peroxidase followed by drop-wise H_2O_2 addition, consumed the nitronate and peaks due to acetone, acetoxime, DMDNB and 2-nitropropane formation become apparent in the NMR spectrum (time=30 min). Finally, at the reaction's completion (t=65 min), no nitronate remained. DMDNB was obtained in approximately 30 % yield from these reactions. Spiking the reaction mixture with authentic acetone, acetoxime, and 2-nitropropane samples identified the reaction products. In the case of 2-nitropropane, a broad singlet was observed at 1.55 ppm for the methyl signals instead of the expected doublet due to deuterium incorporation from the D_2O at 2-nitropropane's 2-position during reprotonation.

Figure 7 shows a graphical representation of the NMR results over time. Because DMDNB was not soluble in the aqueous media, only a small signal was observed by NMR for DMDNB. Therefore, the relative amount of DMDNB shown in figure 7 was calculated by subtracting the amount of acetone, acetoxime and 2-nitropropane from the starting amount of propane-2-nitronate used in the reaction.

Interestingly, comparison of the isolated DMDNB yields obtained from the D_2O pH 5.0 runs (*ca.* 30 % yield) to those obtained from the pH 5.0 KOAc buffered runs (20-25 % yield) showed a slight but repeatable yield improvement. It is not readily apparent why removing the acetate buffer from the reaction media improves the DMDNB yield, although literature reports suggest acetate can bind to the enzyme at lower pH (2-4) and affect the rate at which Compound **I** forms.[15]

Of concern is the discrepancy between the isolated DMDNB yields (*ca.* 30 %) and the amount of DMDNB determined by NMR (*ca.* 48 %). The discrepancy was initially thought to arise from the diglyme volatilizing and being lost during the reaction.To check this possibility, 2-methoxyethyl ether, a less volatile molecule, was tried in place of diglyme. Evaluation of reaction products using 2-methoxyethyl ether instead of diglyme as the internal standard gave results similar to those observed when diglyme was used.

It was apparent from the propane-2-nitronate HRP-catalyzed reaction that the derived byproducts, i.e. acetone, acetoxime, and 2-nitropropane, were similar to those obtained from the well-known Nef reaction of propane-2-nitronate, although, different product ratios were obtained. In a typical Nef reaction, the nitronate anion, generated by treatment of primary or secondary nitro compound with base, is hydrolyzed to its corresponding aldehyde or ketone upon treatment with strong acid, usually sulfuric acid.[16] Work by Armand has shown 0.1 M propane-2-nitronate gives 2-nitropropane (85 %), acetone (7-8 %), and acetoxime (7-8 %) at pH 5.0 in buffer at 21 °C.[17] Upon raising the pH to 5.4, only 2-nitropropane, obtained from reprotonation, was observed.[17] The acetone and acetoxime formed during the HRP-catalyzed reactions suggest the Nef reaction may be occurring concurrently during the reaction. To check if acetone, acetoxime, and 2-nitropropane are formed via the Nef reaction, and to minimize or eliminate these byproducts, the pH of the HRP-catalyzed reaction was raised at the expense of HRP activity.

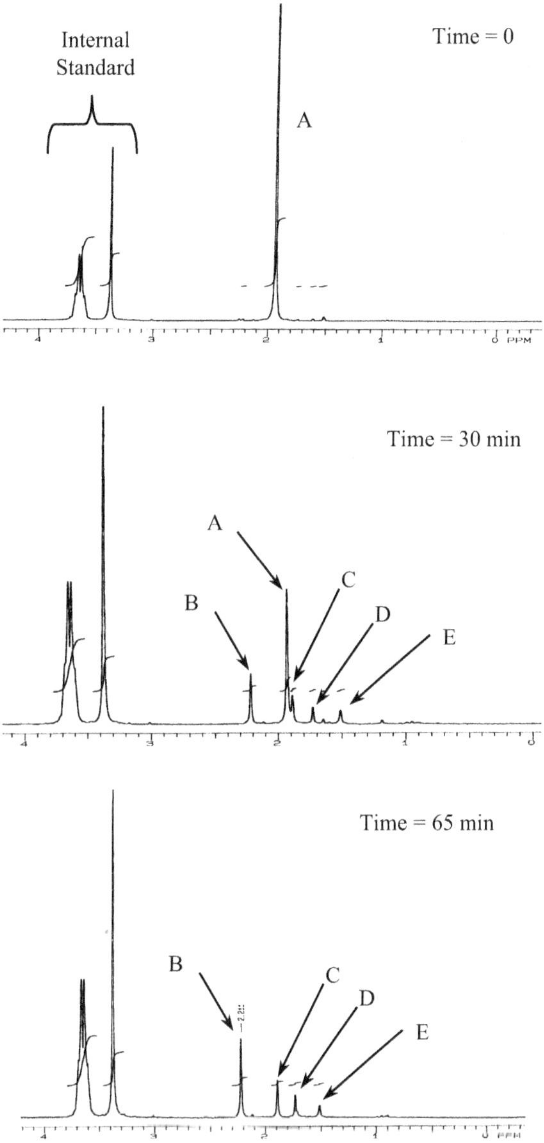

Figure 6. ***Typical ^{1}H NMR Reaction Time Course For The HRP-Catalyzed Oxidation Of Propane-2-Nitronate At Ph 5.0 In D_2O; A) Propane-2-Nitronate; B) Acetone; C) Acetoxime; D) DMDNB; E) 2-Nitropropane***

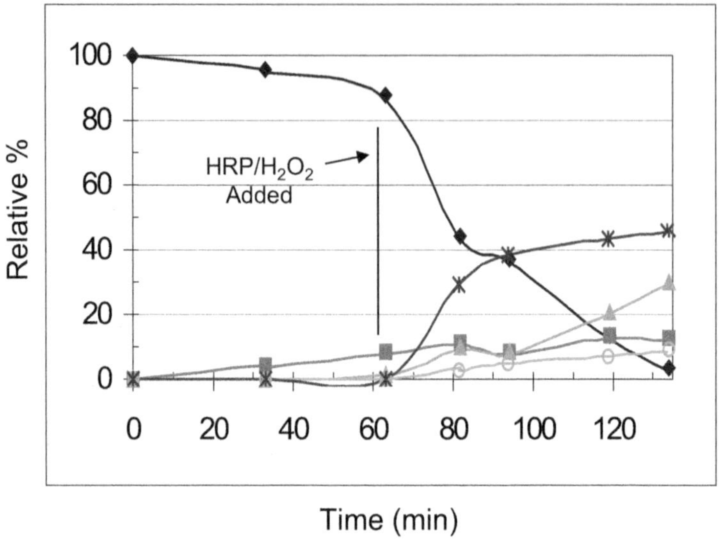

Figure 7. ***Time Course For HRP-Catalyzed Propane-2-Nitronate Oxidation At Ph 5.0; . = Propane-2-Nitronate;. DMDNB; ▲ Acetone; O Acetoxime; ■ 2-Nitropropane***

Accordingly, the propane-2-nitronate HRP-catalyzed reactions were rerun at pH 6.3 in D_2O using a pH controller to maintain the pH. The reaction course was followed by ^1H NMR in a manner similar to the runs at pH 5.0 and the same products (acetone, acetoxime, 2-nitropropane and DMDNB) were formed at pH 6.3 as those produced at pH 5.0. No new products were observed. Figure 8 shows the graphical representation of the NMR results over time. As in the case of the pH 5.0 runs, DMDNB was not soluble in the aqueous media and only a small signal for DMDNB was observed in the NMR spectrum. Therefore, the relative amount of DMDNB shown in figure 8 was calculated by subtracting the amount of acetone, acetoxime and 2-nitropropane from the starting amount of propane-2-nitronate used in the reaction. As can be seen, the NMR results indicate DMDNB is formed in approximately 73 % (recall, by NMR DMDNB was formed in approximately 48 % at pH 5.0, figure 6). Importantly, the isolated DMDNB yields obtained by running the reaction at pH 6.3 were 50-55 %. Increasing the pH (greater than pH 6.0) dramatically improved the DMDNB yield compared to previous attempts to synthesize DMDNB at lower pH. The relative amounts of acetoxime and 2-nitropropane were also lowered when compared to the pH 5.0 runs. In a separate experiment, raising the pH to approximately 7.0 and running the reaction gave DMDNB yields similar to the pH 6.3 runs, although, the reaction seemed to run more sluggishly. Unfortunately, in the pH 6.3 runs acetone is still formed in substantial amounts and represents the major byproduct formed during the reaction. In these runs, acetone formed by the Nef reaction should be negligible because at pH 6.3 the Nef reaction should not readily take place.

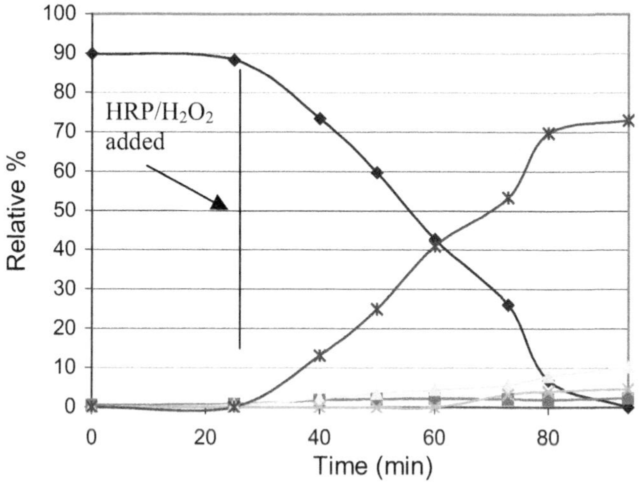

Figure 8. ***Time Course For The HRP-Catalyzed Propane-2-Nitronate Oxidation At Ph 6.3 Determined By ${}^{1}H$ NMR. . = Propane-2-Nitronate;. = DMDNB; ▲ = Acetone; O = Acetoxime; ■ = 2-Nitropropane***

Controls, in which no HRP was added, were run to examine the stability of the reaction products to the reaction conditions. Accordingly, acetone, 2-nitropropane, acetoxime, DMDNB and the internal standard were dissolved in a mixed solvent system of D_2O/acetonitrile (60:40) at concentrations similar to the concentrations run in the actual peroxidase catalyzed reactions. The stirred solutions were subjected to reaction conditions at pH 7.3 in a manner similar to the way the peroxidase-catalyzed reactions were typically run. A pH of 7.3 was used in these experiments because the pH was quite sensitive to the addition of small amounts of acid making pH adjustment difficult. Once a pH near the desired pH was obtained it was left alone. During the control runs, H_2O_2 was added slowly by syringe pump. Aliquots were removed from the reaction solution over time, and subjected to NMR analysis. As shown in figure 9, no appreciable loss of acetone, 2-nitropropane, acetoxime or DMDNB was observed over the time period examined. Not surprisingly, because of acetone's volatility, analysis at extended times (18-24 h) showed acetone was slowly being lost over this time period.

The acetone formed in these HRP-catalyzed reactions must be coming from another type of side reaction. Two possibilities for acetone formation readily come to mind. First, there may be a catalase impurity present in the HRP preparations obtained from the Sigma Chemical Company. Catalase, an enzyme with a turnover approaching diffusion-controlled limits, converts H_2O_2 into O_2 and H_2O as shown.

$$\text{Catalase} + 2\,H_2O_2 \rightarrow O_2 + 2\,H_2O$$

Because the turnover of the catalase is very large, a small catalase impurity will generate considerable amounts of O_2. Once O_2 enters into the reaction media, the 2-nitropropyl radicals generated by the HRP can readily scavenge the O_2 and enter into a chain reaction with the

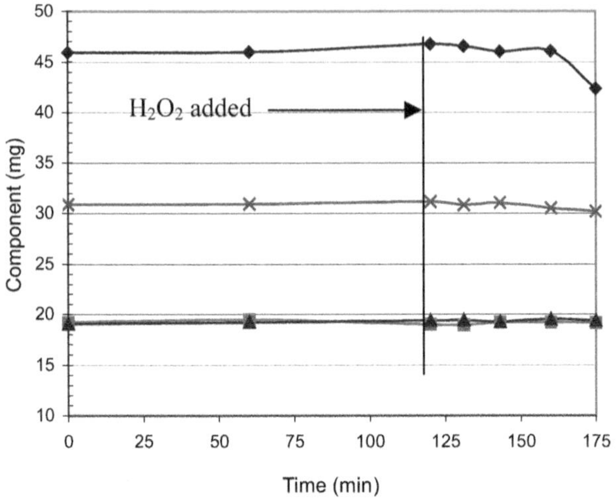

Figure 9. ***Stability Of DMDNB, Acetone, Acetoxime And 2-Nitropropane To Simulated Reaction Conditions At Ph 7.3; . = DMDNB; X = 2-Nitropropane; ▲ = Acetoxime; ■ = Acetone***

oxygen to give a peroxyl radical intermediate that ultimately decomposes to acetone.[18] This mechanism has also been suggested by Porter and Bright to explain the aerobic oxidation of 2-nitropropane by HRP to give acetone.[10a] Experiments were attempted to determine if catalase was present in the HRP preparations by trying to follow the absorbance loss at 215 nm of a known amount of H_2O_2 upon addition of HRP to the reaction cuvette. Unfortunately, the results of these experiments were not readily interpreted.

Chloroperoxidase a well studied peroxidase is known to possess substantial catalase activity.[9] One method researchers use to minimize the catalase activity present in CPO catalyzed reactions is to use an alkyl hydroperoxide such as *t*-butyl hydroperoxide in place of H_2O_2. Chloroperoxidase can utilize *t*-butyl hydroperoxide, although not as efficiently as H_2O_2, while catalase cannot utilize this akyl hydroperoxide as a substrate. To examine if this substitution could be used to minimize any catalase activity that may be present in HRP-catalyzed reactions, H_2O_2 was replaced by *t*-butyl hydroperoxide. When *t*-butyl hydroperoxide was added to the reaction media at pH 6.3 under reaction conditions similar to previous conditions used with H_2O_2, no reaction was observed. Apparently, HRP cannot utilize *t*-butyl hydroperoxide as a substrate to generate Compound **I** that subsequently utilizes the propane-2-nitronate.

When soybean peroxidase was used in place of HRP and the reaction run at pH 6.3, results similar to the HRP runs at pH 6.3 were obtained. Importantly, acetone was formed in nearly identical amounts as found for the HRP reactions. This suggests the catalase impurity in

SBP and HRP are present in nearly equal amounts assuming the catalase turnover is similar. Since the peroxidases are from different sources and purified differently, one would expect the catalase levels found in these peroxidase preparations to be different. The fact that SBP and HRP are purified differently yet give similar amounts of acetone in their reaction with propane-2-nitronate argues against the presence of a catalase impurity that leads to acetone in these reactions.

A second more likely scenario is that there may be an unrecognized reaction between propane-2-nitronate and H_2O_2. Olah and coworkers have shown primary and secondary nitro compounds can be transformed into their corresponding aldehyde and ketones by the action of H_2O_2 in the presence of potassium carbonate.[19] It appears the potassium carbonate is used to generate the nitronate anion *in situ* which is then converted into the corresponding carbony compound by reaction with H_2O_2. Unfortunately, during the course of this work, the control experiment to check propane-2-nitronate's stability towards H_2O_2 was not performed. It is likely the preformed propane-2-nitronate, contained in the reaction solution at a much greater concentration than the HRP, reacts with some of the H_2O_2 added to the reaction. If this reaction is in fact found to occur, methods will need to be devised to minimize the contact time between propane-2-nitronate and H_2O_2. This crucial control experiment needs to be performed.

As in the case of the pH 5.0 runs, the pH 6.3 HRP reactions also showed a discrepancy between the isolated DMDNB yields (50-55 %) and the NMR results (73 %). It is plausible the difference between the NMR and actual isolated yields arises from the inability to accurately follow the DMDNB formation by NMR coupled with the inaccuracies associated with manipulating the aliquots taken for the NMR analyses during the reaction. To quantify DMDNB more accurately by NMR, a mixed solvent system was used in place of D_2O that kept the DMDNB in solution during the HRP-catalyzed reactions. A mixture of D_2O and deuterated acetonitrile (MeCN) in a 60:40 ratio was found to dissolve roughly half of the DMDNB typically produced during a HRP-catalyzed run. Because this solvent mixture could not dissolve all the DMDNB produced during the reaction, and to minimize perturbations to the enzyme due to large amounts of MeCN, the starting concentration of propane-2-nitronate was halved to keep the DMDNB produced during the reaction dissolved. A series of reactions at pH 5.5-6.0 were performed using this solvent system and figure 10 shows graphically a typical time course as determined by NMR. Since the DMDNB stayed in solution for these runs, an isolated yield of DMDNB was not determined.

As can be seen, the amount of DMDNB determined from these runs is lower than the DMDNB actually isolated in the D_2O runs. It is difficult to make direct comparisons between these runs because changing the solvent system may affect how the propane-2-nitronate is utilized by HRP as a substrate, and also how the substrate interacts with the reaction media. Evidence that changing the solvent system does appear to change the reaction course was verified by the appearance of a new signal at 2.18 ppm observed in the NMR spectra. This small signal has not been identified yet, but appears repeatable in the D_2O/MeCN runs but not in the D_2O or the KOAc buffered runs.

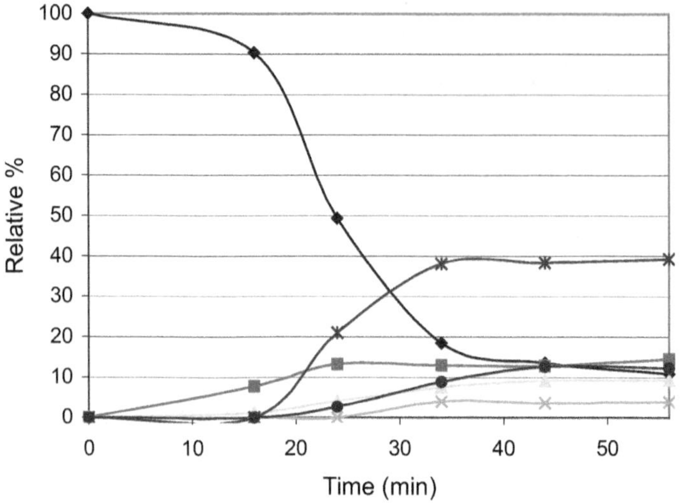

Figure 10. ***Time Course For The HRP-Catalyzed Propane-2-Nitronate Oxidation At Ph 5.5-6.0;. = DMDNB; ▲ = Acetone; O = Acetoxime; ■ = 2-Nitropropane***

Because of the inherent problems associated with manipulating the aliquots during the analyses, attempts to follow an entire HRP-catalyzed reaction in a sealed NMR tube were tried. A 10 mm NMR tube was utilized to increase the products and starting material concentrations in the tube and to improve the signal to noise levels. Unfortunately, the NMR available for these experiments was only equipped with a non-tunable ^{13}C NMR 10 mm probe. Obtaining useful ^{1}H NMR spectra directly with this probe was unworkable, therefore, attempts were made to collect ^{1}H data through the NMR's decoupler channel using this probe. Unfortunately, this technique, coupled with the low concentration of materials present in solution and the inability to control the pH during the reaction, gave a low signal to noise ratio and meaningful data interpretation was not possible.

When SBP replaced HRP in the reactions, slightly lower yields of DMDNB (45-50 %) were obtained. The acetone, acetoxime, and 2-nitropropane byproducts were also obtained. There is no evidence to suggest that other byproducts are formed when SBP is used in place of HRP. In general, SBP served well as a replacement for HRP in these reactions. Soybean peroxidase is an attractive alternative to HRP because it can be obtained in large amounts cheaply from soybean hulls, a byproduct of the food industry. In our work, the activity of the SBP obtained from Sigma tended to be lower than HRP. To obtain similar reaction times and yields to those observed in the HRP-catalyzed reaction more SBP was utilized per run. Otherwise, no other differences between the HRP and SBP-catalyzed reactions were observed.

SUMMARY AND CONCLUSIONS

This report constitutes an initial study into the environmentlally green synthesis of energetic materials using enzymes as biocatalysts. In particular, DMDNB (2,3-dimethyl-2,3-dinitrobutane) a detection agent used to detect plastic explosives, was prepared in nearly 55 % yield using peroxidases such as HRP, hydrogen peroxide, a clean oxidant used at low concentrations, 2-nitropropane, and biotechnology principles. The reaction is run in water and the optimal pH for the reaction is between 6.0 and 7.0. The DMDNB produced by this reaction sequence precipitates from the reaction solution during the reaction and is easily collected at the reaction's completion by vacuum filtration. Soybean peroxidase, an attractive peroxidase due to the economics of its preparation, has also been shown to give DMDNB in 45-50 % yields under similar conditions.

Ideally, the only byproduct formed during this reaction is water. The use of ^{1}H NMR in this study has shown that other byproducts, acetone, acetoxime, and 2-nitropropane, are also generated during the reaction. The 2-nitropropane and acetoxime are formed at low levels, and theoretically the 2-nitropropane, produced by reprotonation of the corresponding nitronate anion, can be recovered and recycled for use in subsequent reactions. Acetone is the major byproduct generated by the reaction, and attempts to minimize its formation have met with limited success thus far. Future work aimed at minimizing acetone formation should focus on the potential reaction between propane-2-nitronate and H_2O_2 in the reaction media. Minimizing the unwanted side reactions that produce acetone should allow the propane-2-nitronate substrate to be better utilized by the peroxidase, thereby, improving the DMDNB yield even further. Ultimately, it is envisioned that this work will help establish this technology within the Navy, and its transition to industry should help reduce costs to the DoD.

The wide array of reactions peroxidases can catalyze make them ideal candidates to perform transformations needed to synthesize other energetic ingredients. For example, it may be possible to use peroxidases to synthesize HNS and DNPOH.[20] Collaborations established between NSWC-IHD, academia and industry are not only examining the feasibility of using peroxidases to synthesize these targets, but also developing more robust peroxidase variants that will extend their usefulness. The future innovative approaches by these researchers should lead to better methods to synthesize energetic materials in an environmentally sound way.

EXPERIMENTAL

Spectrophotometric spectra experiments were recorded performed using a Cary 100 Bio ultraviolet spectrophotometer manufactured by Varian. 1H NMR and 13C NMR spectra were recorded with a Varian XL-200 spectrometer. Infrared spectra were recorded with a Perkin-Elmer model 1600 series FT-IR spectrometer (cm-1) as neat films on NaCl plates or as solids in pressed KBr pellets. A Cole Parmer pH controller (Catalog number E-05652-10, model 501-3400) with a Cole Parmer pH electrode (Catalog number 5992-20) was used for pH control. Elemental analyses were performed by Galbraith Laboratories, Inc., Knoxville, TN. Melting points were obtained on a Thomas-Hoover apparatus and are uncorrected. Column chromatography was performed using Silica gel 60 with dichloromethane as eluent unless otherwise stated. TLC visualization was by UV light or I_2.

Chloroperoxidase, horseradish peroxidase and soybean peroxidase were obtained from Sigma Chemical Co. (St. Louis, MO) as solid powders. Hydrogen peroxide (30 % solution in water) and 2,3-dimethyl-2,3-dinitrobutane and were obtained from Aldrich Chemical Co. (Milwaukee, WI). All other chemicals employed were of the highest purity commercially available.

Horseradish Peroxidase Catalyzed DMDNB Synthesis In 0.1 M KOAc Ph 5.0

Propane-2-nitronate is preformed by mixing 2-nitropropane (857 mg; 9.618 mmol) with 1.0 M KOH (10.6 mL; 10.6 mmol) and stirring the mixture at rt for approximately 2-3 h until homogeneous. Next, a solution of horseradish peroxidase (45 mg; 1.3 umoles) in KOAc buffer (0.2 M, pH 5.0, 100 ml, degassed with N_2 approximately 2 h) is added to the nitronate solution. The pH of the stirred solution is adjusted with concentrated acetic acid and the pH is kept between 5.0 and 5.2 during the course of the reaction. The H_2O_2 (195 mM, 24 mL, 4.8 mmol) is then added at once and the solution is stirred for 4 h. A solid precipitated out of solution and was collected by filtration to give a whitish green solid shown to be DMDNB. Many runs were performed in this manner and the yields ranged from 110-170 mg (13-20 %). ^{1}H NMR (200 MHz, $CDCl_3$): δ 1.73 (s, 12H), ^{13}C NMR (50 MHz, $CDCl_3$): δ 23.02, 91.39. IR (KBr): 1538 (NO_2). Anal. Calcd for $C_6H_{12}N_2O_4$: C, 40.91; H, 6.87; N, 15.90. Found C, 40.76; H, 7.13; N, 15.89.

Small Scale Horseradish Peroxidase-Catalyzed DMDNB Synthesis In D_2O Ph 6.3

In a 15 mL three-necked flask equipped with a stir bar, pH probe and two septa, a mixture of 2-nitropropane (96.9 mg; 1.09 mmol), 2.32 M KOH (490 uL; 1.14 mmol) and 2-methoxyethyl ether (64.7 mg; 0.480 mmol; internal standard for ^{1}H NMR) is stirred at rt for approximately 4 h to preform the corresponding nitronate anion. Next, the nitronate solution is diluted with 10 mL D_2O degassed by bubbling N_2 through the D_2O for approximately 30 min. Horseradish peroxidase (20.3 mg; 0.59 moles) is then added and the solution is stirred to dissolve the HRP. The pH of the stirred solution is adjusted to 6.0, and the pH is maintained between 6.3 and 5.8 by adding HCl (8 % v/v) dropwise over the reaction course via syringe pump operated by a pH controller. H_2O_2 (289 mM, 1.25 mL, 0.36 mmol) is added drop-wise at 2.1 mL/h via syringe pump. Upon H_2O_2 addition, a white silty solid begins to precipitate from solution. The reaction course is followed over time using ^{1}H NMR and products formed are quantified using the 2-methoxyethyl ether internal standard signals. The reaction was stopped after 1h 20 min when the nitronate was deemed consumed by NMR analysis. The mixture is then filtered through a Büchner funnel under weak vacuum using a paper filter. The crude whitish green solid is air dried in a vacuum desicator to give 44.3 mg (46 %). ^{1}H NMR (200 MHz, $CDCl_3$): δ 1.73 (s, 12H), ^{13}C NMR (50 MHz, $CDCl_3$): δ 23.02, 91.39. IR (KBr): 1538 (NO_2).

Large Scale Horseradish Peroxidase-Catalyzed DMDNB Synthesis In H_2O Ph 6.3

In a 100 mL three-necked flask, a mixture of 2-nitropropane (604 mg; 6.78 mmol) and 2.32 M KOH (3.0 mL; 6.98 mmol) is stirred at rt for approximately 3 h until homogeneous. Next, the nitronate solution is diluted with 60 mL H_2O and then horseradish peroxidase (90.6 mg; 2.6 umol) is dissolved in the stirred nitronate solution. The pH 12 solution is adjusted to 6.1 using a small amount of concentrated HCl. The pH is then maintained between 6.3 and 6.0 by adding HCl (8 % v/v) dropwise over the reaction course via syringe pump operated by a pH controller. H_2O_2 (270 mM, 7.0 mL, 1.89 mmol) is added dropwise at 4.4 mL/h via syringe pump. Upon H_2O_2 addition, a white silty solid begins to precipitate from solution. The H_2O_2 addition is continued 95 min after which the pH became difficult to control. This signaled nitronate consumption to be nearly complete and the reaction was stopped. The mixture was then filtered through a Büchner funnel under weak vacuum using a paper filter. The crude whitish green solid is air dried in a vacuum desicator to give 309 mg (52 %). Several recrystallizations from CH_2Cl_2/hexane give white crystals melting at 210-212 °C dec. (Aldrich 214-215 °C dec.). ^{1}H NMR (200 MHz, $CDCl_3$): δ 1.73 (s, 12H), ^{13}C NMR (50 MHz, $CDCl_3$): δ 23.02, 91.39. IR (KBr): 1538 (NO_2). Anal. Calcd for $C_6H_{12}N_2O_4$: C, 40.91; H, 6.87; N, 15.90. Found C, 40.76; H, 7.13; N, 15.89.

Soybean Peroxidase Catalyzed DMDNB Synthesis In D_2O Ph 5.0-5.5

The same general procedure used to run the small scale HRP-catalyzed DMDNB synthesis was used to run the reactions with soybean peroxidase. Accordingly, in a 15 mL three-necked flask equipped with a stir bar, pH probe and two septa, a mixture of 2-nitropropane (98.6 mg; 1.07 mmol), 2.34 M KOH (490 uL; 1.15 mmol) and 1,2-dimethoxy ethane (61 mg; 0..68 mmol; internal standard for ^{1}H NMR) is stirred at rt for approximately 4 h to preform the corresponding nitronate anion. Next, the nitronate solution is diluted with 10 mL D_2O degassed by bubbling N_2 through the D_2O for approximately 30 min. Soybean peroxidase (25.6 mg; 0.69 umoles) is then added and the solution is stirred to dissolve the SBP.

The pH of the stirred solution is adjusted to 5.5, and the pH is maintained between 5.0 and 5.5 by adding HCl (8 % v/v) dropwise over the reaction course via syringe pump operated by a pH controller. H_2O_2 (212 mM, 1.0 mL, 0.21 mmol) is added dropwise at 1.3 mL/h via syringe pump. Upon H_2O_2 addition, a white silty solid begins to precipitate from solution. The reaction course is followed over time using ^{1}H NMR and products formed are quantified using the 1,2- dimethoxy ethane internal standard signals. The reaction was stopped after 46 min when the nitronate was deemed consumed by NMR analysis. The mixture is then filtered through a Büchner funnel under weak vacuum using a paper filter. The crude whitish green solid is air dried in a vacuum desicator to give 23.5 mg (24 %).

Determination Of H_2O_2 Concentration Using HRP And Potassium Iodide

The following procedure represents the general procedure used to determine H_2O_2 concentration. A stock HRP solution was prepared by dissolving horseradish peroxidase (4.2

mg; 121 nmoles) in 10 mL of distilled deionized water. To a quartz cuvette was added 0.1 M pH 5.0 potassium acetate buffer (2975 uL) that contained 10 mM potassium iodide. Freshly prepared 225 mM stock H_2O_2 was prepared by dissolving 2300 uL of 30 % H_2O_2 in 100 mL of water. The stock solution was diluted 40 fold (25 uL of stock H_2O_2 into 975 uL H_2O) and 12 uL (68 nmoles) was added to the cuvette which was then placed into the spectrophotometer and allowed to equilibrate to 25 °C. The reaction was initiated by adding HRP stock (15 uL) and the absorbance of I_3^- (ε_{353}=2.62 x 10^4 $M^{-1}cm^{-1}$) formed during the reaction was followed spectrophotometrically at 353 nm over 2 min. The average absorbance (0.518) was then used to calculate the stock H_2O_2 concentration. The H_2O_2 concentration was found to be 198 mM.

Spectrophotometric Determination Of H_2O_2 Concentration At 215 Nm

This procedure represents a general method to check H_2O_2 concentration spectrophotometrically. Freshly prepared 225 mM stock H_2O_2 was prepared by dissolving 230 uL of 30 % H_2O_2 in 10 mL of water. To a quartz cuvette was added H_2O (2900 uL) and the freshly prepared H_2O_2 stock (100 uL). The cuvette was then placed into the spectrophotometer, allowed to equilibrate to 25 °C, and the absorbance at 215 nm was recorded three times and averaged. The absorbance was then used in Beer's law to calculate the H_2O_2 concentration (molar absorptivity of H_2O_2 at 215 nm = 100 $M^{-1}cm^{-1}$). The H_2O_2 concentration determined in this manner typically ranged from 250 to 280 mM.

Reprotonation Of Potassium Propane-2-Nitronate In Potassium Acetate Buffer (Ph 5.0)

Propane-2-nitronate reprotonation rates were determined spectrophotometrically (230 nm) at 25 °C under anaerobic conditions in several different KOAc buffer concentrations (25-100 mM, pH 5.0). A typical run is outlined as follows. To a cuvette containing 100 mM pH 5.0 KOAc (2970 uL) was added propane-2-nitronate (10 mM, 30 uL). The cuvette contents were mixed, and the decrease in propane-2-nitronate's absorbance at 230 nm was recorded over time ($_{230}$= 9400 M^{-1} cm^{-1} in 25 mM KOH). The decrease in rate was fitted to a first order rate equation using the instrument's software package. The reprotonation rate for this experiment was found to be 0.0044 min^{-1} with a half-life of 157 min. Similar experiments were performed at pH 6.0 and a half-life of 11.5 h was found.

REFERENCES

1. (a) Farber K. in "*Biotransformations in Organic Chemistry 3rd edition*", Springer, Heidlberg 1997. (b) Uhlig, Helmut. in "*Industrial Enzymes and Their Applications*", John Wiley and Sons, NY, 1998. (c) Poppe, Laszlo, Novak, Lajos in "*Selective Biocatalysis, A Synthetic Approach*", VCH, Weinheim, Germany, 1992.
2. Alcohol, Tobacco and Firearms Explosives Symposium, September 1995.
3. Public Law 104-132 "Antiterrorism and Effective Death Penalty Act of 1996." April 24, 1996.
4. Private communication from Marc Magdinec, Naval Surface Warfare Center, Indian Head Division, Indian Head, MD.
5. (a) Limatibul, S.; Watson, J. W. *J. Org. Chem.* **1972**, *37*, 4491-4492. (b) Wawzonek, S; Su, T.-Y. *J. Electrochem. Soc.* **1973** *120*, 745-747.
6. Hardy, R. J.; Lindsay, W. S.; Mees, G. C. British Patent 997,826, 1965.

7. For mechanistic details of these reactions see: (a) Russell, G. A.; Jawdosuik, M.; Makosza, M. *J. Am. Chem. Soc.* **1979**, *101*, 2355-2359. (b) Russell, G. A.; Metcalfe, A. R. *J. Am. Chem. Soc.* **1979**, *101*, 2359-2362. (c) Russell, G. A.; Danen, C. *J. Am. Chem. Soc.* **1966**, *88*, 5663-5665. (d) Bordwell, F. G.; Bausch, M. J. *J. Am. Chem. Soc.* **1986**, *108*, 1985-1988. (e) Bowman, W. R.; Suleiman, I. A.-K.; Symons, M. C. R. *J. Chem. Soc., Perkin Trans. I* **1986**, 555-565.
8. Bowman, W. R.; Symons, M. C. R. *J. Chem. Soc., Perkin Trans 2*, **1983**, 25.
9. For a review of peroxidases see: (a) Dunford, Brian. in "*Heme Peroxidases*", John Wiley and Sons, NY, 1999. (b) Everse, J., Everse, K. E. and Grisham, M. B., Eds. "*Peroxidases in Chemistry and Biology*" 1991, CRC Press, Boca Raton, FL, Vol. I and II.
10. (a) Porter, D. J.; Bright, H. J. *J. Biol. Chem.* **1983**, *258*, 9913-9924. (b) Porter, D. J. T.; Bright, H. J. *J. Biol. Chem.* **1977**, *252*, 4361-4370.
11. Dawson, J. H. *Science* **1988**, *240*, 433-439.
12. Reprotonation experiments run by J. Kenar, Naval Surface Warfare Center, Indian Head, MD.
13. Nielsen, A. T. in *The chemistry of the nitro and nitroso groups* Feuer, H. Ed., 1969, John Wiley and Sons, NY, pp. 349-486.
14. Independent work under a related 6.1 ONR program is focusing on developing evolved enzymes to address peroxidase stability.
15. (a) Araiso, T.; Dunford, H. B. *Biochem. Biophys. Res. Commun.* **1980**, *94*, 1177-1182. (b) Dolman, D.; Newell, G. A.; Thurlow, M. D.; Dunford, H. B. *Can. J. Biochem.* **1975**, *53*, 495-501.
16. March, J. in "*Advanced Organic Chemistry, Reactions, Mechanisms, and Structure*", John Wiley and Sons, NY, 1985, 3rd edition.
17. Armand, J. *Bull. Soc. Chim. France* **1965**, 3246.
18. Russell, G. *J. Amer. Chem. Soc.* **1954**, *76*, 1595-1600.
19. Olah, G. A.; Arvanaghi, M.; Vankar, Y. D.; Prakash, G. K. S. *Synthesis*, **1980**, 662-663.
20. Kenar, J. A. in "*Fiscal Year 1998 Independent Research Program Report*" Report number IHSP 99-435, Indian Head Division, Naval Surface Warfare Center, Indian Head, MD, 1999, May, 5-17.

POLYMERIZATION OF ENERGETIC OXETANES IN SUPERCRITICAL CO_2

SRI International, Menlo Park, CA

INTRODUCTION

Polymers of 2,2'-substituted oxetanes are the components of useful binder materials for rocket propellants. Block polymers made from single monomers are joined together to produce thermoplastic/elastic block co-polymers (TPEs) that serve as the actual binders. Usually, the polymerization is performed in solvents in which the resulting polymer is soluble, such as methylene chloride or toluene. The aim of this work is to explore supercritical carbon dioxide as an alternative solvent for polymerization. Carbon dioxide is considered to be an environmentally benign material because using it does not result in additional green house gas emissions, or volatile organic emissions. Some work in this area has already been conducted:

- Studies by Aerojet showed that polymerization of energetic oxetanes are possible in liquid CO_2. However, this work yielded a significant percentage of cyclic oligomers, a broad distribution of molecular weights, polymeric material that was not soluble in methylene chloride or toluene, and incomplete conversion even after long reaction times. Polymerizations also required far more catalyst than is required in organic solvent systems.
- Polymerization of NMMO in liquid CO_2 catalyzed by gaseous BF_3 was reported by the Naval Surface Warfare Center. Subsequent fractionation of the product using supercritical CO_2 produced 3 fractions of different molecular weight.

Supercritical CO_2 was expected to be a better solvent than liquid CO_2 because its solvent properties could be adjusted over a wide range by changing temperature and pressure and by adding co-solvents.

SUMMARY

SRI conducted Polymerizations of 2-azidomethyl, 2'-methyloxetane (AMMO) and 2,2' bis-azidomethyloxetane (BAMO) in a reactor with super critical-CO2. A systematic study of solubility in super critical-CO2 was not conducted before these experiments were initiated because the focus of the effort was to determine first if polymerization occurred. A systematic study of solubility would have required too much of the project resources and precluded any significant study of polymerization. Some preliminary solubility experiments seemed to indicate that the components of the system were soluble under supercritical conditions. The

Polymerizations were catalyzed by BF_3•THF, and 1,4 butanediol was used to control the molecular weight. At the end of the reaction period, the reactor was cooled with ice to lower the vapor pressure and then vented. The residue was analyzed by NMR to establish the percent monomer conversion and the molecular weight of the polymer. Molecular weights of some polymeric material were also measured by Thiokol using gel permeation chromatography. When polymerization occurred, we obtained polymers of molecular weight of Mw = 5000 – 13000 and narrow molecular weight distribution from over night reaction times of 16 - 18 hours. High conversions were obtained, and changes in molecular weight correlated with ratio of monomer to 1,4 butanediol as was expected.

We began a program to test the feasibility of preparing block co-polymers in supercritical CO_2 by first preparing polyAMMO and then polymerizing BAMO onto both ends. The first step in the study was a series of kinetic experiments with AMMO to determine whether 16 -18 hour reaction periods were required, or if a shorter time period could be used. In these experiments the polymerizations were quenched with MeOH/NaOH while the reactor was still under supercritical conditions to establish the reaction times unambiguously. No polymerization was observed, regardless of reaction time. This suggests that all AMMO polymerizations that were observed occurred only in the neat condition after the CO_2 had been vented. This is consistent with several observations that polymerization continued on the bench top after the autoclave had been vented and opened to observe the results. An examination of the data for BAMO does not show the same clear pattern – polymerization occurred once on the bench top when the autoclave contents were extracted with methanol, but for all other experiments the polymer had formed by the time the autoclave was opened, or it did not form at all.

Additional work was recommended by SRI to clarify the situation, including:

- A kinetic study of BAMO polymerization to determine if polymerization occurs under supercritical conditions or at some point during the pressure let down operation.
- A systematic study of the solubility of AMMO monomer and BAMO monomer in super critical-CO_2.
- A systematic study of the solubility of AMMO and BAMO polymer in super critical-CO_2. The study should include additional gas phase components (such as N_2O) to increase solubility.
- An investigation of the impact of higher concentrations of BF_3 and an investigation of other catalyst systems under conditions where polymer and monomer are soluble.

ONR personnel at Indian Head suggested that it would be valuable to find out if the AMMO and BAMO that was made at SRI in the supercritical CO_2 reactor behaved the same as AMMO and BAMO made by a conventional solvent based process by attempting to form AMMO-BAMO block copolymers with the SRI material.

SRI finished the research program by conducting block polymerization tests and attempting to make some of the recommended solubility measurements. There were insufficient time and resources to investigate the kinetics of BAMO polymerization or catalyst concentrations and system.

Block copolymers were made utilizing a procedure developed by Thiokol. The unreacted hydroxide end groups on AMMO and BAMO polymers were reacted with toluene 2,4 diisocyanate to create reactive polymers terminated by an isocyanate group. 1,4 butandiol was used to link the terminal isocyanates to form a block copolymer. When this procedure was used with polyAMMO (Mw_W = 5,834) and polyBAMO (Mw_W = 8,360) made by Thiokol using a solvent process, the result was a block copolymer with a molecular weight of 75,370. When this procedure was used on polyAMMO (Mw_W = 7,400) and polyBAMO (Mw_W = 12,000) made by SRI using its supercritical CO_2 process, the result was a block copolymer with a molecular weight of 14,780. Although the molecular weights of the starting materials are different, the difference in the block copolymers was larger than expected. This result could indicate that the polymers made by the two processes are chemically distinct in a manner that is not apparent using an NMR, but it could also indicate that the starting materials have different impurities and water contents. A generalized recipe was used which may have provided too large an excess of diisocyanate. And finally, since very small amounts of SRI's material were available, the block copolymerization procedure could only be conducted once, and the failure might represent part of the learning curve. Additional study of the copolymerization is required.

Solubility measurements were made in a redesigned apparatus that provided temperature control for all of the high pressure components, smaller tubing to minimize the disturbance to the system when samples were taken, and a six port valve with a 2 mL sample loop to remove material from the autoclave.

EXPERIMENTAL PROCEDURES

General

O-ring material for valves was made from EP because Viton or Buna-N was found to weaken within hours. Copper or Teflon was used as gasket material for the autoclave. In early experiments, Teflon gaskets were used at temperatures below 40°C and copper at higher temperatures where Teflon caused leaking. At the end of the program copper gaskets were used for all experiments, regardless of temperature.

Early Solubility Measurements

The autoclave was loaded with 0.4 g of polymer. The sample volume on the top of the autoclave was kept isolated by closing valve V1 (See Figure 1). The autoclave was cooled in an ice bath, connected to the CO2 supply cylinder, and loaded with 30 g of CO2 using condensation as the driving force for material transfer. The autoclave was placed in a temperature bath at 40°C and developed a pressure of ~1500 psi. The sample volume remained at room temperature because it was not immersed in the temperature bath. After 50 minutes of equilibration with the magnetic stir bar turning, valve V1 was opened to capture a sample, and it was our assumption that any dissolved material contained in the 2 ml volume of fluid that was admitted would immediately precipitate on its walls. The autoclave was removed from the temperature bath to bring the system below supercritical conditions, and the pressure of the system was then released slowly over a period of an hour by opening valve V2. The released gas was bubbled through a methylene chloride bath to capture any polymer that might be entrained or dissolved in the high pressure gas phase. An additional volume of methylene chloride was used to rinse the inside of the tubing that the gas had passed through to pick up any polymer that might have precipitated.

For a single experiment where N_2O was tested as a material to increase polymer solubility under supercritical conditions, 15 g of N_2O and 16 g of CO_2 were added to the autoclave before it was immersed in a 37°C temperature bath. The system pressure after temperature equilibrium was established was ~1500 psi.

Polymerization

Each batch of BAMO monomer was purified on a basic alumina column. The column was assembled and the bed filled with methylene chloride. The BAMO was placed on the column neat and eluted with additional amounts of methylene chloride.

- Approximately 50 g of alumina were used to treat 50 g of monomer.
- 200 to 300 mL of methylene chloride was used. The methylene chloride was assumed to be dry because it was obtained from a freshly opened bottle of reagent grade material. It was used as is – no attempt was made to measure the water content or remove residual water that may have been present. The alumina column was expected to remove any trace amounts of water that may have been present.
- A fresh batch of alumina was obtained at the beginning of the project – it was stored in a sealed container under nitrogen after opening. A new column was prepared from this material every time purification was performed. The alumina column was pretreated under a vacuum for 1 hour before being used.

After purification, the monomer solution was evaporated under a vacuum with a liquid nitrogen trap to remove the methylene chloride. The purified monomer was stored in a Schlenk flask under a nitrogen atmosphere to prevent exposure to oxygen and water. The Schlenk flask was sealed with a septum and retained on the laboratory bench. Monomer for a polymerization experiment was removed by penetrating the septum with a hypodermic syringe. The first batch of purified BAMO monomer (batch A) was material several years old that had been produced by SRI for another program. It was repurified partway through this experimental program to test a theory that the old material had impurities that were preventing polymerization, and that these materials had not been removed by the first purification. Batch A was used until it had been consumed. The second batch of purified monomer (batch B) was from the same SRI stock. It was repurified at the end of the program by filtration to test a theory that a fine particulate of zeolite from the purification was preventing polymerization.

The AMMO monomer used was material several years old that had been produced by SRI for another program. Two batches were removed from this supply (Batch C and Batch D). Each batch of AMMO monomer that was used was purified by vacuum distillation. After purification, the material was stored in a Schlenk flask under a nitrogen atmosphere to prevent exposure to oxygen and water. The Schlenk flask was sealed with a septum and retained on the laboratory bench. Monomer for a polymerization experiment was removed by penetrating the septum with a syringe with a 6-inch needle.

1,4-Butanediol and BF_3•THF, which had been purchased from Aldrich, were stored in glass inside of a drybox under a nitrogen atmosphere. Prior to storage the butanediol was dried by dissolving 1% by weight of sodium in it and then evaporating it under vacuum. The purified material was collected in a trap cooled by liquid nitrogen. The appropriate amounts of the diol and the BF_3•THF were premixed on the day when an experiment was conducted, 5 minutes was allowed for any chemical reactions to occur, and the mix was charged to the sample volume that had been removed from the autoclave. The valve on the sample volume was closed, and the sample volume was removed from the glove box.

The experimental apparatus is shown in Figure 1. It was based on a 25 mL autoclave with a single penetration through the top. All fittings were 1/4 inch stainless tubing and Swagelock fittings. An electronic pressure gauge was included to permit continuous recording of pressure. A pressure relief valve was provided to protect the system from rapid reactions. Valve V2 connected the system to the carbon dioxide supply, which contained an hplc pump to provide the pressures required. A 2 mL sample chamber was attached by means of valve V1. The sample chamber was used to introduce catalyst, to sample the reactor contents when solubility measurements were made, and to add a quenching agent when kinetic studies were conducted. Prior to a polymerization attempt, all parts of the autoclave except pressure transducer were heated for at least an hour at 100°C to drive off water and solvents. Next the autoclave was assembled with a cap instead of the sample volume and valve V1. The system was tested for any leakage by applying CO_2 pressure (900 psi), and then the cap was removed and a slow flow of CO_2 initiated. Monomer was placed in autoclave through the open top using the syringe. The sample volume, which had been filled with catalyst, was connected to autoclave at the top with valve V1 in the closed position. The apparatus was placed in the temperature controlled water bath and pressurized using a pump connected to a CO_2 tank with siphon. A heating tape was wrapped around the tubing and other parts that were outside the water bath. The power input to the tape was set to maintain the exposed parts 1-2 degrees above temperature of the water bath so that any temperature differential would favor material being found in the cooler autoclave, rather than in the tubing. The temperature between the heating tape and the heated parts was tested at several points with a hand held thermocouple to confirm the power setting. When the temperature and pressure stabilized at the values required by the experimental plan, the valve of the sample volume (V1) was opened to allow the catalyst mixture to drain into the CO_2 and monomer mixture to start the reaction. V1 was closed again. Valve V2 was closed to isolate the reactor from the pump, and the connecting line from the pump was removed. Most reactions were started in the mid-afternoon, and allowed to run until the next morning, thereby achieving reaction times of 16 – 18 hours. At the end of the reaction time the water bath was exchanged with an ice bath to cool the autoclave. 5 minutes of cooling brought the temperature below the supercritical point for CO_2, and SRI's assumption was that polymer and any remaining monomer either formed a separate condensed phase, or were retained in the condensed CO_2 phase that formed. After cooling, the CO_2 gas phase was vented over a period of 15-20 min by opening valve V2, and then the autoclave was opened. The contents of the autoclave were treated with one of the workup procedures, and NMR samples were taken from the product.

After a BAMO polymerization attempt the reactor contents were treated by four different procedures that evolved as the program proceeded. During the extraction operations that were part of some of these procedures, no insoluble residues were observed in the reactor.

- In the early experiments they were extracted from the reactor with methylene chloride and quenched with water. The water layer was discarded and the methylene chloride layer was evaporated to dryness in a rotovap
- A simplified procedure used a methanol quench, which solublized the contents of the reactor. The methanol solution was evaporated to dryness in a rotovap. When it was observed that polymerization continued on the bench top with material that had been treated with methanol, the methanol quench was eliminated.
- A methylene chloride extraction followed by a quench with water saturated with sodium carbonate was substituted for the methanol quench.
- In the last experiments no workup was applied to the material in the autoclave.

After an AMMO polymerization attempt, the contents of the reactor were sampled directly without any workup.

KINETIC EXPERIMENTS WITH AMMO

A kinetic experiment was attempted first by cooling the reactor and venting it after a chosen reaction time had elapsed. During this procedure it was observed that polymerization continued on the bench top after the autoclave was opened. This meant that any correlation of extent of polymerization versus reaction time would be confounded by the time it took to cool and vent the reactor and handle the sample. It was determined that a chemical quenching step would be required to establish a definite reaction time. The sample volume on the top of the autoclave that had been used to add the catalyst mixture was also used to add a methanol solution saturated with sodium hydroxide to the autoclave. After the catalyst had been added, the sample valve (V1) was closed, and the cap was removed from the sample volume. This operation vented a small fraction of the reactor contents. The methanol solution was added to the sample volume, and the cap was replaced. When the chosen reaction time had elapsed, valve V1 was opened to dump the methanol solution into the autoclave. The pressure in the system fell by less than 100 psi when the valve was turned.

Block Copolymerization Of AMMO And BAMO

Block copolymers were made utilizing a procedure developed by Thiokol. The unreacted hydroxide end groups on AMMO and BAMO polymers were reacted with toluene 2,4 diisocyanate (TDI) to create reactive polymers terminated by an isocyanate group. Diphenyltin dichloride (DTD) was used as a catalyst. 1,4 butanediol was used to link the terminal isocyanates to form a block copolymer.

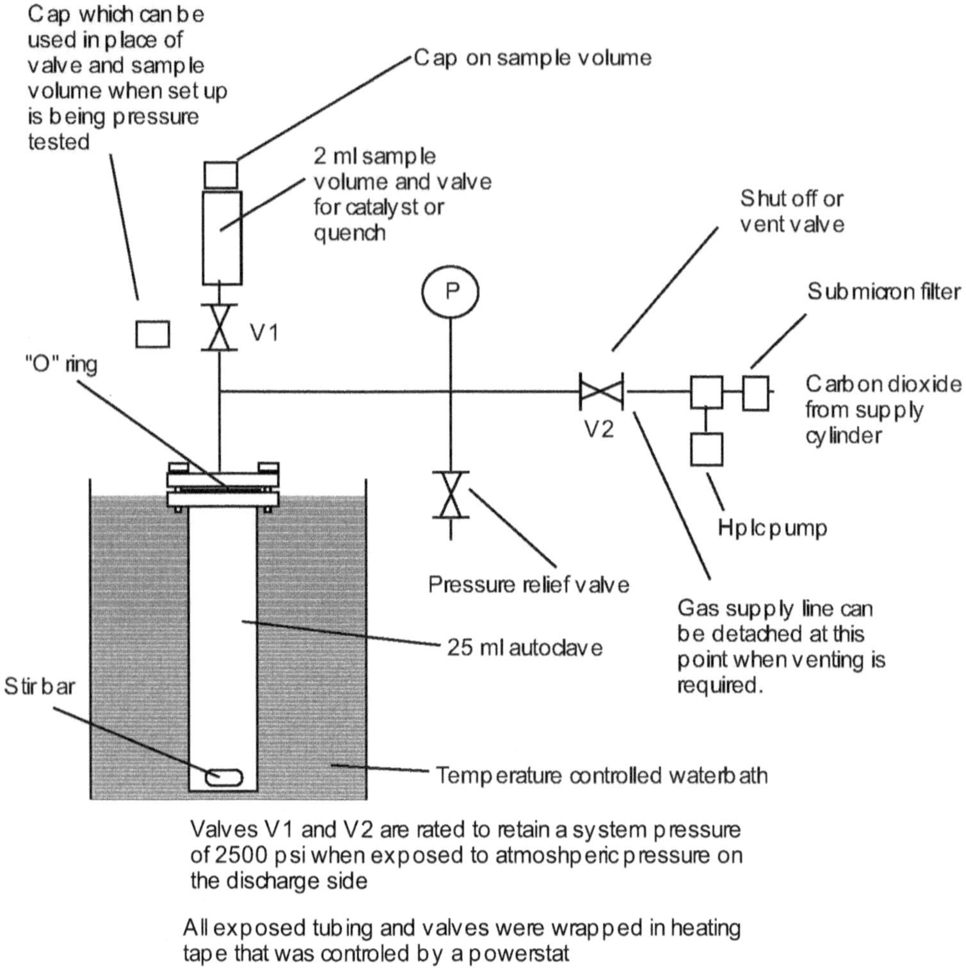

Figure 1: ***Apparatus for Polymerization Experiments***

Before any reactions were conducted, all materials were dried to remove water that would interfere with the coupling reaction. Poly AMMO and poly BAMO samples were dried under high vacuum (< 0.025 mm Hg) for 24 hrs before use. The chloroform solvent and 1,4-butanediol coupling agent were stored over 4Å molecular sieves for 24 hrs. All the reactions conducted under dry argon.

In the first step, 1.275 g poly AMMO was dissolved in 2 mL $CHCl_3$. 64.8 μL of TDI and 1.5 mg of DTD were added while the mixture was stirred. In a separate flask, 0.428 g poly BAMO was dissolved in 2 mL $CHCl_3$ and 34.6 μL TDI and 1.5 mg DTD were added while the mixture was stirred. Both reaction flasks were fitted with condensers and then heated in an oil bath to 45 °C for 24 hrs to allow the isocyanate reaction to go to completion.

In the second step, the two reaction flasks were combined and 30.6 μL of 1,4-butanediol was added. Heating was continued at 45 °C for 36 hrs. At this point, the reaction was poured into 5 mL of stirring methanol to isolate the block copolymer as an oily precipitate. Methanol is a non-solvent for the block copolymer. The precipitate was isolated by decanting. The precipitate was then washed several times with methanol under conditions of vigorous stirring until a granular product could be isolated. SRI was unable to isolate a granular product using this procedure, and modified it to include several washes with 3:1 $CHCl_3$: methanol until a mostly granular product resulted. The granular product was isolated by filtration and dried under vacuum. At this small scale isolation of the product was difficult and the polymer remained somewhat tacky after drying.

Additional Solubility Measurements At The End Of The Program

The experimental apparatus was modified as shown in Figure 2. The major improvements included:

- Eliminating the sampling volume at the top of the reactor
- Installing a 6 port sampling valve so that catalyst, and/or quenching solution could be added directly, and samples of the reactor contents could be removed easily.
- Substituting fine bore tubing wherever possible to minimize the dead volume included in a sample.
- Installing a fine frit in the line between the reactor and the sampling valve to remove any particulate material that might be flushed out of the reactor during sampling
- Installing a thermocouple to measure the temperature inside the reactor
- Replacing the water bath with an air bath. Most of the piping at the top of the reactor was included in the air bath, which made it unnecessary to try to control

the piping temperature with heating tape. The air bath contains a heat source, a temperature controller, and a fan.

ANALYSIS

^{1}H NMR spectroscopy was used to analyze the BAMO and AMMO polymerization reaction mixtures in terms of monomer conversion, and molecular weight. Product isolated from the autoclave was dissolved in $CDCl_3$. Trifluoroaceticanhydride (TFAA) was added to the $CDCl_3$ solution to facilitate the molecular weight determination. TFAA reacted with CH_2OH end groups to form CH_2O-TFAA end groups, and this shifted the frequency of the signal enough to be read easily. The molecular weight by NMR was calculated from the ratio of the methylene signals in the polymer chain (3.3-3.5 ppm for both polyBAMO (corresponds to 8H) and polyAMMO (corresponds to 6H)) versus the end group signals (4.29 ppm for polyBAMO; 4.24 ppm for polyAMMO (corresponds to 4H)). Thiokol performed low temperature GPC analysis with THF as mobile phase on some selected samples of polyBAMO and polyAMMO to provide an independent estimate of the molecular weight and to determine the molecular weight distribution. All block copolymers were analyzed by GPC at Thiokol.

RESULTS

Solubility

A limited number of polymer solubility experiments were conducted before polymerization was attempted. When the autoclave was opened after the pressure had been released to atmospheric, the polymer was observed to have changed state from a single lump to a thin coating on the walls of the autoclave and a fine powder at the bottom. No material was found in the sample volume, the methylene chloride bubbler, or the tubing. When N_2O was added to the system (in a single experiment), a small amount of polymer was detected in the sample volume. The results were interpreted to mean that polymer did dissolve, but that the sampling procedure using the sample volume was flawed, in that it disturbed the system too much and/or did not obtain a representative sample. Additional experiments using a better sampling procedure and an autoclave with a window were not conducted because they would have consumed most of the limited time and budget that had been allocated to the project, and the real project objective was to achieve polymerization.

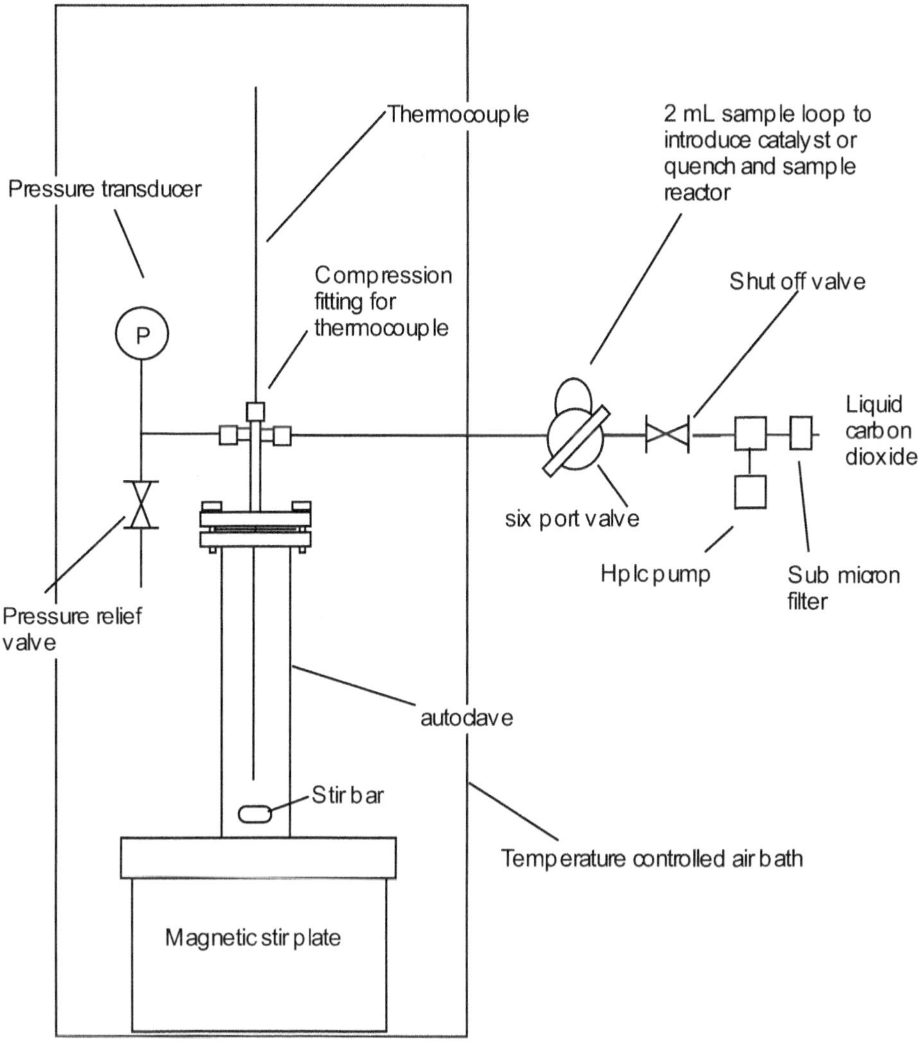

Figure 2. ***Improved Experimental Apparatus with Temperature Control of all High Pressure Parts and a 6 Port Sampling Valve***

Polymerization

Table 1 and 2 summarize the results of our work in the period of March-May 1999. Table 1 includes all of the experiments performed with BAMO monomer. The experiment number refers to the page in the laboratory notebook where the experiment is described. Table 2 includes experiments that were performed with AMMO monomer, including kinetic experiments where the reaction was quenched by addition of 0.5 ml MeOH/NaOH at the end of the reaction time before the water bath was exchanged for the ice bath.

BAMO Polymerization

The results with BAMO are mixed. The first polymerization failed, but that was assumed to be part of the shake down process. In the experiments that followed, several different methods were used to limit the amount of polymerization that could occur after the autoclave was opened. In experiment 7, water, which is known to destroy the activity of the BF_3 catalyst, was used to quench the reaction after the autoclave contents had been dissolved in methylene chloride.

Table 1. ***BAMO Polymerization Experiments***

Experiment	T (°C)	Pressure (psi)	monomer	Monomer: C_4: BF_4	Conv. %	Mw (NMR)	Mw (GPC)	Polydispersity (GPC)	Comment
13100-6	36-37	1380 - 1385	A	22: 1: 0.55	0				Contents of opened autoclave were quenched with methylene chloride and water
13100-7	38-39	1425 - 1680	A	28: 1: 0.49	97	8245	12000	1.92	Contents of opened autoclave were quenched with methylene chloride and water
13100-9	32-33	2200 - 2600	A	20: 1: 0.51	>97	3120	6100	1.66	Contents of opened autoclave were quenched with methanol. 20% polymerization observed when methanol evaporated, complete polymerization one hour later

13100-10	39-47	1370 - 1990	A repurified	20: 1: 0.57	97	3380	5800	1.48	Polymer observed in autoclave when opened. Contents extracted with methylene chloride and sodium carbonate saturated water
13100-11	35	1510 - 1310	A repurified	25: 1: 0.66	0				System may have leaked. Contents extracted with methylene chloride and sodium carbonate saturated water
13100-12	38-25-39	2752 - 1795	A repurified	21: 1: 0.57	0				Temperature control lost overnite, reactor reheated in morning for 2 hrs. Contents extracted with methylene chloride and sodium carbonate saturated water

Table 1. ***BAMO Polymerization Experiments (concluded)***

Experiment	T (°C)	Pressure (psi)	monomer	Monomer : C_4: BF_4	Conv. %	Mw (NMR)	Mw (GPC)	Polydispersity (GPC)	Comment
13100-13			A repurified	22: 1: 0.57	0				Contents extracted with methylene chloride and sodium carbonate saturated water
13100-14	39-47	1540 - 1440	A repurified	26: 1: 0.57	83	7160	10780		Polymer observed in autoclave when opened. Solid washed with water and methanol
13100-15	44-42	1300 - 1290	A repurified	24: 1: 0.57	>97	7640	9070		Polymer observed in autoclave when opened. No workup
13100-17	36	2034 - 1613	A repurified	27: 1: 0.57	0				No workup
13100-24	42	2000 - 2290	A repurified	46: 1: 1.14	0				No workup
13100-25	41	2580 - 2607	A repurified	47: 1: 1.11	0				No workup. No polymerization observed on bench
13100-26	41	2320 - 2370	B	44: 1: 1.14	0				No workup. No polymerization observed on bench
13100-27	40-39	1690 - 1710	B	47: 1: 1.11	0				No workup. No polymerization observed on bench
13100-28	38-37	1400 - 1350	B refiltered	54: 1: 1.07	0				No workup. No polymerization observed on bench. Hot spot found in tubing with yellow residue – deactivated catalyst?

Table 2. ***AMMO Polymerization Experiments***

Experiment	T (°C)	Pressure (psi)	Monomer	Monomer : C_4: BF_4	Conv. %	Mw (NMR)	Mw (GPC)	Polydispersity (GPC)	Comment
13100-16	35-40	2800 - >3000	C	28: 1: 0.57	91	3080	7420	2.17	Polymerization occurred on bench top in 4-5 hrs
13100-18	36	2650 - 2226	C	33: 1: 0.56	0				
13100-19	42	1650 - 1150	C	32: 1: 0.57	46				Polymer observed in autoclavewhen it was opened
13100-20	38	2884 - 2300	C	30: 1: 0.59	NA				Polymerization occurred on bench top in 24 hrs
13100-21	41-40	2050 - 2780	C Redistilled	28: 1: 0.76	93	2680	4760	1.47	Polymer observed in autoclave when it was opened
13100-22	41	1600 - 1945	C Redistilled	46: 1: 0.92	87	3920	7400	1.87	Polymer observed in autoclave when it was opened
13100-23	41	1950 - 1030	C Redistilled	61: 1: 1.02	84	5660	12950	2.15	System leaked. Polymer observed in autoclave when it was opened

13100-33	39	2200 - 2350	D	14.1: 1: 0.83	28	2.5 hrs in reactor, unquenched, polymerization completed on bench top
13100-34	38-39	2250 - 2260	D	16.7: 1: 0.83	0	3 hrs in reactor, quenched with methanol/NaOH
13100-35	38-39	2230 - 2360	D	16.7: 1: 0.83	0	14 hrs in reactor, quenched with methanol/NaOH
13100-36	38-39	2200 - 2350	D	14.1: 1: 0.82	0	16.5 hrs in reactor, quenched with methanol/NaOH
13100-37	38-39	2200 - 2145	D	14.1: 1: 0.82	100	16 hrs in reactor, unquenched, exotherm during handling

Experiment 8 is not reported because a leak necessitated opening the autoclave, replacing a gasket, sealing the autoclave, and repressurizing. No polymer was formed, but it was assumed to be because water or air exposure destroyed the catalyst. In experiment 9, methanol was used as a quench, but it did not prove effective, since some additional polymerization occurred on the bench after the methanol was evaporated to facilitate NMR analysis. Experiment 9 shows polymerization taking place after the CO_2 pressure was released. SRI's interpretation at the time, however, was that the bulk of the polymerization occurred at pressure, and that the failure to polymerize completely under the high-pressure conditions was due to an impurity that slowed the polymerization rate. The monomer was repurified after experiment 9.

Experiment 10 yielded a polymerization that was completed before the autoclave was opened. Experiment 11, which suffered a pressure leak some time in the night, yielded no polymerization, and the failure was believed to be due to the low pressure for most of the overnight reaction period. The low pressure would have yielded a low-density fluid inside the reactor and limited the solubility of the reactants or polymer. Experiment 12, which suffered a failure of the temperature control system, yielded no polymerization, and the failure was assumed to be because the low temperature suppressed reaction rates. If we consider that experiment 10 demonstrated that the monomer was pure enough to polymerize, these two failures and the failure of experiment 13 as well argue against the concept that polymerization occurred in the autoclave only after CO_2 had been released.

Experiments14 and 15 again yielded polymerization, with no evidence of polymerization occurring after the autoclave was opened. Experiments 17, 24, and 25 showed no polymerization under experimental conditions close to ones that had yielded polymerization previously. A new batch of BAMO monomer then had to be prepared because batch A had been used up. The new batch did not polymerize either. On the last experiment, batch B was refiltered to provide assurance that a fine particulate residue of zeolite from the purification was not interfering with the chemistry. After the experiment, a hot spot was found in the tubing leading to the autoclave. The hot spot was found by way of a yellow residue, and it was assumed that the residue represented destroyed catalyst. SRI's presumption at the time was that this hot spot had been present during all of the experiments following number 15 and was the cause of the failure to polymerize.

When polymerization occurred, poly(BAMO) was formed with molecular weight ranging from Mw = 5800-12000 g/mol with a polydispersity index (PDI) of 1.48-1.92 (polydispersity index = Mw/Mn where Mw is the weight average molecular weight, and Mn is the number average molecular weight) at conditions of 1300-2000 psi carbon dioxide pressure and a temperature range of 38-47°C. The PDI increased with higher molecular weight. Molecular weights correlated well with the ratio of monomer to1,4 butanediol.

Most unsuccessful BAMO polymerization runs could not be explained by specific reaction conditions such as temperature or pressure, or product purity. A plot of polymerization outcome with respect to temperature and pressure showed no pattern. Additional polymerization experiments, including ones with in situ quenching, such as was used with the final AMMO experiments, will be required to clarify matters. These experiments should be conducted in an apparatus that does not have the potential for hot spots. Systematic solubility studies of monomer and polymer should also be conducted to clarify matters.

AMMO Polymerization

The results with AMMO are mixed. In experiment 16 the polymerization occurred slowly on the bench top. In experiment 18, which was conducted under similar conditions, no polymerization occurred at all, in or out of the autoclave. In experiment 19 polymerization was incomplete, and in experiment 20 it was slow. If we accept the argument made later in the analysis of results that polymerization occurred only after CO_2 had been released from the system, one explanation for experiments 16-20 is that an impurity in the monomer slowed the polymerization reaction. The monomer was repurified for experiments 21 through 23, and almost complete polymerization was observed when the autoclave was opened. It is unfortunate for understanding the system that the amount of catalyst used was also increased for these experiments.

When poly(AMMO) was formed, it had molecular weights in the range of Mw = 4760 to 12950. The trend in molecular weight correlates well with the ratio of monomer to 1,4 butanediol. As in the case of BAMO polymerization, the polydispersity index increases with increasing molecular weight from PDI=1.47 to 2.15.

In the next step we investigated the overall kinetics of AMMO polymerization in preparation for experiments where BAMO would be added to the ends of complete polyAMMO. In experiment 33, which was conducted with only a short reaction time and without quenching, it was apparent that polymerization occurred after the autoclave was opened. This meant that polymerization could be occurring during NMR analysis. This made it necessary to develop the quenching procedure. Experiments 34, 35, and 36, which were conducted under the same conditions as experiment 33, made it clear that if the reaction was quenched before the autoclave was opened, no polymer would be found, regardless of the reaction time. Polymerization occurred some time during the 20-min handling time that was required for cooling and venting – presumable near the end when CO_2 concentrations were low. Additional studies will be required to determine if the failure to polymerize under supercritical conditions is a solubility problem, or a catalyst activity problem. Catalyst activity could be investigated by testing higher concentrations of BF_3 and different catalyst materials.

Block Copolymerization

Block copolymerizations of polyAMMO and polyBAMO samples were performed in our laboratory using a procedure supplied by Thiokol. We performed the block polymerizations on samples of polyAMMO and polyBAMO that were made at Thiokol (solvent process) and on samples that were made at SRI (supercritical CO_2).

Table 3 shows the molecular weights of the starting polymers and the resulting block copolymers. Molecular weights were determined by GPC at Thiokol. Clearly, there are differences between block polymers derived from the Thiokol and SRI polymers. We believe that one explanation for this difference is the nature of the synthesis procedure used. The procedure as practiced does not isolate the reactive polymers that are terminated by an isocyanate group from unreacted diisocyante or unreacted polymer. In the linking step, unreacted excess diisocyanate can react directly with the butanediol and prevent it from linking the block polymers. The amounts of toluene diisocyanate (TDI) used in the procedure was not adjusted to reflect the specific molecular weights of the starting polyAMMO and polyBAMO. We may have used way too much toluene diisocyanate relative to the amount of -OH groups present on the starting polymers. Based on the molecular weights reported below we calculated that excess TDI was used in both cases but the larger excess was used in the case of the SRI derived polymers.

The TDI-butanediol polymers formed from excess diisocynate could account for the difficult isolations observed. The block copolymer produced by the procedure using SRI produced polymer could not be isolated as a truly granular product. It remained a tacky material, and was tackier than block copolymer produced by the procedure using Thiokol produced polymer. The difficult isolations themselves could be responsible for the apparent differences in the block polymers. SRI had to modify the isolation procedure for both starting materials to include repeated washing with a 3:1 $CHCl_3$: methanol solution. Chloroform is a solvent for the block copolymer, and repeated washing with this solution would remove some block copolymer from the isolated product, thus changing its molecular weight. Other possible reasons for the different results with Thiokol and SRI produced material include differences in the impurities in the polymers, and differences in water content that will adversely effect block polymerization. Further work will be required to determine if these or other reasons are responsible for the differences in molecular weights of these polymers.

Limited supplies of AMMO and BAMO made by supercritical CO2 polymerization were available, and the block copolymerization procedure was completed only once. It may be the case that the results simply reflect being low on the learning curve.

Table 3. Results of Block Copolymerization

Sample	Origin	M_n	M_w	M_w/M_n
PolyAMMO	SRI		7,400	
PolyBAMO	SRI		12,000	
PolyAMMO	Thiokol		5,834	
PolyBAMO	Thiokol		8,360	
$(pAMMO\text{-}pBAMO)_n$	SRI	9,260	14,780	1.60
$(pAMMO\text{-}pBAMO)_n$	Thiokol	17,730	75,370	4.25

Additional Solubility Measurements

We attempted to measure the solubility of BAMO in super critical CO_2 by making modifications to our existing apparatus. Our procedure for measuring solubility was to load the temperature controlled reactor with a quantity of BAMO and then fill it with CO_2 until the desired pressure was reached, typically about 3000 psi at 35° C. The CO_2/BAMO solution was stirred and after about 10 min the 2.0 ml sample loop was open to the system and filled with CO_2 /BAMO solution. Subsequent careful venting of the sample loop was used to isolate the BAMO in solution.

Unfortunately, we experienced some problems with our system that prevented us from obtaining useful solubility data. One issue was that we experienced problems with our HPLC pump during this time in the project. Frequently we were unable to achieve pressures that corresponded to a supercritical state. We were unable to effect permanent repairs to the pump within the time frame that we had to make these solubility measurements. A second problem was system design. When the sample loop was open at high pressure there was too much of a volume change in the system - despite our efforts to minimize this. This resulted in large pressure changes while sampling. For example, one sampling cycle could cause the pressure to drop from 2800 psi to 1800 psi or lower. Thus the actual sampling conditions were not known, but they certainly were not made at supercritical pressures. An additional difficulty resulted from the dead volume in the pressure gauge and pressure relief valve at the top of the reactor. CO2 in this dead volume probably was not in equilibrium with the BAMO that had been placed in the reactor, but it was the first CO2 out of the reactor when the sample valve was actuated. Future solubility measurements must use a redesigned system that addresses the above problems. Systems are commercially available to complete this work.

SYNTHESIS OF ENERGETIC MONOMERS USING SUPERCRITICAL FLUID REACTION MEDIA

GenCorp Aerojet, Sacramento, CA

INTRODUCTION

Aerojet recently completed a two year SERDP (Strategic Environmental Research and Development Program) funded contract, N00014-96-C-0255, to investigate and demonstrate the cationic induced polymerization of cyclic ethers (oxetanes) using supercritical carbon dioxide as a replacement solvent for methylene chloride. We successfully synthesized a series of Homo and copolymers which, on analysis, proved to be equal or superior to those produced in the ODS (ozone depleting substances) methylene chloride. This program is intended to expand on that program and in partnership with NOS/Indian Head, we will transfer technology to the Navy with the intent of scaling up to a larger reactor. This program will merge the innovative technology of Aerojet with the scale up expertise of the Navy.

Background

A number of energetic oxetane polymers were developed at Aerojet to be used as binders in PEP (Propellants, Explosives and Pyrotechnic) compositions. These polyoxetanes contain azido, nitrate ester, difluoroamino and nitramine pendant groups. They are either chemically cured telechelic polymers or random copolymers with hydroxyl or other reactive groups or energetic thermoplastic elastomers (ETPEs). ETPEs are tri-block or multi-arm star block copolymers with an amorphous center block (core) and crystalline end blocks (arms). ETPE oxetane binders offer a number of advantages over chemically cured binders since they can be re-melted. The melting transition of the crystalline blocks can be tailored to suit the application. These polyoxetanes are obtained by cationic ring-opening polymerizations using appropriate initiators.

In a previous program, funded through SERDP and administered by the Office of Naval Research, we reduced to practice the synthesis of monomers and polymers under environmentally acceptable conditions. It was the overall objective of this program to demonstrate the manufacture of polymeric materials using clean manufacturing technology from initial design through end-of-service recycle. This technology eliminated the use of potentially hazardous solvents and chemicals, reduce or eliminate waste streams, recycle when feasible and reduce explosive hazards. It was intended to reduce to practice a general manufacturing technique requiring a single plant design, thus allowing a wide variety of materials to be produced, eliminating large capital expenditure.

In the first phase of the program, we developed improved methods for synthesis of oxetane monomers which are used to make polyoxetane energetic binders from commercially available starting materials. Original methods of monomer synthesis used toxic, hazardous and environmentally undesirable reactants and solvents, such as thionyl chloride, pyridine, hydrobromic acid, DMSO and dimethylformamide. We developed new synthetic methods that do not use these chemicals and that produce reduced amounts of waste streams, some of them aqueous that can be discharged into sewer. We demonstrated these synthetic methods in two pilot plants built at Aerojet.

In the second phase of the program, we proved viability of oxetane polymerization in liquid carbon dioxide solvent. Liquid carbon dioxide replaced toxic and environmentally undesirable solvents, such as methylene chloride and it can be easily recycled. Initial polymerization work demonstrated that the catalyst BF_3.THF was effective with oxetane monomers and found suitable for polymerizations leading to homopolymers and random co-polymers.

In the last phase of the program, we designed, built and operated an apparatus with one-liter reactor and collected data on polymerization of energetic oxetanes in liquid carbon dioxide solvent. The program covered by this report was built upon this fundamental concept and took the synthesis through initial scale up and then ultimately to block linking a crystalline and amorphous polymer to form a simple Thermoplastic Elastomer.

RESULTS AND DISCUSSION

Synthesis of Oxetane Monomers

Oxetane monomers are not currently commercially available and must be therefore prepared from commercially available materials. We have prepared at Aerojet, on a laboratory scale, a number of energetic polyoxetanes containing azido, nitrate ester and difluoroamino groups. At present, we have produced larger amounts of polyoxetanes based on BAMO, AMMO, BNMO and NMMO. As the requirement for this program is to produce Poly [BAMO] and coPoly [BAMO/NMMO] for further block linking to form the ABn TPE [BAMO-b-BAMO/NMMO] the monomers BAMO and NMMO were prepared by the improved methods described below.

Azide Oxetane Monomers

The original method for synthesis of the BAMO monomer was based on synthesis of 3,3-bis(chloromethyl)oxetane (BCMO) from the commercially available pentaerythritol. BCMO was then converted into BAMO by the reaction of BCMO with sodium azide in hot dimethylformamide (DMF) or dimethylsulfoxide (DMSO).

The reactions are shown below:

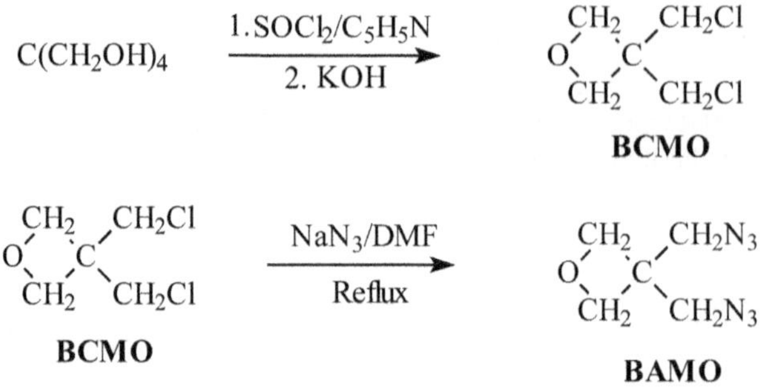

Old Method

The synthesis of BCMO uses flammable and toxic solvent and produces very corrosive fumes of hydrogen chloride. Use of hot DMSO or DMF is also undesirable. Organic waste solutions containing azide are produced and DMSO is known to facilitate transport of toxic substances through skin. DMF is a suspected carcinogen.

The new method that we developed uses commercially available tribromopentaerythritol which is converted into 3,3-bis(bromomethyl)oxetane (BBMO). BBMO is then converted into BAMO by the reaction with sodium azide in water using a phase-transfer catalyst, tetrabutylammonium bromide. The reactions are shown below:

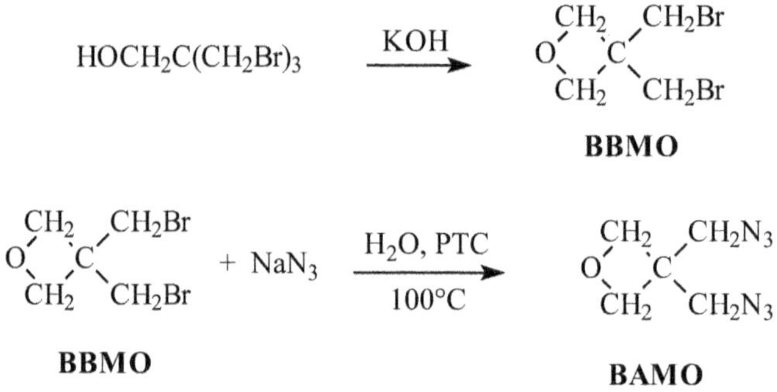

New Method

This method produces BAMO with high yield and in high purity suitable for polymerization. All aqueous waste is treated to kill the residual sodium azide waste solution. Four pounds of BAMO were produced by the new method for subsequent polymerization.

Nitrate Ester Oxetane Monomers

The original method for synthesis of nitrate ester oxetanes, such as 3-nitratomethyl-3-methyloxetane (NMMO) and 3,3-bis(nitratomethyl)oxetane (BNMO) was carried out as a batch process in methylene chloride. Acetyl nitrate is generated in a flask from acetic anhydride and nitric acid. This is a hazardous substance, which decomposes at elevated temperatures. The reaction mixture must be therefore kept at low temperatures. When scaled up, there would be large quantities of this material present.

We developed an improved method for synthesis of nitrate ester oxetanes in a continuous reactor pilot plant. This process can be used for synthesis of various nitrate esters, such as 3-nitratomethyl-3-methyl-oxetane (NMMO) or 3,3-bis(nitratomethyl)oxetane (BNMO). The reaction scheme for synthesis of NMMO from commercially available 3-hydroxymethyl-3-methyloxetane (HMMO) is shown below:

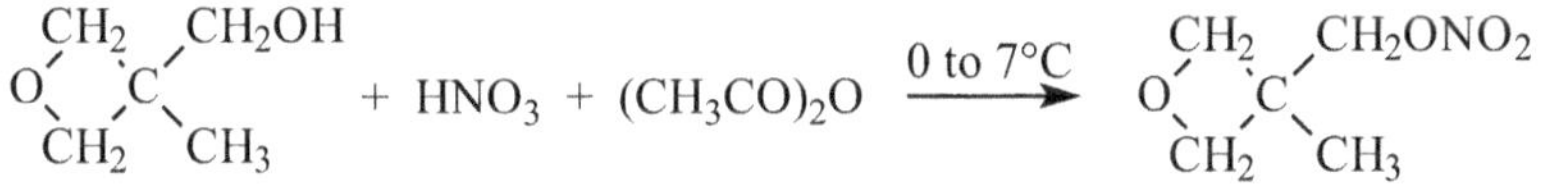

HMMO **NMMO**

Two pounds of NMMO monomer were produced by this method for subsequent polymerization

Oxetane Homo and Co-Polymerization

Since the early eighties, we have polymerized a number of energetic and inert oxetanes to high polymers of well-defined structure with high yields. However, all polymerizations required large amounts of toxic and environmentally undesirable solvents. Methylene chloride is the most commonly used solvent in these reactions. Methylene chloride is toxic and is also a suspected carcinogen. The permissible vapor concentrations at workplaces were recently drastically reduced. It was imperative to find an alternate process in which methylene chloride would not be used.

Liquid carbon dioxide solvent was chosen as a polymerization media to replace methylene chloride. In addition to being environmentally responsible, carbon dioxide exhibits many solvent properties, such as low polarizability, low viscosity, high density and high diffusion rates, that provides an excellent polymerization medium.

Polymerizations of oxetanes in methylene chloride are usually performed at temperatures below ambient. Under those conditions, side-reactions are suppressed and polymer characteristics are improved. Oxetane polymerizations at elevated temperatures usually produce large amounts of low molecular-weight oligomers. In particular, oxetane polymerizations with bis-cumyl chloride are usually performed in methylene chloride in the -70 to -90°C range. Inspection of the phase diagrams of carbon dioxide reveals that the triple point of CO_2 is -55°C and the critical temperature is +31°C. This means that CO_2 freezes to solid at -55°C and can not be used as a solvent below these temperatures. On the other hand, CO_2 can not exist as a supercritical liquid at temperatures below +31°C. Supercritical fluids exhibit some desirable properties as polymerization solvents. However, it appeared that temperatures above +31°C might not be suitable for oxetane polymerization. We therefore concentrated on sub-critical liquid carbon dioxide as a solvent

Polymerization of Energetic Oxetanes with BF_3-THF Initiator

We have used initiators based on boron trifluoride for some time at Aerojet for synthesis of homopolymers and random copolymers from energetic oxetane monomers. Polymerization is performed at a temperature of approximately 0°C. These initiators are not suitable for synthesis of block copolymers, however. An example of homopolymer formation is shown below for poly-BAMO. An example of random copolymer formation is shown below for poly-BAMO-co-AMMO.

Homopolymers:

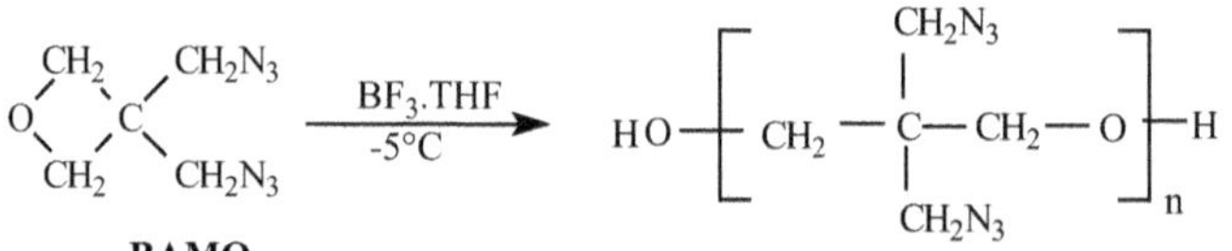

BAMO **poly-BAMO**

Random Co-Polymers:

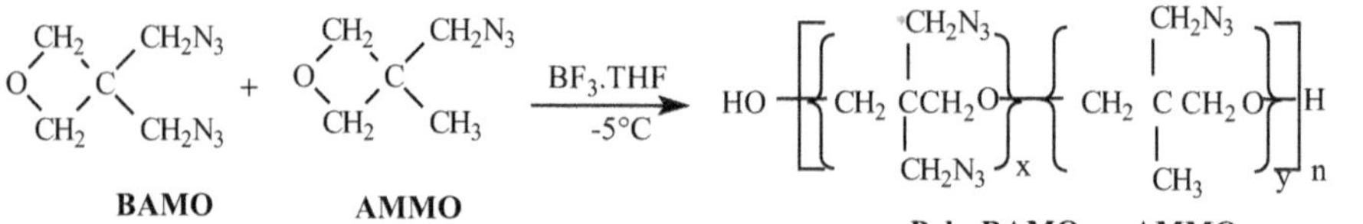

BAMO **AMMO** **Poly-BAMO-co-AMMO**

The BF_3-THF initiator was developed in Aerojet laboratories for oxetane polymerization. Prior to this development, we used the adduct of boron trifluoride and diethyl ether. However, this adduct is very hygroscopic and it has a short shelf life. The adduct of BF_3 and tetrahydrofuran (THF) is much more stable and less hygroscopic. An additional benefit is that replacing BF_3-etherate by BF_3-THF improves the polymer functionality. Therefore BF_3.THF was used for this work to form polyether glycols that will be further chair extended to form a TPE.

Liquid Carbon Dioxide Reactor Configuration

In the previous contract we designed and built a one-liter reactor system that produced 25-gram batches of a wide range of energetic and inert Polyether glycols. As the final design provided useful materials we continued its use for this work.

A line drawing is shown below:

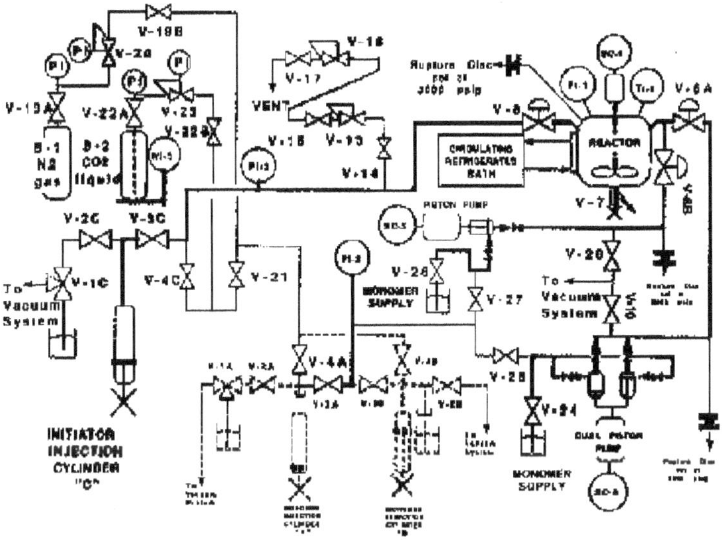

Figure 1. ***One-liter Liquid CO_2 Reactor and Support Equipment***

Initial work concentrated on finding the maximum loading possible in this reactor. It became apparent that the magnetic drive couldn't stir greater than 60-gram batches. As this was a limiting factor to scaling up further without going to a direct drive we decided that this batch size would allow us to synthesize sufficient materials to fulfill our program goals.

After two 60 gram batches of poly(BAMO) were produced an incident occurred that demonstrated a fundamental fault in the reactor design which had not been apparent. During the addition of the monomer BAMO through the first dual piston pump an over pressurization occurred that ruptured the feed line into the reactor. After an exhaustive investigation it appeared that the most plausible cause was attributed to the monomer feed pump losing pressure and allowing active catalyst to flow out of the reactor into the feed pump in direct contact with neat monomer. This caused a rapid polymerization of the monomer and over pressured the line. The above design did not have a mechanism to maintain a positive pressure outside the reactor.

As a result of this incident and taking into account the most plausible cause we redesigned the reactor to have a failsafe design so that the feed system will always be at a higher pressure than the reactor. This is achieved by introducing twin injection vessels (one for monomer and one for water quench) that are maintained at 100 psig above the reactor pressure via a differential pressure regulator. Under these conditions flow can only occur into the reactor.

A line drawing of the improved design is shown below:

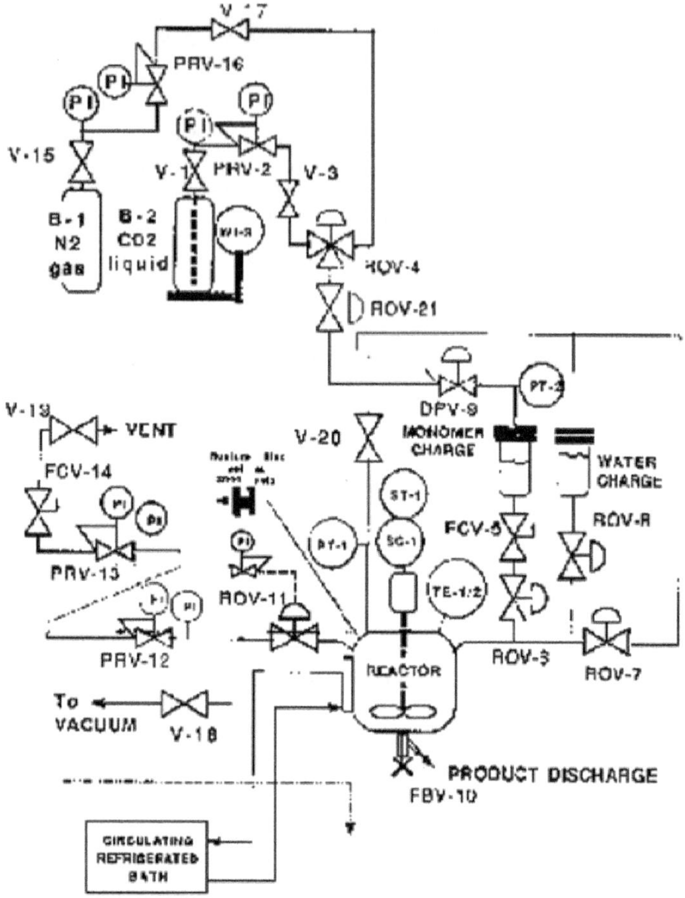

Figure 2. ***Modified Reactor Design and Support Equipment***

Approximately 35 runs were made in this set up and no problems were experienced. This configuration was used, unaltered, for the remainder of the program. The only future modification would include a greater stirrer torque capability allowing higher reactor loading to be achieved.

Polymer Synthesis

Two Polyether glycols were synthesized, poly[BAMO] and Poly 50/50 [BAMO/NMMO]. Initially we planned to produce one pound of Poly[BAMO} and two pounds of poly[BAMO/NMMO] for subsequent block linking. This was change in concurrence with the Navy to one half pound of poly[BAMO] and one pound of poly{BAMO/NMMO]. Each polymer was produced in 60-gram batches and combined to produce a single batch of each polymer.

The basic method employed for polymerization was the same for all monomers, as described below.

Reactant Weights

- ***60 grams monomer***
- ***5 grams catalyst***
- ***50 grams water***
- ***600 grams Liquid Carbon dioxide***

- ***The reactor was thoroughly cleaned and dried before each run by sweeping with nitrogen followed by evacuating at elevated temperature overnight.***
- ***The injection vessels flow rates were calibrated by filling with monomer (monomer vessel) or water (water vessel) and opening the needle valves to achieve 10ml/min for monomer and 100 ml/min water.***
- ***The reactor was evacuated to >50mmHG.***
- ***Monomer and quench solutions were added to injector vessels.***
- ***Nitrogen was introduced to the total system and the pressure differential regulator set at 100 psig***
- ***The reactor was brought to atmospheric pressure.***
- ***The catalyst was introduced through an injector port.***
- ***The reactor was filled with 600 grams of liquid carbon dioxide ~600 psig at 0^oC.***
- ***With stirring the monomer was introduced.***
- ***The polymerization was allowed to continue for two hours.***
- ***The reaction was quenched.***
- ***The reactor vented to atmospheric pressure.***
- ***The polymer was removed from the reactor.***

Overall yields throughout both polymer synthesis were in excess of 85% (by mass balance).

Polymer Characterization

The resultant polymers were characterized by Gel Permeation Chromatography (GPC) using a Waters GPC system.

The GPC was calibrated using polystyrene standards. As such, we refer to the molecular weight as the "apparent Molecular Weight", as the molecular volume of polystyrene is considerably smaller the straight chain polyether and as such we expect the actual molecular weight of these polymers to be much lower.

The calibration was achieved by injecting known polystyrene standards and producing a calibration curve based on a cubic curve fit with a CC of 0.99998.

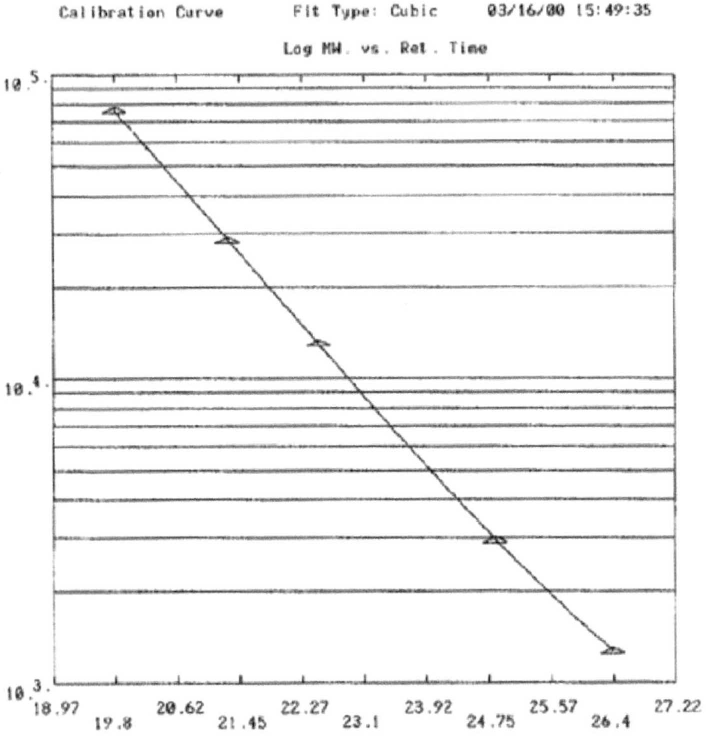

Figure 3. Polystyrene Calibration Curve

Poly [BAMO] GPC Analysis

The polymer was dissolved in Tetrahydrofuran and injected in to the GPC. The analysis trace is shown below.

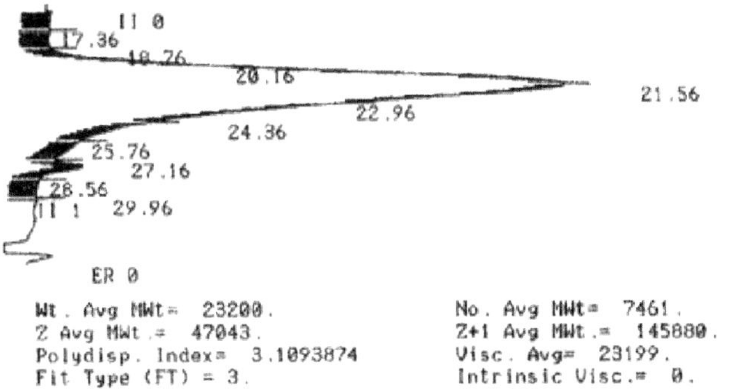

Figure 4. Poly[BAMO/NMMO] GPC Analysis

As the GPC traces indicated the polymers were well formed with very little cyclic tetramer impurity. They we considered suitable for block linking.

Polymer Block Linking

It was proposed to use a block linking technique for this contract as in the previous contract the normal "quasi-living polymerization" catalyst proved to be insoluble in liquid Carbon Dioxide.

The technique uses the reaction of the Polyether glycol with Toluene Diisocyanate (TDI) in Dimethylformamide (DMF) at elevated temperature using dibutyl tin dilaurate (DBTDL) as a catalyst. No blocking of the isocyanate with a Diol such as butanediol was found necessary

We first blocked the BN-5 by dissolving an equivalent weight of TDI in DMF followed by a drop of DBTDL. The Polyol was dissolved in DMF to give a 25%w/w solution and then added slowly to the isocyanate solution with stirring at room temperature. After complete addition the combined solutions were heated to 55°C for sixteen hours. The solution was allowed to cool to room temperature then quench by adding slowly to a stirred 10 fold excess of methanol. The precipitated polymer was dried and analyzed by GPC.

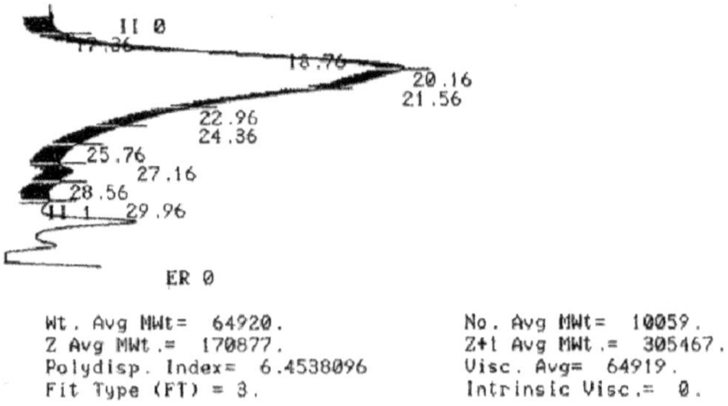

Figure 5. BN-5 Block Linked Polymer

Clearly the original polymer had been doubled in molecular weight as expected from the reactant stoichiometry. We were satisfied that the technique would produce a simple block polymer.

The experiment was repeated taking half of the poly [BN-5], end capping with a full equivalent of TDI then adding an equivalent of Poly [BAMO]. After heating the solution for sixteen hour we quenched in methanol and isolated the precipitated polymer.

GPC analysis is shown below.

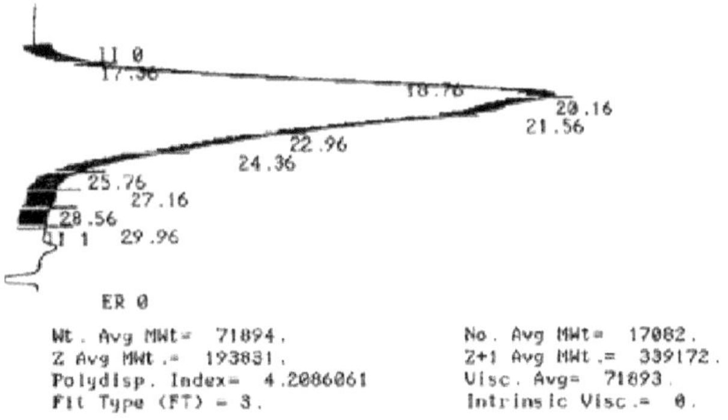

Figure 6. AB_n Thermoplastic Elastomer [BAMO]-b-[BAMO/NMMO]

Further analysis was not made, but clearly the GPC analysis indicates that blocking occurred by the increase in apparent molecular weight.

Samples of each of these polymers will be shipped to the Navy for internal evaluation.

CONCLUSIONS

This program demonstrated environmentally sensible methods for manufacture of energetic oxetane binders for use in PEP compositions. The processes that we developed eliminated a number of toxic and hazardous solvents and reagents that were previously used in laboratory-scale syntheses. Waste streams were considerably reduced and the liquid carbon dioxide solvent that was used in the polymerization can be easily recycled. Replacing batch nitration substantially reduced explosive hazards by the continuous process. Pilot plants that we operated allowed a variety of materials to be produced there. This reduced the capital expenditures.

We successfully reproduced the polymerization technique developed on the previous SERDP contract and increased the reactor loading threefold. The inability to stir the polymerization at higher loading was limited by the power of the magnetic drive on the reactor. The polymerization becomes an emulsion polymerization at a critical molecular weight when the polymer precipitates out of solution due to insolubility. Although this precipitation apparently does not limit the molecular weight, the viscosity increases rapidly. The scale up of the Polyether glycols Poly[BAMO] and Poly[BAMO/NMMO] appear to be suitable for scale up to large reactors without further experimentation.

The synthesis of Thermoplastic Elastomers by step-wise formation is still limited by the absence of a suitable, soluble catalyst. Further work needs to be initiated to identify a dicarbocation that is soluble in liquid carbon dioxide. However we successfully block-linked the Polyols [BAMO] and [BAMO/NMMO] through an isocyanate route to an apparent molecular weight that was expected for an AB_n blocked polymer.

GLOSSARY

AMMO	3-Azidomethyl-3-methyloxetane
BAMO	3,3-Bis(azidomethyl)oxetane
BBMO	3,3-Bis(bromomethyl)oxetane
BEMO	3,3-Bis(ethoxymethyl)oxetane
BCMO	3,3-Bis(chloromethyl)oxetane
BMMO	3-Bromomethyl-3-methyloxetane
BNMO	3,3-Bis(nitratomethyl)oxetane
DBTDL	Dibutyl Tin Dilaurate
DMF	N,N-Dimethylformamide
DMSO	Dimethyl sulfoxide
EMMO	3-Ethoxymethyl-3-methyloxetane
ETPE	Energetic Thermoplastic Elastomer
GPC	Gel Permeation Chromatography
HMMO	3-Hydroxymethyl-3-methyloxetane
LCO_2	Liquid Carbon Dioxide

NMMO	3-Nitratomethyl-3-methyloxetane
PEP	Propellants, Explosives and Pyrotechnic Compositions
PTC	Phase Transfer Catalyst
SCO_2	Supercritical Carbon Dioxide
THF	Tetrahydrofuran
TPE	Thermoplastic Elastomer

NITRATION IN LIQUID AND SUPERCRITICAL CARBON DIOXIDE

Indian Head Division, Naval Surface Warfare Center Indian Head Division, Indian Head, MD

INTRODUCTION

The objective of the program was to develop environmentally acceptable synthetic methodologies to prepare energetic ingredients (i.e., oxidizers, plasticizers, and energetic monomers) using liquid or supercritical fluid carbon dioxide.

The majority of the energetic ingredients comprise two main classes of compounds, nitrate esters and nitramines. The conventional synthesis of these materials involves the nitration of alcohols to obtain nitrate esters, or nitration of amines (or amine derivatives) to give nitramines.

O-Nitration of alcohols to afford nitrate esters (e.g., NG, TMETN, BTTN, NC, GN, etc.) is carried out classically with nitric acid alone, or more frequently with mixed acid (mixtures of nitric and sulfuric acid).[1] Likewise, N-nitration of polyamines, and protected amides, to give nitramines (e.g., RDX, HMX, and CL-20), as well as nitration of aromatics to give poly-nitroaromatics (e.g., TNT, picric acid, tetryl, etc.), employs the use of mixed acids or other nitrating agents.[1] Selective nitration of ingredients with acid sensitive moieties, such as glycidol, is commonly performed with dinitrogen pentoxide (N_2O_5), or nitronium salts,[2] at low temperature in halogenated solvents. The latter processes all require the use of class II ODS solvents (e.g., methylene chloride). This is because halogenated solvents are resistant to attack from nitronium cation. However, when these solvents become contaminated with energetic materials it makes their recycling or disposal difficult. Therefore, we focused efforts on exploring the feasibility of nitrating alcohols and amines in liquid or supercritical carbon dioxide.

RESULTS AND DISCUSSION

The nitration reactions were typically carried out at 0 to –10 °C in liquid carbon dioxide using dinitrogen pentoxide as the nitrating reagent. Nitration reactions of several model compounds were reported previously by scientists at DERA.[2]

With the successful nitration of model compounds reported, we focused our investigations to the synthesis of other energetic compounds. Based on the nitration conditions we had previously developed for the nitration of cellulose to cellulose nitrate, using dinitrogen pentoxide in liquid carbon dioxide, we converted cyclodextrin to cyclodextrin nitrate in 30% yield. The elemental analysis showed the up to 12 % N incorporation in the molecule which translates in to approximately 22 hydroxyl sites, out of possible 24, nitrated. We also had several samples of cyclodextrin attached to a polymer chain. These poly-CD samples were nitrated in liquid carbon dioxide using dinitrogen pentoxide. These reactions are heterogeneous in nature since the polymer is insoluble in carbon dioxide. However, the nitration reactions proceeded under these conditions to give a 12-15% yield of the nitrated product, poly-CDN (polymer bound cyclodextrin nitrate) with high degree of nitration (as evidenced by elemental analysis). However, extensive chain degradation occurred.

The nitration of cyclic alcohols, such as, glycidol and 3-hydroxymethyl-3-methyloxetane (HMMO) gave the corresponding glycidol nitrate and 3-nitratomethyl-3-methyloxetane (NMMO) in 46 and 67 % yields, respectively. The yields for the nitrated products for the cyclic alcohols are low due to concurrent ring-opening oligomerization of the monomer and/or product. In our attempt to nitrate 1,2,4-butanetriol (BT) to obtain the corresponding 1,2,4-butanetriol trinitrate (BTTN), we encountered difficulties due to the highly viscous nature of the substrate. It was difficult to inject the substrate, via an HPLC pump, in to the reactor containing dinitrogen pentoxide dissolved in liquid carbon dioxide. The dilution of the substrate BT by an organic solvent was necessary to lower the viscosity of the mixture so that the substrate could be introduced into the reactor.

The investigation of the nitration of an amines to nitramines was undertaken. We investigated the nitration of piperidine, a cyclic secondary amine, as a model compound. The nitration of piperidine using dinitrogen pentoxide in liquid carbon dioxide did not succeed under various temperature (-10 °C to +25 °C) and pressure (800 to 1500 psi) conditions. This was not surprising since dinitrogen pentoxide is not a strong enough nitration reagent for amines under these non-polar, aprotic conditions. The nitration of amines and amine derivatives are typically carried out using nitric acid/ acetic anhydride mixture. We therefore investigated the nitration of piperidine using nitric acid/ acetic anhydride mixture in liquid carbon dioxide. The nitration reaction was carried out by adding piperidine (via an HPLC pump) into a reactor containing nitric acid in liquid carbon dioxide. The acetic anhydride was then added, via an HPLC pump, into the reactor and the mixture was allowed to stir for 3 hrs. The yield of 1-nitropiperidine was low (ca. 10%) when the reaction was carried out at room temperature. The yield of the 1-nitropiperidine improved to 40 % when the temperature of the reaction was raised to 40-45 °C. The nitration of hexamine to obtain RDX was attempted using nitric acid/acetic anhydride mixture in carbon dioxide at 45 °C. However, a very complicated product mixture was obtained with very low yield.

SUMMARY AND CONCLUSION

In conclusion, the O-nitration of alcohols to obtain corresponding nitrate esters using dinitrogen pentoxide dissolved in liquid carbon dioxide was accomplished. The nitration reaction of an amine (piperidine) was also achieved using nitric acid/acetic anhydride mixture in carbon dioxide. Nitration of other amines (or an amine derivatives, such as, N-acetyl amines) needs further investigations to obtain energetic nitramines.

EXPERIMENTAL

The description of the apparatus used in nitration of organic compounds in liquid carbon dioxide using N_2O_5 has been reported previously.[4]

General Method for Nitration of Alcohols

Solid Substrates

Solid dinitrogen pentoxide (0.11 mol) was placed in a cooled pressure reactor at -20 °C. Cellulose or cyclodextrin (0.1 mol) was placed in a separate cooled reactor and the reactor was cooled to 0 to –10 °C . The liquid carbon dioxide was slowly passed through the reactor with N_2O_5, and allowed to flow into the reactor with the alcohol. Dinitrogen pentoxide dissolved in the flowing carbon dioxide and was allowed to react with the alcohol in the main reactor. After ca. 3 h, the main reactor was completely vented and the solid product washed out with water. The product was washed several times with water until the washings were neutral and then dried in an oven at 60 °C. The products were usually obtained in ca. 30% yield. All nitrated were subjected to elemental analysis.

Liquid Substrates

Dinitrogen Pentoxide (typically 0.11 mol) was placed in a cooled pressure reactor at –20 °C containing liquid carbon dioxide. Liquid alcohol (0.1 mol) was added to this reactor using HPLC pump. The mixture was allowed to stir for the requisite time and then vented. The product was removed using an organic solvent and then organic layer was washed with water, sodium carbonate and water again. After drying over anhydrous magnesium sulfate, the solvent was removed on rotary evaporator affording 40-67% liquid nitrate ester. The analysis of the product was done by NMR and comparison with an authentic samples.

Nitration of Piperidine

The method employed was similar to the one reported for the nitration of liquid alcohols described above. In place of alcohol, an amine was added to the reactor using the HPLC pump. The product was analyzed by NMR and was compared with an authentic sample.

REFERENCES

1. Olah, G. A., Malhotra, R. and Narang, S. C., "Nitration – Methods and Meachanisms" Organic Nitro Chemistry Series, **1989**, VCH Publishers, Inc.

2. (a) Millar, R.W.; Colclough, M.E.; Desai, H.; Golding, P.; Honey, P.J.; Paul, N.C.; Sanderson, A.J.; Stewart, M.J., "Novel Syntheses of Energetic Materials Using Dinitrogen Pentoxide"; *Nitration Recent Laboratory and Industrial Development* ; Albright, L.F.; Carr, R.V.C.; Schmitt, R.J., Eds.; ACS Symposium Series 623, 1996, pp 104-121. (b) Fischer, J.W., "The Chemistry of Dinitrogen Pentoxide," *Nitro Compounds: Recent Advances in Synthesis and Chemistry*; Feuer, H.; Nielson, A.T., Eds.; VCH Publishers, Inc., NY, 1990; pp 267-365.

3. Caesar, G.V.; Goldfrank, M., "Nitration of Starches with Dinitrogen Pentoxide in Presence of Sodium Fluoride", *J. Amer.Chem. Soc.*, 1946, 68, 372.

4. Nauflett, G.W., Farncomb, R.E., Patent Disclosure Navy Case No. 79,009.

HYDROGENATION OF CL-20 PRECURSORS IN SUPERCRITICAL SOLVENT MEDIA

Naval Surface Warfare Center Indian Head Division, Indian Head, MD

INTRODUCTION

The production of CL-20 in the U.S. by propellant companies has been carried out by processes which begin with HBIW, a hexaazapolycyclic in which the amine nitrogens are stabilized by attached benzyl groups ($PhCH_2$); with the overall formula of $(PhCH_2)_6C_6N_6$. Nitration to CL-20, $(NO_2)_6C_6N_6$, must be preceded by partial replacement of the benzyl groups with acetyl groups by a hydrogenolysis done in the presence of acetic anhydride, to produce TAIW $(Ac)_4C_6N_6(PhCH_2)_2$. The remaining benzyl groups are removed in a further hydrogenolysis in the presence of formic acid (Thiokol) to produce TADF $(Ac)_4C_6N_6(CHO)_2$. They can also be removed in a reaction with N_2O_4 (Aerojet) to produce $(Ac)_4C_6N_6(NO)_2$. At this stage the ring system is stable to the conditions of nitration and a complete nitrolysis of all protecting groups is done in a final step to give CL-20. In the ultimate simplification of this process the HBIW would be nitrated directly, but this is not possible with current nitration technology.

The hydrogenolysis method is suited to the transition from HBIW to a protected substrate for nitration because it is a reducing medium. HBIW is sensitive to acids, electrophiles, and oxidizing reagents. A significant improvement in the production of CL-20 could be achieved by converting HBIW in a one-step hydrogenolysis to a fully acylated substrate. A hydrogenolysis in which the benzyl groups are removed, but not all the amine sites are subsequently acylated, may also produce a substrate suitable for nitration.

The extension of the hydrogenolysis reaction to the removal of the remaining two benzyl groups in TAIW, by use of a supercritical fluid medium in the hydrogenation reaction, was the goal of this investigation. A driving force for this reaction would be the approximate order of magnitude increase in the effective concentration of H_2 expected in $scCO_2$. Hydrogen, which is only slightly soluble in organic solvents, is miscible in supercritical fluids. The challenge is to be able to take advantage of this in a medium such as $scCO_2$, which in solvent characteristics, is like carbon tetrachloride. The TAIW substrate has low solubility even in polar solvents. This could negate the expected benefits of the increased concentration of the H_2. The use of co-solvents to overcome this obstacle of low solubility of some organics in supercritical fluid, heterogeneous catalytic reactions has so far not been demonstrated according to a recent review[1]. From prior work we knew, however, that in hot acetic acid, TAIW does have some solubility. Possible candidate co-solvents would thus include acetic acid, acetic anhydride, and acetic acid/HBr solutions. The presence of a small amount of the hydrohalic acid would generate small equilibrium amounts of acetyl halide from its reaction with acetic anhydride, which can facilitate the benzyl group hydrogenolysis. The HBr content was anticipated to also aid in the dissolution of the TAIW into the $scCO_2$.

RESULTS AND DISCUSSION

When the hydrogenolysis of the benzyl groups (Bn) in HBIW is done in the presence of acetic anhydride as solvent, there is an initial removal of these protecting groups at the five-membered ring nitrogen sites, followed by acetylation to produce TAIW. This generation of four amide bonds results in a decreased solubility of the isowurtzitane ring system.

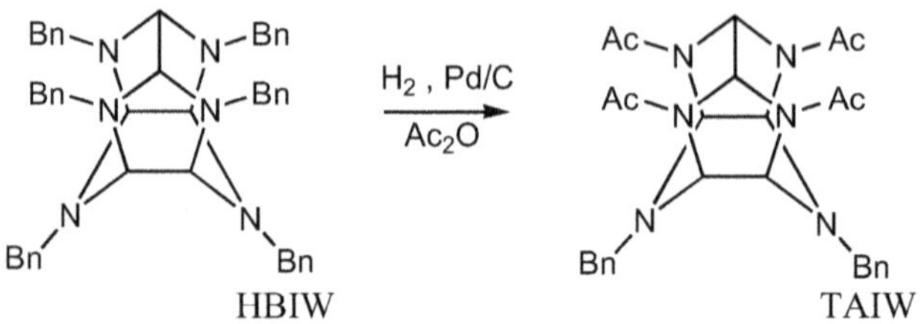

In the $scCO_2$ system used in this study, the low solubility of TAIW was recognized as an obstacle to achieving complete hydrogenolysis. Some initial solubility tests of both HBIW and TAIW in a syringe pump, supercritical fluid apparatus (pressurized by a screw driven piston) were therefore done to determine the effects of co-solvents. Solubility was checked by pressure and temperature variations to induce cloud points, and by isolating aliquots of the solution through a sampling port. Under a condition of 4400 psi and 34 °C, the aliquot sample showed no detectable amount of TAIW. With 3% acetic acid as co-solvent and conditions of 4000 psi and 34 °C, there was also no indication of dissolved material. In the case of HBIW, conditions of neat CO_2 and 1% acetic acid resulted in no dissolution at pressures up to 5700 psi and 37 °C. At 5150 psi and 36 °C, a 2% co-solvent level of DMF resulted in a 0.01% solution of HBIW. The solubility of a potential model compound for this reaction, 1-benzyl-4-acetylpiperazine, was also investigated at 4075 psi and 35 °C. This liquid benzylamine appeared to be much more soluble, but its percent solubility was not precisely determined.

The poor solubility of the TAIW was addressed in the hydrogenation experiments by use of acetic anhydride and acetic acid as co-solvents, as well as the use of temperatures up to 50 °C and pressures up to 4000 psi in a one-liter autoclave. Following the addition of the TAIW and co-solvents, the autoclave was pressurized to 600-900 psi with CO_2. The H_2 was then pumped in to a final pressure of 1400-1500 psi. The temperature of the autoclave was ramped up and held at the set temperature for reaction periods of 5-6 hours, with the final pressure determined by the autogenous pressurization from heating. The contents of the vessel were recovered by suspension in ethanol and aspiration from the vessel. Filtration gave a partitioning of unreacted TAIW between the filtrate and as a deposit on the recovered catalyst (Pd on charcoal). Products of the hydrogenolysis were predominately in the ethanol filtrate. The TAIW was recovered from the catalyst by extraction into hot acetic acid. Analysis of these fractions by HPLC showed evidence for only trace amounts of HAIW, the hexaacetyl product produced by lysis of the two benzyl groups and subsequent acetylation. Efficient acetylation following hydrogenolysis of HBIW occurs in an organic solvent mixture of acetic anhydride and DMF to produce TAIW. In the autoclave reactions described here, the acetylating agent is much reduced in concentration, and the products might reasonably include the un-acylated diamine TADA.

In addition to the concentration effect, the amines of TADA would be less basic and

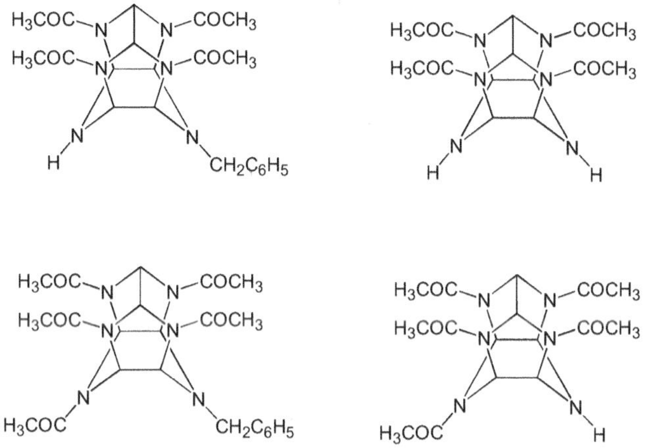

might be hindered sterically due to the boat conformation of the six-membered ring of the isowurtzitane. The non-polar $scCO_2$ medium would also be less favorable for acetylation. An acetylation reaction would also compete with the known formation of carbamic acids from amines and CO_2 in the liquified or supercritical state. The mono-benzylated products shown above are also possible, but if conditions are suitable for removal of one of these protective groups, complete lysis of the groups is more likely.

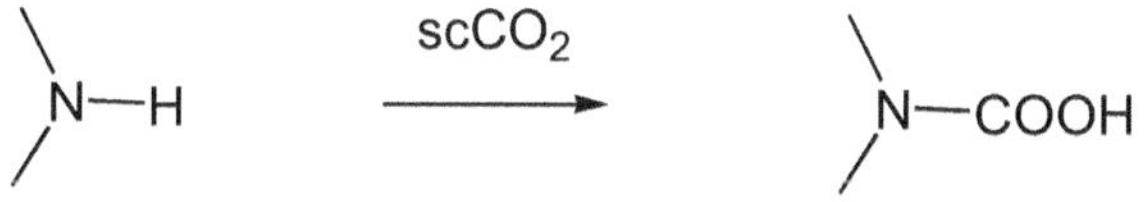

In parallel to the lack of HAIW formed in the $scCO_2$ hydrogenolysis of TAIW in the presence of acetic anhydride, the formylation of TADA formed in the $scCO_2$ hydrogenolysis of TAIW in the presence of formic acid was incomplete, and the previously unknown monoformyl product was isolated instead of the tetracetyldiformyl isowurtzitane (TADF) product. TADF is the major product when the reaction is done in formic acid solvent at ambient P/T.

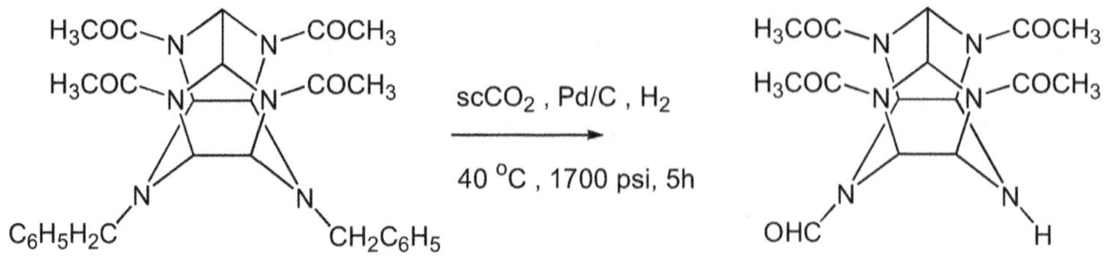

The HOLC analysis of the ethanol and acetic acid extracts of a series of hydrogenations in $scCO_2$ showed recoveries of TAIW in the range of 80-85% for the reactions done in the presence of acetic acid, HBr, and acetic anhydride. The high conversion of TAIW in the case of the hydrogenolysis done in the presence of formic acid may relate to its effectiveness as a co-solvent in dissolving the TAIW, as well as its effect of activating the N-Bn bonds of the TAIW by protonation. The HPLC analysis of the series of reactions with low conversions show a product peak with retention between that of HAIW and the monoformyl-tetraacetyl product, which would fit for the pentaacetyl-monoamine product, plus a long-retention peak due to TADA. The major product peak is TADA, due to inefficient acetylation as discussed above.

SUMMARY AND CONCLUSIONS

The hydrogenolysis of the benzyl groups in the hexaazapolycyclic TAIW was demonstrated in $scCO_2$. The products, which were the diamine TADA and the pentaacetyl monoamine, based on the HPLC analysis, were produced in low yield. The very polar amide, TAIW, had low solubility in $scCO_2$. The advantage of providing a large increase in accessibility to H_2 by use of the supercritical carbon dioxide medium was countered by the low solubility of the substrate in this medium. Some mitigation of this may have been achieved with use of CO_2 -soluble catalysts, such as Pd with phosphine ligands, but this was considered to be outside the scope of a program to explore a practical hydrogenolysis procedure using "green" chemistry techniques. The use of co-solvents was explored and, while there was little benefit from the use of acetic acid, formic acid resulted in over 90% conversion to the monoformyl amine, $(Ac)_4C_6N_6(CHO)H$. The feasibility of a CL-20 synthesis based on this supercritical fluid process was thus demonstrated.

REFERENCES

1. A. Baiker, "Supercritical Fluids in Heterogeneous Catalysis", Chem. Rev. **1999**, *99*, 453-473.

ACKNOWLEDGEMENTS

The Authors would like to thank the following individuals: Mr. Marc S. Magdinec and Mr. Charles R. Painter (NSWCIHD) for programmatic coordination of the SERDP program, Mr. Steven E. Mitchell (NSWCIHD) for conceptual and editorial input into the gun propellant modeling effort; and Drs. Richard S. Miller, Harold Bright, and James M. Short (NSWCIHD detail) of ONR for their technical and program managerial support. The authors also wish to thank Mr. Bradley P. Smith and Dr. Jeffery A. Marqusee of the SERDP Office.

ENERGETIC MATERIALS – SYNTHESIS AND DEVELOPMENT OF INSENSITIVE AND GREEN COMPOUNDS

Portugal,A., Campos, J., Simões,P., Plaksin,I., Pedroso,L.

Laboratory of Energetics and Detonics, Faculty of Sciences and Technology of University of Coimbra, 3030 Pinhal de Marrocos, Coimbra, PORTUGAL

ABSTRACT

The synthesis and development of new energetic compounds is a complex process that requires a global perspective. Performance, the key property for new product development in the past, is no longer paramount and safety and environment impact are factors that have to be taken at equal terms.

From the idea to the application of new energetic materials several steps have to be followed. They are: theoretical prevision of properties, compound synthesis, formulation, engine design, qualification and finally application. In each step several assessments are required namely, performance, safety, storage, cost and environment impact. Each step is revaluated at anytime by the new knowledge acquired during the other phases of the product/process development. Therefore, is not surprising that a long time is necessary for the acceptance of new energetic materials. History is full of examples in which the path from idea to application had to be interrupted...

In this work, we present an overview of our experience in the LEDAP in the development of energetic materials. The selected compound was 2,4 – dinitramine – 6 – oxy – s – triazine (DNAM), an energetic molecule that can be used in explosive formulations as well as in the propellant field. The discussion of the decision making process along the product development process is described, from a global point of view.

P. Castelo Branco et al. (eds.),
Defense Industries: Science and Technology Related to Security, 209–220.

Energetic Materials: Synthesis and Development of Insensitive and Green Compounds

Oporto, 30 October 2001

António Portugal
José Campos
Pedro Simões
Igor Plaksin
Luís Miguel Pedroso

Some History

- Roger Bacon (13th century)

  Black Powder —(1 century)→ Gun Powder

- Nitrocellulose (19th century) [1833-1846]
- Sobrero – Nitroglycerine – $C_3H_5N_3O_9$ [1847]
- Wilbrand – TNT – $C_7H_5N_3O_6$ [1863]
- Nobel → Scientist and Industrial

  Dynamite
  \+
  Gel explosives (NG + Nitrocellulose)

Some History

- Van't Hoff (1909)
- Plets (1935) – proposes to relate chemical groups with energetic characteristics for energetic compounds (explosophores)

Military Explosives

- Between the world wars – military explosives based on TNT (TNT+AN) → Amatol
- Second world war – RDX (hexogen, $C_3H_6N_6O_6$) – Composition B (RDX, TNT, wax)
  Cyclotol

PBX – Idea of a Polymeric Matrix + Filler

Some History

- Group trinitromethyl (-$C(NO_2)_3$) – performance (1950, 1965)
- Kamlet (1959) – Rotation of C-NO_2 bound
- Group (-$CF(NO_2)_2$) – energetic oxidisers
- Group azide (-N_3) → Primary explosives
  → Energetic binders – GAP (poli-glycidyl azide)

Some History

Looking for thermal stability

- Poliaromatics polinitrated

- Heterocycles with nitro and pycric groups

- Aromatics with adjacent NO_2 and NH_2 groups – TATB (1,3,5-triamine- 2,4,6 trinitro benzene)

Some History

- In the last 30 years: thousands of new molecules have been synthesised
  interest on NO_2 groups and nitrocompounds

But how many are in use today?

Strategic Research vs. Applied Synthesis Research

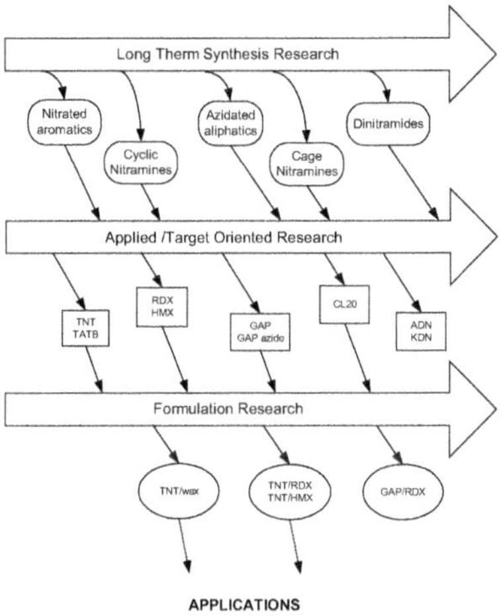

Adapt. from NATO Meeting 1994

The route from idea to service

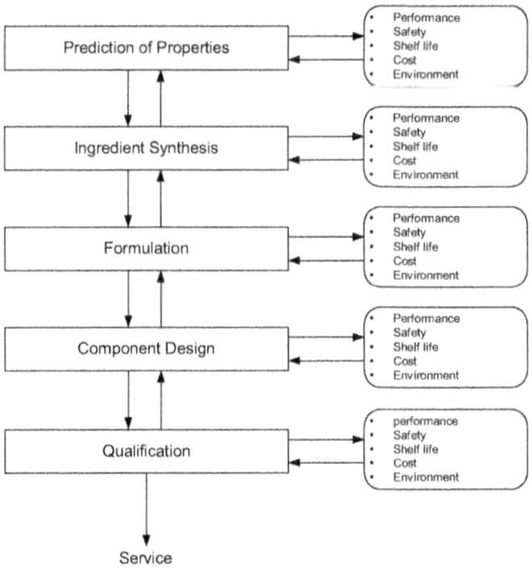

Adapt. from Sanderson et al., 1994

The governing factors for development of energetic materials

1. A new material will not be used unless it offers at least comparable performance to materials it might replace.
2. A new material with safety characteristics worse than a compound such as PETN will not be used (except possibly within explosive trains of fusing systems).
3. A new material will not be used unless it will have a reasonable shelf life in a finished article, e.g., 10 years or more.
4. A new material containing a compound with a production cost significantly greater than TATB will not be used unless a significant cost-benefit can be demonstrated or it is the only suitable/allowed material for an application.
5. A new material will not be used if there are concerns about its effect on its manufacturing or the global environment.

NATO 2001 – Oporto Meeting

Steps for the Design and Synthesis of New Energetic Compounds

Adapt. from Sanderson et al., 1994

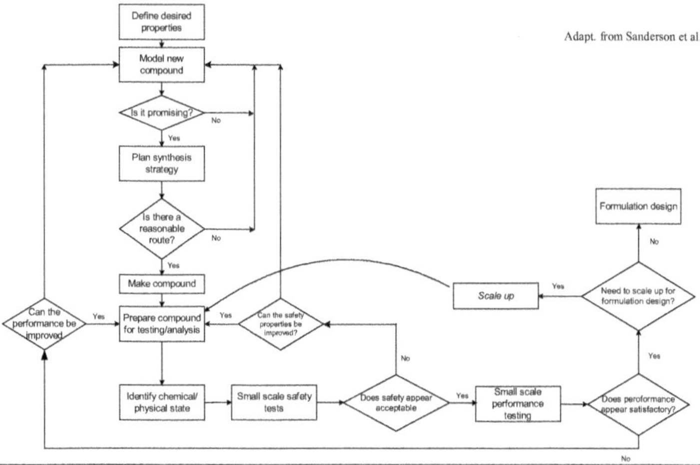

NATO 2001 – Oporto Meeting

Essential Chemical and Physical Analysis of New Compounds

Measurement	Type of measurement		
	Chemical charact.	Purity	Physical charact.
Heat of formation	v		
Functionality	v		
Mass spectrum	v	v	
Acidity	v	v	
Elemental analysis	v	v	
UV spectra	v	v	
Solvent content		v	
Multinuclear nmr	v	v	
Hplc/gpc		v	
IR spectra	v	v	v
Decomposition temperature	v	v	v
Powder and single crystal x-ray analysis	v		v
Crystal size and morphology		v	v
Melting point/boiling point/phase transitions (DTA/DSC - obligatory qualification tests in some countries)		v	v
Density (flotation)			v
Vapour pressure			v

Adapt. from Sanderson et al., 1994

NATO 2001 – Oporto Meeting

Objectives

To study nitration reactions in order to:

- Improve the knowledge in this area
- Better control of the reaction
- Synthesis of nitrated products to be used in energetic formulations and/or propellants

NATO 2001 – Oporto Meeting

Selection of the System to be Studied

Po:

Melamine
+
Nitration Agent

- ➢Melamine is available at low prices
- ➢Previous experience on DNAM synthesis

Nitration Agents

Nitration Agents

Nitric Acid + Acidic Mixtures
- ➢Classic procedure
- ➢Safety precautions
- ➢Necessary reaction in liquid phase

Nitronium Salts ($NO_2^+ BF_4^-$)
- ➢Reaction in liquid phase
- ➢Appropriate solvents required

Nitrogen Oxides (N_2O_5, N_2O_4)
- ➢Complex equipment with cryogenic facilities
- ➢Gaseous phase
- ➢Stability of compounds questionable

State of Art

Reaction:

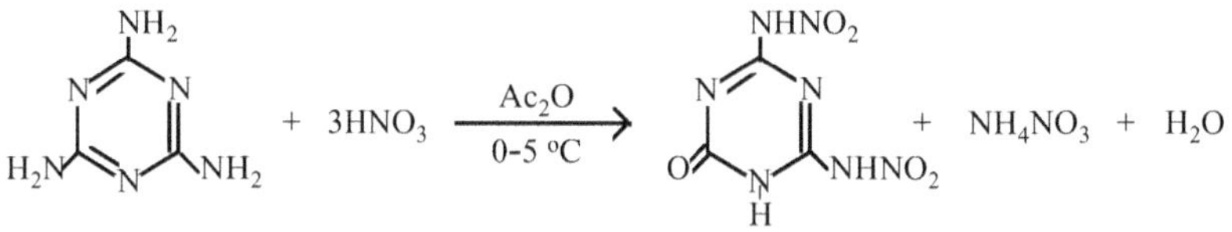

Atkinson, E. R., 1951,The Nitration of Melamine and Triacetylmelamine. *J. Am. Chem. Soc.* **73**, pp.4443-4444.

Cason, J., 1947, The Nitration of Melamine and Triacetylmelamine. *J. Am. Chem. Soc.* **69**, pp. 495-498.

State of Art

IR Spectrum

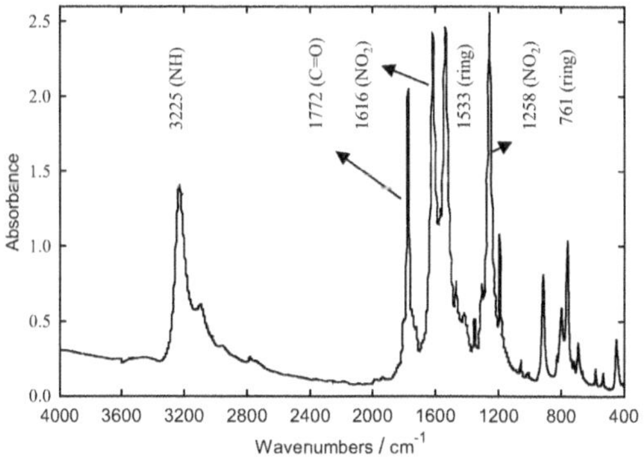

State of Art

Granulometry

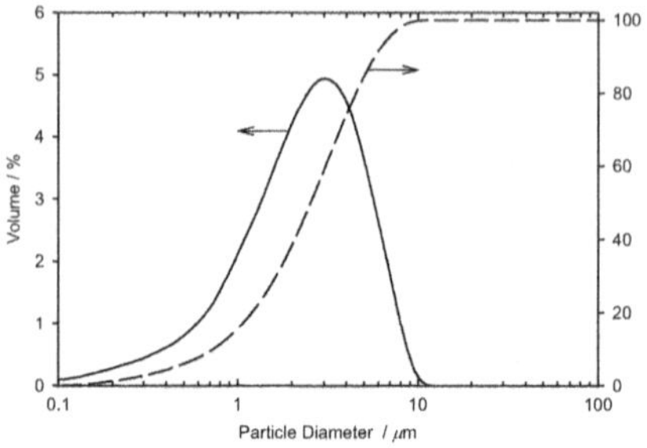

➢Laser diffraction technique (Coulter LS130)
➢Monomodal, narrow distribution: d_{50}=2.5 μm

NATO 2001 – Oporto Meeting

State of Art

Density: 1.945 (Gas picnometry)
(Theoretical value by Stine's algorithm 1.916)

Heat of combustion: 1508.3 $J \cdot mol^{-1}$ (Bomb calorimeter)

Impact sensitivity (BAM): > 50.5 J

Friction sensitivity: > 360 N

Vacuum stability: 0.257 $cm^3 \cdot g^{-1}$ (40h @ 100ºC)

NATO 2001 – Oporto Meeting

State of Art

Thermal stability

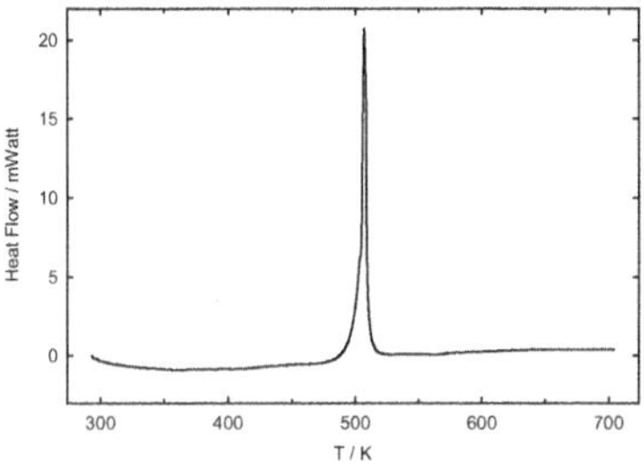

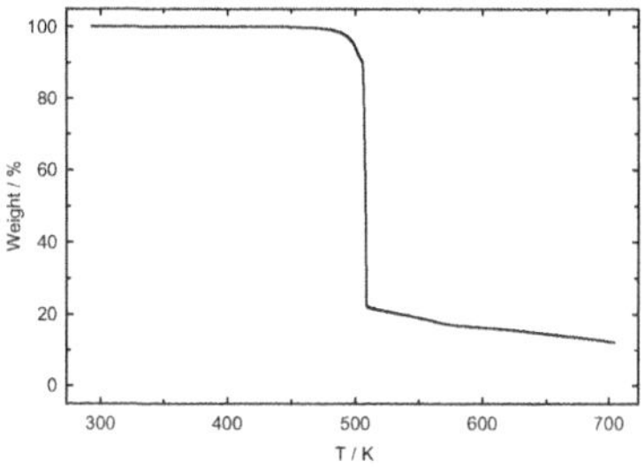

- ➢Stable product
- ➢Decomposition temperature: > 223°C

Reasons of Choice

➢It is a chlorine-free material

➢It presents a fairly attractive density

➢The particle size distribution may be advantageous in terms of processability, of the higher solid load charges that might be reached, and of the burning characteristics

➢It is thermally stable over a wide temperature range

Reasons of Choice

➢It presents low sensitivity both to impact and friction; the vacuum stability test (performed under standard conditions by default) also revealed a stable material

➢The *s*-triazine nucleolus of the molecule is highly resistant to cleavage and, at high temperatures and pressures, is prone to lead to cyclic azines (fused *s*-triazine rings); these are known as having good characteristics as burn-rate modifiers

NATO 2001 – Oporto Meeting

Future Work

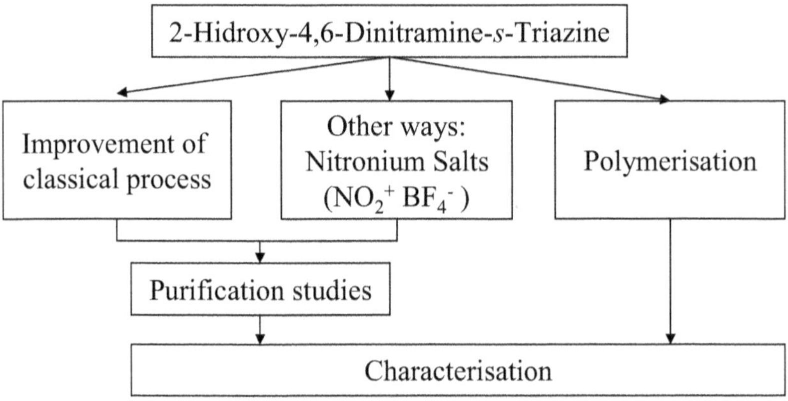

NATO 2001 – Oporto Meeting

Replacement of Toxic and Ecotoxic Components of Military Smokes for Screening

Dr. Uwe Krone

Nico Pyrotechnik, Hanns-Jürgen Diederichs GmbH & Co. KG
P.O. Box 1227, D-22943 TRITTAU, GERMANY
Phone: +49 4154 805 181; Fax: +49 4154 2451; e-mail: u.krone@nico-pyro.de

ABSTRACT:

Pyrotechnic smokes for screening, as referred to in this paper are solid/liquid aerosols being generated by pyrochemical reactions (chem. formation of the active aerosol component) as well as by explosive dissemination (dispersion of already preshaped aerosol components by HE or pyrotechnic bursting charges).
HC-smoke, WP- or RP-smoke, fog oil or artificial smokes based on dissemination of titanium tetrachloride are well known formulations of the past century. Their environmental impact and their toxicity are also well known.
During the last 20 years efforts were made in various NATO countries, not only to replace these hazardous smokes but also to improve the screening capacity and cover also the longer wavelengths like near and far IR and the millimeter wave section of the electromagnetic spectrum (EMS).
NICO has developed two new types of pyrotechnic smoke formulations, one for training purposes (KM-smoke) and the other (NG-smoke) for multispectral applications (VIS, IR, MMW). Both types were subject to environmental compatibility tests according to the OECD-guidelines. The tests were carried out by the Fraunhofer Institute of Environmental Chemistry and Ecotoxicology (IUCT). Test results, pyrotechnic smoke formulations and screening performance of these new smokes are discussed.

AGENDA:

Pyrotechnic Smokes for Screening (Definition and Types)
The Environmental Impact of Pyrotechnic Smokes
New Aims for New Smoke Formulations
The Answers
1. KM-Smoke
2. NG-Smoke
Conclusions

Pyrotechnic Smokes for Screening (Definition and Types)
Pyrotechnic smokes for screening, as referred to in this paper are solid/liquid aerosols being generated by pyrochemical reactions (chem. formation of the active aerosol component) as well as by explosive dissemination (dispersion of already preshaped aerosol components by HE or pyrotechnic bursting charges).
HC-smoke, NT, WP- or RP-smoke, or smokes based on dissemination of titanium tetrachloride are well known formulations of the past century. So are their environmental

P. Castelo Branco et al. (eds.),
Defense Industries: Science and Technology Related to Security, 221–238

impact and their toxicity. These smokes are mainly effective in the visual (VIS) part of the electromagnetic spectrum (EMS) with some screening power in the nearest IR (up to 2 µm wavelength) and the exception of phosphorous based formulations which may cover at appropriate conditions also parts of the longer IR wavelengths.
Particulate smokes, aerosols formed out of brass or aluminium flakes, of graphite powder or of graphite fibers and other chaff-like material are capable of screening the whole IR-sector of the EMS and the higher frequencies of the millimeter wave radar (MMW-RADAR).

The Environmental Impact of Pyrotechnic Smokes

With regard to the chemical nature of the aerosol forming substances of the above mentioned classic visual smokes it is obvious, that follow-on products like HCl and polychloro-aromatic hydrocarbons and derivatives including hexachloro-benzene and polychloro-dibenzo-dioxines and –dibenzo-furanes (HC), ortho-phosphoric acid and phosphanes (WP/RP) have a serious environmental impact. This is especially the case on repeatedly used training grounds, were soil and aquatic ecosystems are widely contaminated with Zn, chlorine compounds and/or phosphanes. The use of brass flakes and fibers of all kind give comparable hazards including inhalation problems for people present in the contaminated area, where through wind or simply footsteps the particles, disseminated on the ground, get airborne again.

New Aims for New Smoke Formulations

Since a number of years non- or lowtoxic smoke has been an urgent requirement of our defense procurement agencies.
Even in an age of increasing weapon sophistication, smoke and obscurants have maintained their place as effective combat multipliers and low-cost countermeasures to modern target acquisition and surveillance.
So, if the tactic intention exists further to use smoke on a possible battlefield, it is only obvious and logical, to keep the armed forces well trained in using artificial smoke screens for tactical, deception or selfprotection purposes and to avoid „colateral" (environmental) damage on the training ground.

1. KM-Smoke

Developed in the mid 80`s, KM was a replacement of mainly HC-smoke in the German army. Intended originally for a training smoke because of its partly poor screening power it is today the main substitute for HC-formulations in hand grenades and AFV-self protection devices.
KM-smoke underwent all ecotoxicology tests according to the OECD-guidelines to prove its environmentally friendly character in soil and aquatic ecosystems. Furthermore acute and subacute inhalation tests with rats and cytotoxic tests with human cell cultures were carried out. KM-smoke passed all tests with good or excellent results, especially when compared to the former used HC-formulations.
According to its nature, an aerosol formed of KCl droplets and MgO particles, the KM-smoke covers only the visual part and the nearest IR-sector of the EMS up to about 1.5 µm. When obscuration in the IR and even MMW-part is necessary, NICO offers another smoke formulation called **NG** which is able to do the job.

2. NG-Smoke

NG-smoke is based like KM on the energy output of burning Mg. Mg is burnt by potassium nitrate and perchlorate and the generated oxydation energy is used to expand the main component of this Composition – intercalated graphite – to multipole like particles of numerous graphite layers connected along the z-axis. Multipole, because spots of dielectric, not conductive MgO and KCl are condensed at random on the surface of the expanded graphite. When exposed to electromagnetic radiation of longer wavelengths, broadband absorption takes place. With a mass extinction coefficient alpha of 0.5 m^2/g the absorption is from the far IR (8 – 12 μm) to the high frequency millimetre wave spectrum (35 – 140 GHz). Also with this smoke ecotoxicological and toxicological tests analogue to those done with the KM-smoke were carried out by the Fraunhofer Institute (FhG-IUCT) under close supervision of the BWB the German procurement agency.

Conclusions

Todays available pyrotechnic smokes are very different from those, used only a decade ago. Besides a wider application range in the EMS their environmental impact is much lower and there are only minor risks for soldiers when exposed to the concentrated aerosol without proper breathing protection, which was a deadly issue when HC was in use and also with RP not always without casualties. The mentioned pyrotechnic KM smoke is fielded now in several NATO armies and NG is on its way. Both will be introduced into Germany´s new Battlefield Training Centre (GÜZ).

NICO - PYROTECHNIK
Welcome

NICO PYROTECHNIK

Replacement of Toxic and Ecotoxic Components of Military Smokes for Screening

...for pyrotechnics

center of competence

1

NICO - PYROTECHNIK
Welcome

NICO PYROTECHNIK

Replacement of Toxic and Ecotoxic Components of Military Smokes for Screening

- **Dr. U. Krone**
  NICO-PYROTECHNIK
  Hanns-Jürgen Diederichs GmbH & Co. KG

  P.O. Box 1227
  D - 22943 Trittau
  Germany
  Phone: +49 41 54 80 51 81
  Fax: +49 41 54 24 51
  u.krone@nico-pyro.de

...for pyrotechnics

center of competence

1a

NICO - PYROTECHNIK
Agenda

- Pyrotechnic Smokes for Screening (Types and Definitions)
- The Environmental Impact of Pyrotechnic Smokes
- New Aims for New Smoke Formulations
- KM-Smoke
- NG-Smoke
- Conclusions

...for pyrotechnic

center of competence

2

NICO - PYROTECHNIK
Definitions

- Smoke for screening is an artificial aerosol of airborne particles, or droplets or a mixture thereof.
- Smoke for screening absorbs, scatters and reflects electromagnetic radiation. The result is always an attenuation of its initial intensity.
- Application of screening smokes today covers the EMS (electro magnetic spectrum) from 100 nm to 5 mm

...for pyrotechnic

center of competence

3

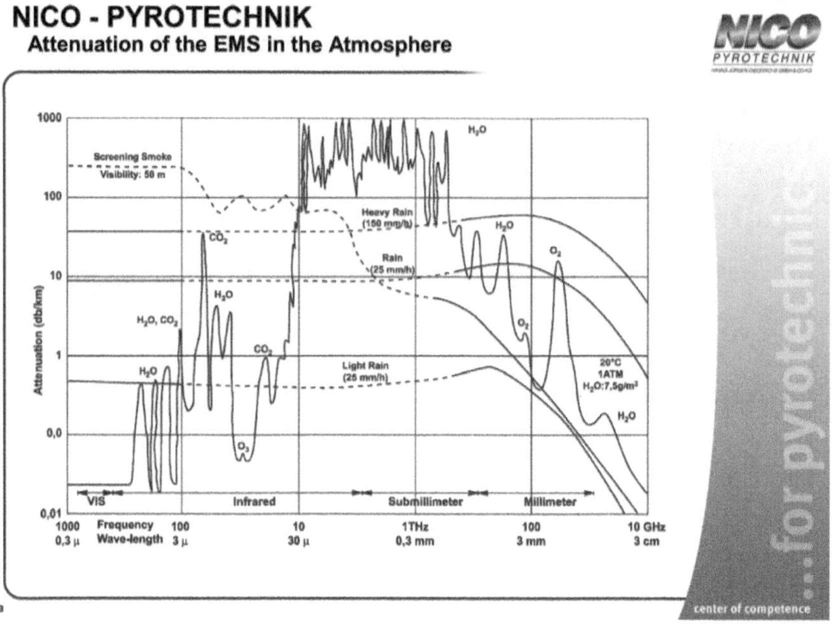
NICO - PYROTECHNIK
Attenuation of the EMS in the Atmosphere
NICO PYROTECHNIK
Attenuation (db/km)
1000
100
10
1
0,0
0,01
Screening Smoke
Visibility: 50 m
Heavy Rain
(150 mm/h)
Rain
(25 mm/h)
Light Rain
(25 mm/h)
20°C
1ATM
VIS
Infrared
Submillimeter
Millimeter
1000 Frequency 100
0,3 μ Wave-length 3 μ
10
30 μ
1THz
0,3 mm
100
3 mm
10 GHz
3 cm
...for pyrotechnic
center of competence

4a

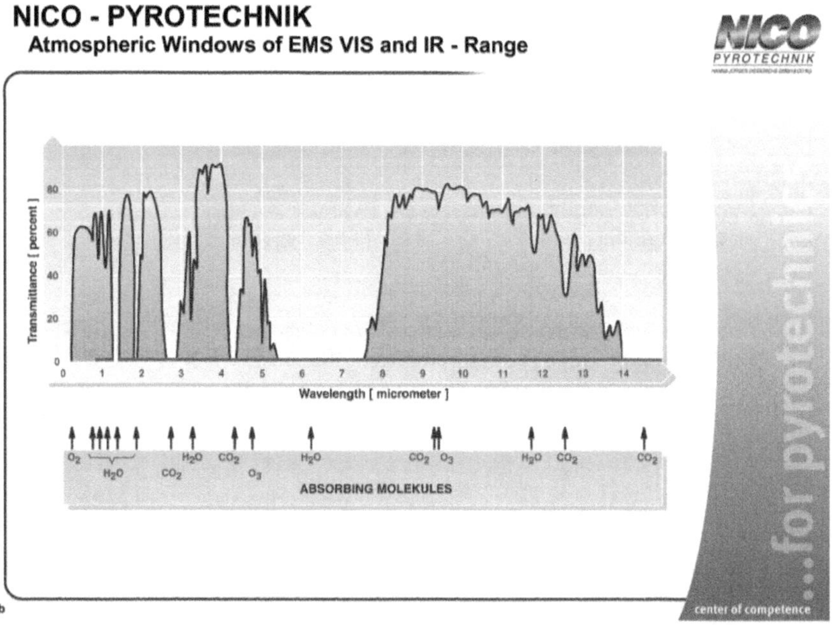
NICO - PYROTECHNIK
Atmospheric Windows of EMS VIS and IR - Range
NICO PYROTECHNIK
Transmittance [percent]
80
60
40
20
0
0 1 2 3 4 5 6 7 8 9 10 11 12 13 14
Wavelength [micrometer]
ABSORBING MOLEKULES
...for pyrotechnic
center of competence

4b

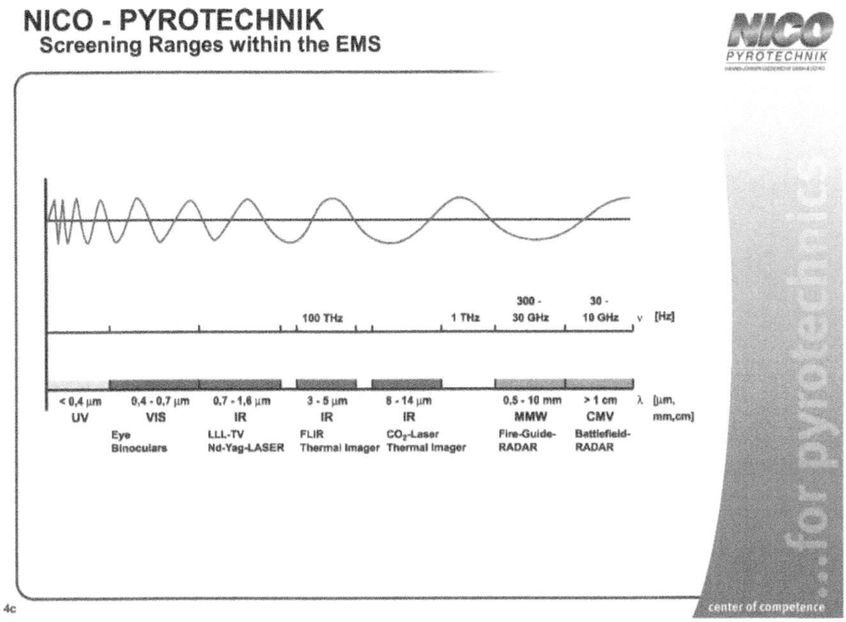
NICO - PYROTECHNIK
Screening Ranges within the EMS
NICO
PYROTECHNIK
300 -
30 GHz
30 -
10 GHz
100 THz
1 THz
ν [Hz]
< 0,4 µm
UV
0,4 - 0,7 µm
VIS
Eye
Binoculars
0,7 - 1,6 µm
IR
LLL-TV
Nd-Yag-LASER
3 - 5 µm
IR
FLIR
Thermal Imager
8 - 14 µm
IR
CO2-Laser
Thermal Imager
0,5 - 10 mm
MMW
Fire-Guide-
RADAR
> 1 cm
CMV
Battlefield-
RADAR
λ [µm,
mm,cm]
...for pyrotechnics
center of competence
4c

NICO - PYROTECHNIK
Example VIS-Smoke
NICO
PYROTECHNIK
...for pyrotechnics
center of competence
5a

NICO - PYROTECHNIK
Example VIS/IR-Smoke

NICO PYROTECHNIK

...for pyrotechnics
center of competence

5b

NICO - PYROTECHNIK
Example IR-Smoke

NICO PYROTECHNIK

...for pyrotechnics
center of competence

5c

NICO - PYROTECHNIK
HC-Smoke - Reactions and Products

$$3\ ZnO + 2\ Al + C_2Cl_6 \longrightarrow 3\ ZnCl_2 + Al_2O_3 + 2\ C$$

$$ZnCl_2 + 6\ H_2O \longrightarrow [Zn(OH)_2(H_2O)_4]Cl_2 \longrightarrow$$

$$\longrightarrow Zn(OH)_2 + 2\ HCl$$

$$Fe_2O_3 + Al + C_6Cl_6 \longrightarrow Fe_2Cl_6 + Al_2O_3 + 6\ C$$

$$Fe_2Cl_6 + 6\ H_2O \longrightarrow \longrightarrow 2\ Fe(OH)_3 + 6\ HCl$$

...for pyrotechnics

center of competence

6a

NICO - PYROTECHNIK
HC-Smoke - By-Products

$n\ C_2Cl_6 \longrightarrow \longrightarrow m\ C_2Cl_6$ Hexachlorobenzene is a main by-product of low temperature burning HC composition.

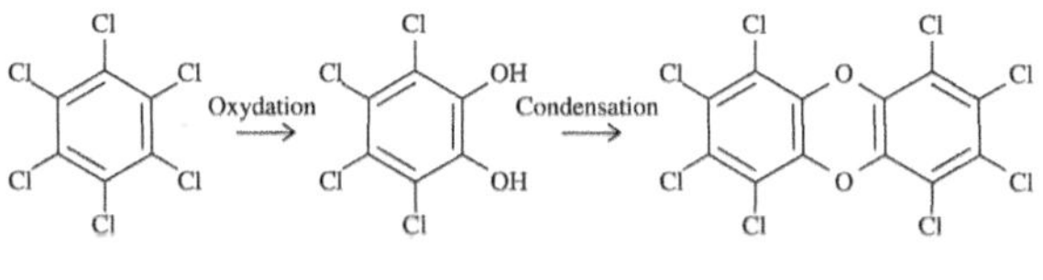

and

The formation of the dibenzo-dioxin or dibenzo-furan system ist obvious and lead to numerous PCDF´s, PCDD´s, TCDF´s andTCDD´s.

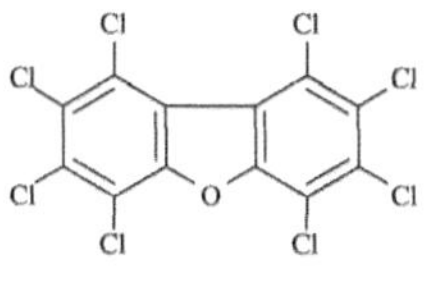

...for pyrotechnics

center of competence

6b

NICO - PYROTECHNIK
NT-Smoke

NICO PYROTECHNIK

$$16\ ZnO + 12\ NH_4ClO_4 + 8\ NH_4Cl + 3\ (C_5H_4Cl_4) + C_{24}H_{38}O_4 \longrightarrow$$

$$[16\ ZnCl_2 \cdot 20\ NH_3 \cdot 35\ H_2O] + 33\ CO + 6\ C$$

6c

...for pyrotechnics

center of competence

NICO - PYROTECHNIK
RP-Smoke

NICO PYROTECHNIK

$$RP \rightarrow P_4 + 5\ O_2 \rightarrow P_4O_{10}\ (+\ H_2O)$$

$$\rightarrow (H_3PO_3)_x\ (+\ H_2O) \rightarrow H_3PO_4$$

$$RP + Mg + H_2O \rightarrow 2H \rightarrow \rightarrow PH_3, P_2H_6, ...$$

6d

...for pyrotechnics

center of competence

NICO - PYROTECHNIK
KM-Smoke - Components

Starter composition	63 % KNO_3
	27 % Mg
Transition charge	47 % KNO_3
	21 % $CaCO_3$
	32 % Mg
Smoke composition	44 % KCl
	27 % KNO_3
	16 % $(NH_2)CO\text{-}N{=}N\text{-}CO(NH_2)$
	8 % Mg
	5 % $KClO_4$

...for pyrotechnics
center of competence

7a

NICO - PYROTECHNIK
KM-Smoke - Reaction and Products

$$9\,Mg + 2\,KNO_3 + KClO_4$$

$$\rightarrow 9\,MgO + K_2O + KCl + N_2 + E$$

$$K_2O \xrightarrow{H_2O,\,CO_2} K_2CO_3$$

44% of additional KCl is evaporized

...for pyrotechnics
center of competence

7b

NICO - PYROTECHNIK
KM-Smoke - Environmental Compatibility Tests

NICO PYROTECHNIK

1. Investigation of the acute and subacute inhalative toxicity in rats
2. Fish, acute toxicity test
3. Toxicity for earthworms, artificial soil test
4. Daphnia sp., 14-day reproduction test including an acute immobilisation test
5. Growth test with higher plants (avena sativa; brassica rapa)
6. Algae, growth inhibition test
7. Bacteria, test for inhibition of oxygen consumption by activated sludge

...for pyrotechnics

center of competence

7c

NICO - PYROTECHNIK
KM-Smoke - Comparative Tests of Different Smokes

NICO PYROTECHNIK

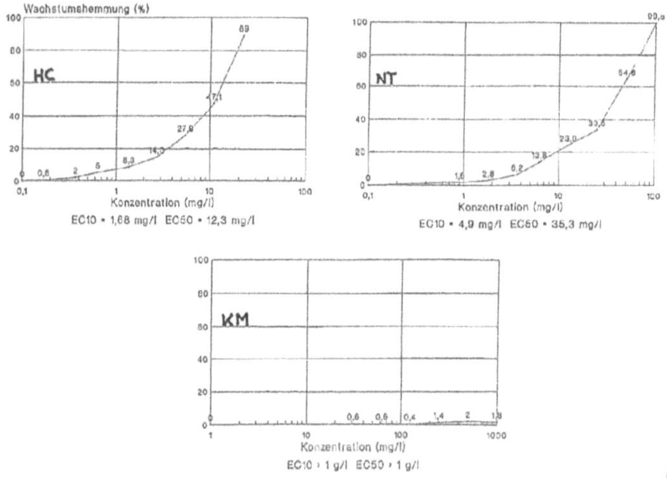

...for pyrotechnics

center of competence

7d

NICO - PYROTECHNIK
KM-Smoke - Test Results

Test Item	KM-Smoke	HC-Smoke
Rats	-	xx
Fish	-	x
Earthworm	-	-
Daphnia	x	x
Avena sativa	-	-
Brassica rapa	-	x
Algae	-	xx
Bacteria	-	x

xxx extremely toxic
xx very toxic
x toxic
- less toxic

...for pyrotechnics
center of competence

7e

NICO - PYROTECHNIK
NG-Smoke - Composition

Intercalated graphite
magnesium
potassium perchlorate
catalyst, burning moderator

...for pyrotechnics
center of competence

8a

NICO - PYROTECHNIK
NG-Smoke - Reaction and Products

$$7\ Mg + 6\ C_nX + KClO_4 + \text{add.}$$
$$\longrightarrow C_n + 4\ MgO + 3\ MgX_2 + +$$

8b

NICO - PYROTECHNIK
NG-Smoke - Graphite Structures

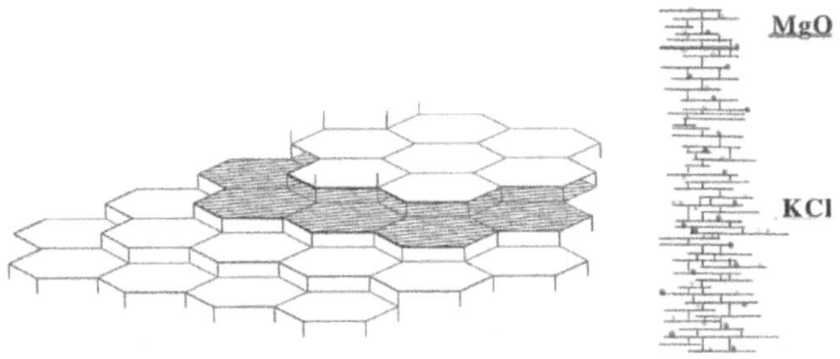

Hexagonal Graphite Structure **Expanded Graphite**

8c

NICO - PYROTECHNIK
NG-Smoke - Graphite Particles

NICO PYROTECHNIK

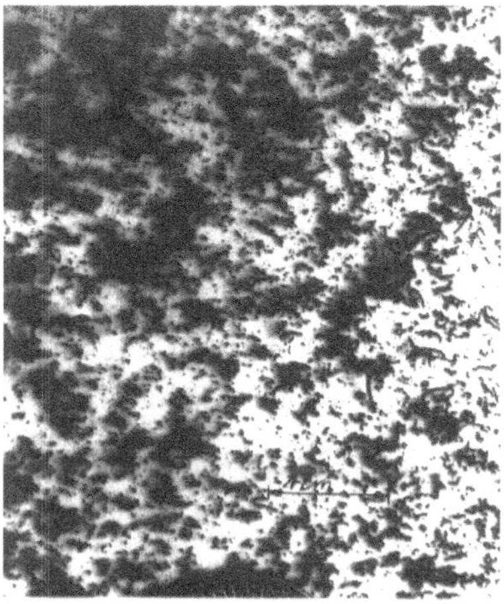

Panorama of a fall out area

8d

...for pyrotechnics

center of competence

NICO - PYROTECHNIK
NG-Smoke - Graphite Particles

NICO PYROTECHNIK

Example of a counting area of
pattern class 40

8e

...for pyrotechnics

center of competence

NICO - PYROTECHNIK

NG-Smoke - Environmental Test Results

NG-Screening Smoke - Toxicity Data - EC_{50}/LC_{50} Values

The concentrations refer to the maximum theoretical substance concentrations.

TEST		Unburnt Composition	Total combustion products
Algae		17	458
Daphnia		> 1000	> 1000
Fish (Zebra)		> 1000	> 1000
bacteria		> 100[1]	> 100[1]
Plants	beans	247	> 1000
	oat	101	> 1000
	turnip	32	> 1000
Plant Germination Test		> 1000	> 1000
Earthworm		> 1000	> 1000
Nematode		> 1000	> 1000
Cytotoxicity		>10.000[2]	8.600[2]
Biodegradability		not tested	not biodegradable

[1] corresponds to highest tested concentration

[2] mean of two tests. For total combustion products the EC_{50} values for growth inhibition (measured as protein content of culture) were > 10.000 and 7.200 mg/l.

...for pyrotechnics

center of competence

8f

NICO - PYROTECHNIK

RADAR - Attenuation, 94 GHz

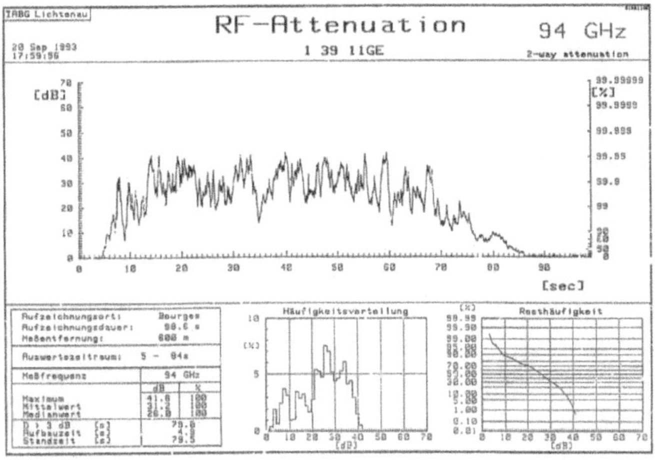

...for pyrotechnics

center of competence

8g

NICO - PYROTECHNIK
RADAR - Attenuation, 35 GHz

NICO PYROTECHNIK

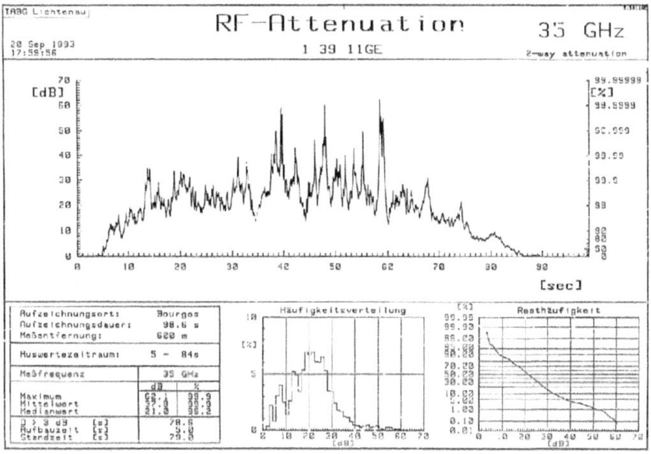

...for pyrotechnic

center of competence

8h

NICO - PYROTECHNIK
RADAR - Attenuation, 16 GHz

NICO PYROTECHNIK

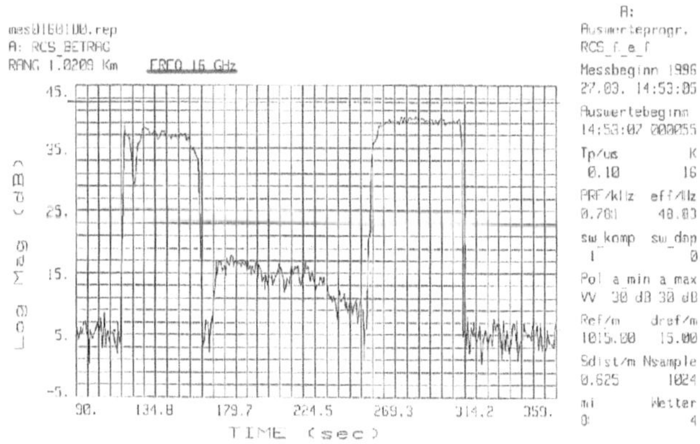

...for pyrotechnic

center of competence

8i

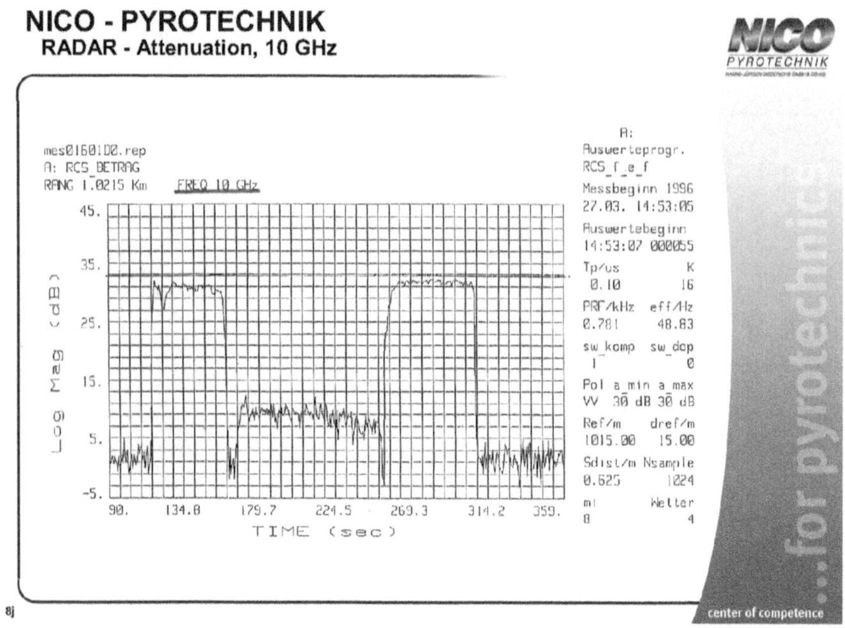
NICO - PYROTECHNIK
RADAR - Attenuation, 10 GHz
NICO
PYROTECHNIK
mes0160100.rep
A: RCS_BETRAG
RANG 1.0215 Km
FREQ 10 GHz
45.
35.
25.
15.
5.
-5.
Log Mag (dB)
90.
134.8
179.7
224.5
269.3
314.2
359.
TIME (sec)
A:
Auswerteprogr.
RCS_f_e_f
Messbeginn 1996
27.03. 14:53:05
Auswertebeginn
14:53:07 000055
Tp/us K
0.10 16
PRF/kHz eff/Hz
0.781 48.83
sw_komp sw_dop
1 0
Pol a_min a_max
VV 30 dB 30 dB
Ref/m dref/m
1015.00 15.00
Sdist/m Nsample
0.625 1024
mi Wetter
8 4
8j
...for pyrotechnics
center of competence

Todays Standardized Survivability Tests

Manfred Held
EADS-TDW, 86523 Schrobenhausen, Germany

Abstract

In the last decades are required so-called insensitive munitions what in practice means less sensitive munitions which are less violent reacting by different attack modes and under strange environmental conditions as fire. One very important part is the requirement that the ammunitions should not more sympathetic or mass detonate. Such behavior would give definitely a much higher safety feature and the related security will be strongly increased.

These tests are more or less required for military use and applications but they give also good features for higher related security aspects. summarized The different test methods are shortly described and their pros and cons are discussed on the "Impact of Conventional Munitions on Environment and Population".

1. Introduction

The original MIL-STD-2105 A (NAVY) from 19th January 1990 was a draft edition which could be definitely discussed and modified. This is mentioned on the front page on the left upper corner (Fig. 1).

This "draft" contains also some proposals for the implementation of experiments which cannot be accepted without any critics. The individually described tests should be shortly presented and their usefulness and importance respectively their weaknesses presented and illustrated in the following part. Described are the following seven test procedures, at which the charge should react to the following

P. Castelo Branco et al. (eds.),
Defense Industries: Science and Technology Related to Security, 239–248

Stimulus	Maximum reaction levels:
Fast cook off (fuel fire test)	Maximum burning
Slow cook off	Maximum burning
Bullet impact	Maximum burning
Fragment impact	Maximum burning
Sympathetic detonation	No mass detonation
Shaped charge jet	No full detonation
Shaped charge spall fragments	No violent burning

Originally these are military requirements. Naturally the different armed forces division *i*n USA have different requirements on this less sensitive or normally called insensitive munitions aspect. The ARMY is mostly interested in the survivability of their armor fighting vehicles, the NAVY in the survivability of their ships and the AIR FORCE in the survivability of their airports respectively that the mutual distances of the ammunitionbunkers are short to the runways, or their mutual distance can be held short. These are certainly important aspects of the survivability and the logistics of the individual branches of the services.

Further it should be emphasized that all the tests have to be performed with the ammunition in the „logistic packages". Certainly these tests can be performed in standard casing or standard warheads to get a screening of the sensitivity of the high explosive charges. But these results belong only to the selected test geometry and to the confinement used per accident. It is known that the high explosive charges are reacting with different degrees of violence in different quantities and especially confinements <2>. Held PEP

2. Tests

20 test items are necessary for the performance of the tests (Fig. 2). At the beginning all test items are inspected by X-ray. Three test items are used for the 28 day temperature and humidity tests. The X-ray inspection is followed by the vibration test, after repeated X-ray inspection the 4 day temperature and humidity tests and

after a third X-ray inspection the 12 m drop test to be carried out. These tests generally corresponds with the MIL 810 tests for ammunitions.

12 items are used for the specific tests of less sensitive munitions for fast cook-off, the slow cook-off, the bullet impact, the fragment impact, the shaped charge jet load. For the sympathetic detonation test at least 5 items are necessary. All these tests are consuming the test items because after each individual load, also if no reaction happens, no repetition with the same item is possible.

2.1 Fast cook-off tests

The test setup for the fast cook-off is schematically presented in Fig. 3. This is a real condition test which can easily happen in practice. In the past it was happening, that ammunition was hanging on a airplane on top of an aircraft carrier and was exposed to a strong fire. The ammunition reacted violently with a strong blast wave and accelerated fragments dangerous for the fire brigade *etc*.

2.2 Slow cook-off test

The so-called slow cook-off test should simulate a slowly increasing temperature in the case of a smouldering heat. The increase of temperature of the ammunition should be 3.3 °C per hour (Fig. 4). The ammunition is heated up uniformly over the total volume until the critical temperature is reached. If the reaction starts it often comes to a very fast and violent reaction with some pressure and fragment load.

The probability is however very small that an ammunition is exposed to such a very slow by increasing temperature rise. Therefore this test should be not a mandatory item. It would be good enough to have a screening test, to know the reaction behavior under this very randomly coming event. The authors agree that this test should be done up to a maximum temperature for example 150 °C but not at temperatures of over 200 °C which are definitely very very rare. On the other hand this slow cook-off test needs some special equipment and is typically expensive.

2.3 Bullet impact test

The sample has to be loaded with a 12.7 mm armor piercing projectile with 2800 ± 200 ft/sec velocity for the bullet impact test. A burst of 3 projectiles with 100 ms ± 10 ms time difference was originally required on the spot with the largest amount of explosives. The observation of the reaction behavior of the sample should be made (Fig. 5) with:

- pressure gauges
- high speed cameras and/or video
- whitness plate

For military purposes this is surely a real and therefore also very useful test.

2.4 Fragment impact test

The sample has to be hit by a fragment of 250 grain respectively 16 gram with a velocity of 8300 ± 300 ft/s which corresponds roughly to 2500 m/s. The sample should be hit once on a spot with the largest amount of explosive and a second new sample on the most sensitive spot

Original the MIL-STD-2105 A is written so that the sample has to be hit with „one" fragment. In USA tests are performed already with at least 3 fragments per test and hit location.

The velocity is measured with light barriers and the hit location on the sample with a screening plate in front of the sample. The reaction behavior of the sample is measured again with pressure gauges and high speed cameras, respectively video recordings (Fig. 6).

The 16 gram heavy fragment with 2500 m/s impact velocity should be realized in a eastern world ammunition. However the shown references in annex.2 demonstrate that fragments with this mass and velocity do not exist in the diagram „possible fragment masses as a function of the fragment velocities".

In USA a detonative fragment generator to accelerate the fragments to velocities of 2500 m/s for one or several fragments with masses of 16 g needed originally an high explosive charge with around 50 kg Composition B and finally about 20 kg Octol with a multi initiation device on the rear side.

The author is of the opinion that fragment hits against ammunitions are always possible. But a realistic fragment threat with realistic masses and velocities should be taken into account and the limit should take 99 % of the threat or maximum 99,9 % of the summarized possible cases and not an extraordinary extreme case for such a test.

Required are cubic fragments. Cubic fragments are realistically used on different warhead types. But as a test method cubic fragments are very critical. If a cubic fragment impacts with a flat surface the threshold velocity is about 2 times lower compared to a hit over the edges or the corner <Held pep >. For tests such different orientations cannot be taken into account statistically.

Another point is that if the fragment is perforating packaging material and casings with velocities above 1500 m/s. Then the fragments are eroding and mushrooming and impacting as semi - spheres on the high explosive. Ammunition should be always tested in the original package and in the original casings. Spherical fragments definitely make more sense for reproducible load and test conditions.

2.5. Sympathetic detonation tests

Sympathetic tests have to be provide with the full original ammunitions in the logistic package. Fig. 7 shows an example for the arrangement of a donor charge to the acceptor charges. Tests with HE shells in paletts have shown that not always the direct neighboring projectile in the nearest distance is detonating sympathetically, but relatively often the diagonally arranged acceptor charge. Over a larger distance the metal casing is accelerated to a higher velocity compared to the shorter distance and lead therefore to a more violent reaction with higher probability.

Because the detonation of the donor charge is strongly damaging the acceptor charges the reaction behavior can only be measured with pressure gauges with high confidence. One useful arrangement of pressure gauges is presented in Fig. 8 where they are oriented in a 90° setup. The total event should be also photographed with high speed cameras.

The sympathetic detonation for example would be avoided in a tank, then such vehicles would be much less vulnerable. The safety distances of ammunition bunkers could be drastically reduced if bombs or ammunition would not sympathetically detonate. Therefore such tests are very useful still if they are also expensive.

2.6 Shaped charge impact test

The sample should be attacked directly with a 81 mm precision shaped charge in a 5.8 inch standoff distance equal to 150 mm (Fig. 9). For the „81 mm shaped charge“ the dimensions, the high explosive filling and the liner are exactly defined (Fig. 10). A drawing of the shaped charge is not presented in the paper. The author added the drawings of this so-called 81 mm precision shaped charge of the BRL from different publications (Fig. 11 and Fig. 12).

Should a sample be tested with a smaller load of a shaped charge jet then it is recommended to use the M-42/M-46 shaped charge. A sketch of this shaped charge is presented in Fig. 13. Surprisingly this small shaped charge does not correspond to the fielded ammunition because the straight 60° liner (Fig. 14) is changed to a trumpet liner with a 3“ radius, corresponding to 76,2 mm (Fig. 15).

The initiation threshold of shaped charge jets is presented by the v^2d criteria where v is the impact velocity of the jet on the high explosive charge and d is the jet diameter. The requirement for tests with shaped charges taken into account the description of jet characteristics and not an exact requirement for the copper material which has to be used and the production method for the liner.

It should be required that the shaped charge is standardized which can be costly produced in the individual countries and which can be easier afforeded for such tests.

Because ammunition can be attacked generally with shaped charges either by terrorists or in a military conflict such requests after „shaped charge resistant" ammunition are understandable. But the tests performed up to now give no realistic inputs.

2.7 Spall impact tests

At the spall impact test the shaped charge is fired in 5.8" standoff against a 1" thick steel plate corresponding to 25.4 mm. The ammunition should be attacked by the spalling secondary fragments of the target plate with 4 fragments per 10 " and total 40 fragments (Fig. 16). This test is surely useful to get the reaction behavior or cartridges in a vehicle, caused by the spalling secondary fragments.

3. Conclusions

Fast cook-off test	Certainly useful
Slow cook-off test	Very costly for complete ammunition and because this load nearly not exists this test is not really necessary and not very useful, at least not mandatory
Bullet impact test	Very useful test
Fragment impact test	16 grams with 8300 ft/s not a realistic combination of fragment mass and velocity; More realistic combinations of fragment mass and velocity should be selected

Sympathetic detonation	Very useful test for reducing the lethality in vehicles, ships and to reduce the safety distances of bunkers (magazines)
Shaped charge impact test	A testing shaped charge should be recommended which can be produced in the individual countries and which corresponds to the real threats
Spall impact test	Useful test, but using a standard shaped charge which is easily and everywhere available

Annex 1

Reaction levels

In the MIL-STD-2105 A (Navy) the different reaction levels are described in great detail with 6 levels (Fig. 17)

Type 1	Detonation reaction
Type 2	Partial detonation reaction
Type 3	Explosion reaction
Type 4	Deflagration reaction
Type 5	Burning reaction
Type 6	Propulsion

The descriptions of the individual reaction levels are relatively long. It is a little troublesome to find out the right reaction level. The author has this visualized in a *chart or VG* (Fig. 18) *with the following different effects at the individual reaction levels (niveau):*

- Pressure history of the blast wave
- Size of propelled fragments
- Crater on the ground

The UK NAVY has made a requirement for the testing of less sensitive ammunition earlier than the MIL-STD-2105 A, whereby the expected threat was mainly terrorist attacks against fielded ammunition (Fig. 19). Also here are presented 6 reaction levels, which are shown in Fig. 20. After a presentation of the author with the remark on the discrepancy in the row of the reaction levels between UK to USA the author achieved a letter with the remark that UK is now using the nomenclature of MIL-STD-2105 A. UK give additionally the maximum pressure values of the individual reaction levels in a defined 6 feet distance (Fig. 21).

Annex II

Threat by Fragments

T.L. Boggs <AGARD-CP-511, 1992> shows in his contribution „Hazards of Energetic Materials and their Relation to Munition Survivability“ the threat of fragments with regard to mass and velocity in one diagram with a rough estimation of their frequency. The marked square shows the 16 gram/2500 m/s fragment which is not in the threat scenario with regard to fragment velocity and mass (Fig. 22). In contrast a very frequently existing fragment has 2050 m/s velocity with 73 g mass. A further critical fragment but only with about 1 % frequency in contrast to the above fragment is the fragment with 430 g mass with the velocity of 1280 m/s. These two fragments present much larger threat and risk to the ammunition compared to the unrealistically fast fragment what is especially demonstrated in the published Figures 8 and 9 after Wagenhals <AGARD - CO-511, 1992> (Fig. 23 and Fig. 24), in which the perforations of steel confinements with different wall thicknesses are shown after 2 different equations (Thor, respectively Fate).

The author <M. Held, "Vulnerability Tests with Fragment Impacts", 29th Int. Annual Conference of ICT, Karlsruhe, Germany, V32:1-12, 1998> recommended EFP fragments with a hemispherical shape to attack the ammunitions on the wanted spots with a few hundred gram high explosive charge compared to the 20 kgs Octol fragment generators <Stotsen, NWC - TM 6411, 1988>

DaimlerChrysler Aerospace
TDW Gesellschaft für verteidigungstechnische Wirksysteme mbH

Fig. 1

TDW • Schrobenhausen

Item Number and Test Sequence

DaimlerChrysler Aerospace
TDW Gesellschaft für verteidigungstechnische Wirksysteme mbH

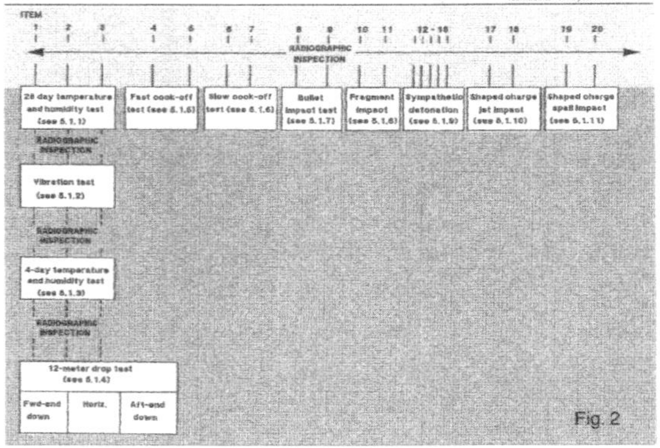

Fig. 2

TDW • Schrobenhausen

Fast Cook-off Test

5.1.5 MIL-STD-2105A (NAVY)

DaimlerChrysler Aerospace
TDW Gesellschaft für verteidigungstechnische Wirksysteme mbH

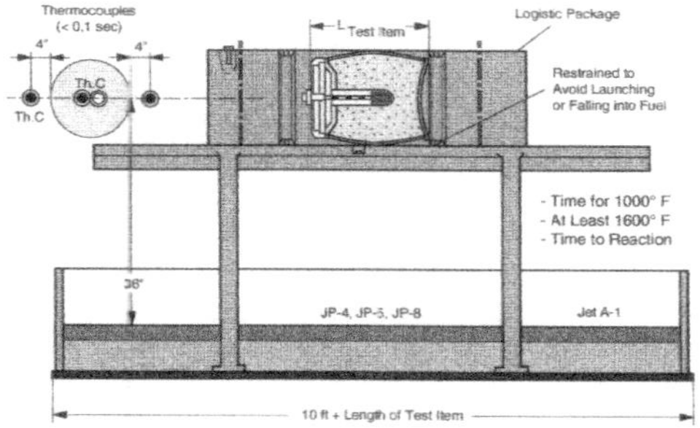

Fig. 3

TDW • Schrobenhausen

Slow Cook-off Test

5.1.6 MIL-STD-2105A (NAVY)

DaimlerChrysler Aerospace
TDW Gesellschaft für verteidigungstechnische Wirksysteme mbH

6° F per Hour ≙ 3,3°C / h

Start at 100° F ≙ 55°C Below Predicted

Reaction Temperature

Continuously Measurement of Temperature

Fig. 4

TDW • Schrobenhausen

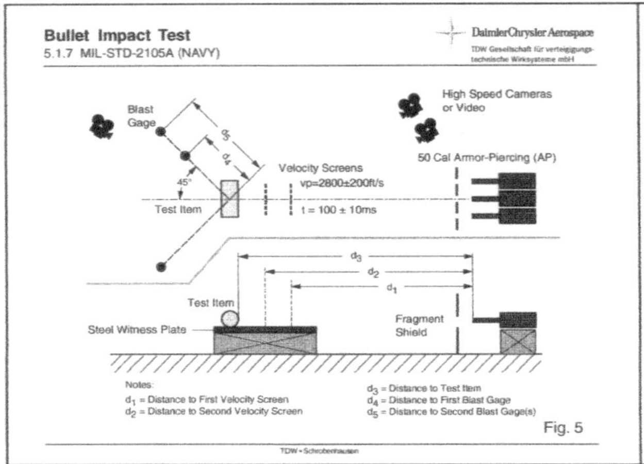

Fig. 5

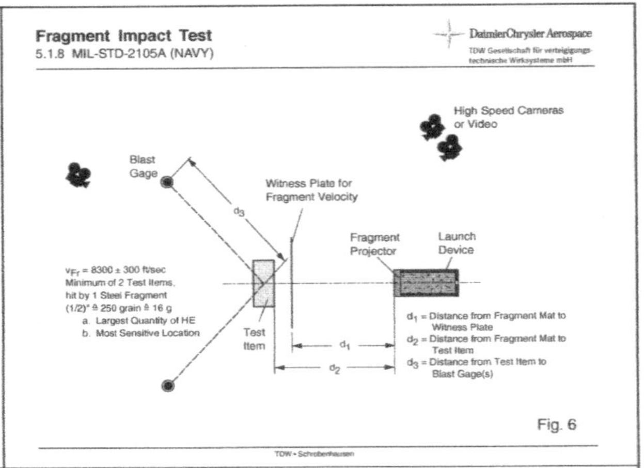

Fig. 6

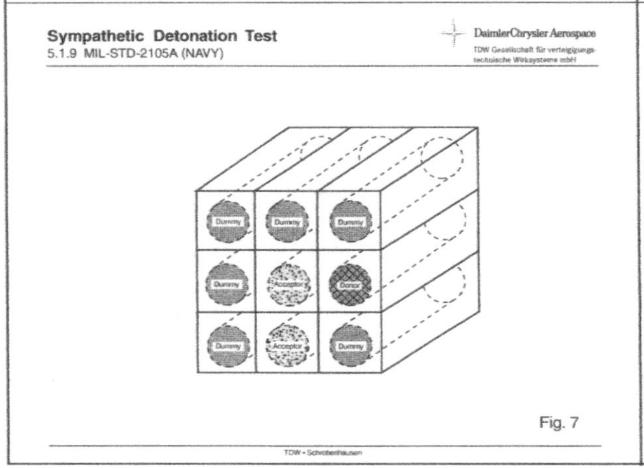

Fig. 7

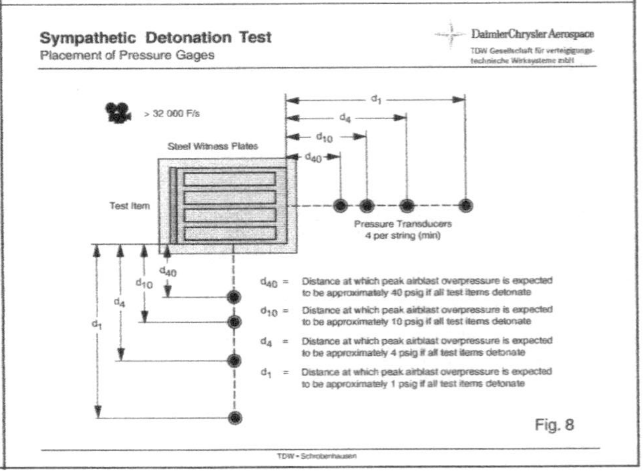

Fig. 8

Shaped Charge Impact Test
5.1.10 MIL-STD-2105A (NAVY)

DaimlerChrysler Aerospace
TDW Gesellschaft für verteidigungstechnische Wirksysteme mbH

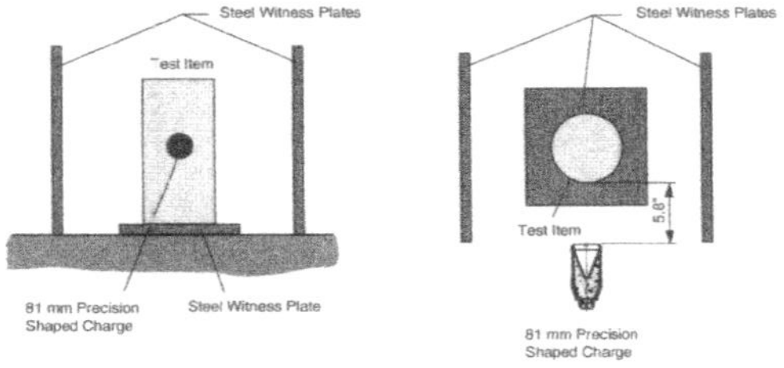

Fig. 9

TDW • Schrobenhausen

Shaped Charge Jet Impact Test
5.1.10 MIL-STD-2105A (NAVY)

DaimlerChrysler Aerospace
TDW Gesellschaft für verteidigungstechnische Wirksysteme mbH

The 81 mm precision shaped charge shall be configured as follows:

Explosive fill:	**1,8 pounds of Composition B conforming to MIL-C-401**		
Cone angle:	**42°**		
Dimensions:	**Height of cone**	=	**3.7 inches**
	Outside diameter	=	**3.2 inches**
	Inside diameter	=	**2.91 inches**
	Wall thickness	=	**0.075 inches**
Liner description:	**Oxygen-free copper conforming to ASTM B152 with a temper of OS025** **Grain size < 50 microns after stress relief** **No shear forming** **Deep drawn anneal**		
Body:	**Standard 90-mm M371E1 recoilles rifle round**		

Fig. 10

TDW • Schrobenhausen

Standard Shaped Charge

DaimlerChrysler Aerospace
TDW Gesellschaft für verteidigungstechnische Wirksysteme mbH

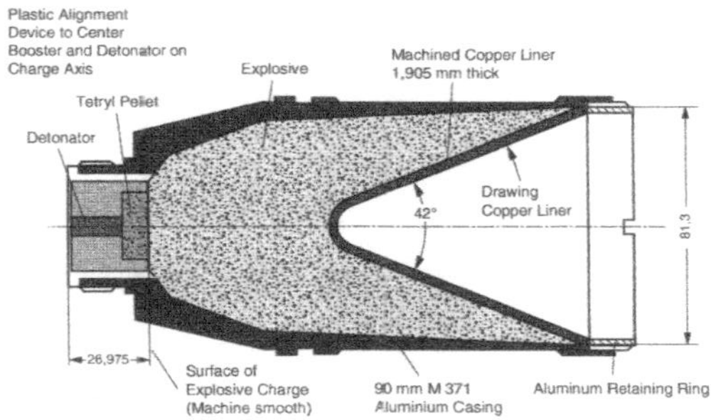

Note:
Liner Thickness not to vary more that .0051mm in any transverse plane and .051mmm in a longitudinal plane

Aseltine (1990)

Fig. 11

TDW • Schrobenhausen

42° Shaped Charge

DaimlerChrysler Aerospace
TDW Gesellschaft für verteidigungstechnische Wirksysteme mbH

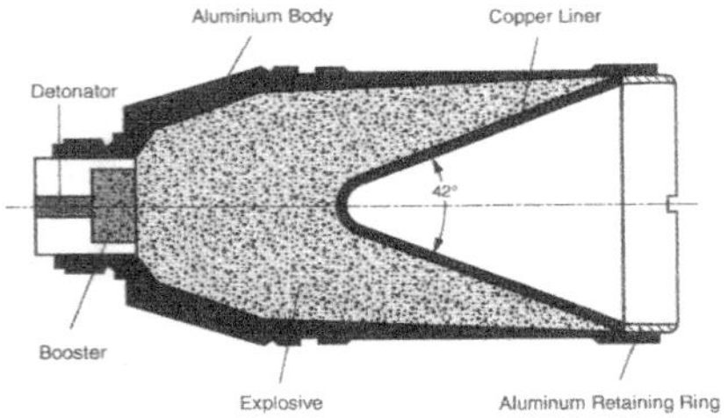

Walters - Fundamental of Shaped Charges 1989

Fig. 12

TDW • Schrobenhausen

Shaped Charge Jet Impact Test
5.1.10.3.2 MIL-STD-2105A (NAVY)

The M 42/M46 grenade shall be configured as follows:

Explosive fill: 30 grams of Composition A-5 conforming to MIL-E-14970

Cone angle: Trumpet with 3" radius

Dimensions:

Height of cone	=	1.3 inches
Outside diameter	=	1.315 inches
Inside diameter	=	1.237 inches
Wall thickness	=	0.075 inches

Liner description: Copper strip, cold-rolled, soft annealed, conforming to QQ-C-576
Electrolytic tough pitch
Grain size < ASTM grain size 8
Non-earring quality with suppressed cube texture

Body: M 42/M46 body load assembly (without fuze)

Fig. 13

Grenade M 42 / M 46

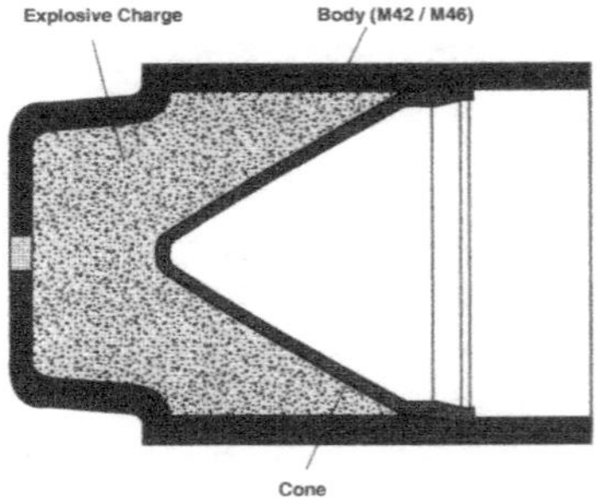

Fig. 14

Grenade M 42 / M 46

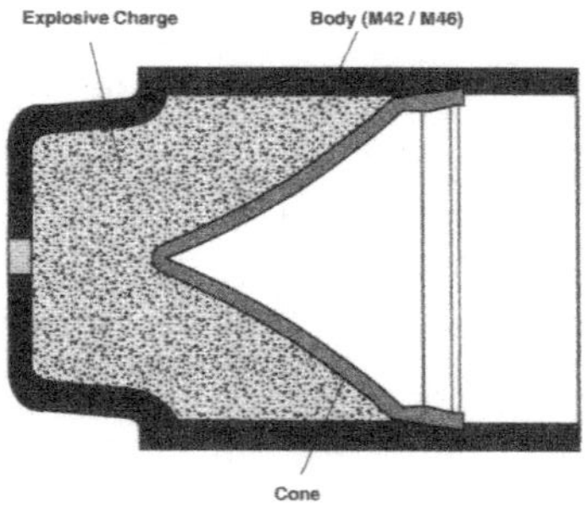

Fig. 15

Spall Impact Test
5.1.11 MIL-STD-2105A (NAVY)

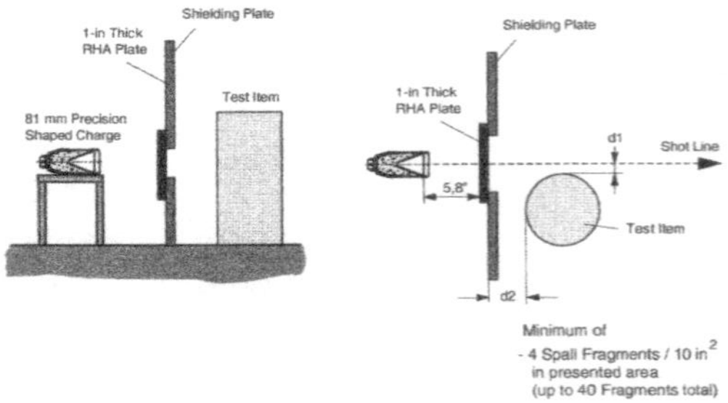

Fig. 16

MIL-STD-2105A (NAVY)

DaimlerChrysler Aerospace
TDW Gesellschaft für verteidigungstechnische Wirksysteme mbH

Explosive reaction levels.

a. Detonation Reaction (Type I). The most violent type of explosive event. A supersonic decomposition reaction propagates through the energetic material to produce an intense shock in the surrounding medium,e.g., air or water, and very rapid plastic deformation of metallic cases followed by extensive fragmentation. All energetic material will be consumed. The effects will include large ground craters for munitions on or close to the ground, holing/plastic flow damage/fragmentation of adjacent metal plates and blast overpressure damage to nearby structures.

b. Partial Detonation Reaction (Type II). The second most violent type of explosive event. Some, but not all of the energetic material reacts as in a detonation. An intense shock is formed; some of the case is broken into small fragments; a ground crater can be produced, adjacent metal plates can be damaged as in a detonation, and there will be blast overpessure damage to nearby structures. A partial detonation can also produce large case fragments as in a violent pressure rupture (brittle fracture). The amount of damage, relative to a full detonation, depends on the portion of material that detonates.

c. Explosion Reaction (Type III). The third most violent type of explosive event. Ignition and rapid burning of the confined energetic material builds up high local pressures leading to violent pressure rupturing of the confining structure. Metal cases are fragmented (brittle fracture) into large pieces that are often thrown long distances. Unreacted and/or burning energetic material is also thrown about. Fire and smoke hazards will exist. Air shock are produced that can cause damage to nearby structures. The blast and high velocity fragments can cause minor ground craters and damage (break-up, tearing, gouging) to adjacent metal plates. Blast pressures are lower than for a detonation.

d. Deflagration Reaction (Type IV). The fourth most violent type of explosive event. Ignition and burning of the confined energetic materials leads to nonviolent pressure release as a result of a low strength case or venting through case closures (leading port/fuze wells, etc.). The case might rupture but does not fragment; closure covers might be expelled, and unburned or burning energetic material might be thrown about and spread the fire. Pressure venting can propel an unsecured test item, causing an additional hazard. No blast or significant fragmentation damage to the surroundings; only heat and smoke damage from the burning energetic material.

e. Burning Reaction (Type V). The least violent type of explosive event. The energetic material ignites and burns, nonpropulsively. The case may open, melt or weaken sufficiently to rupture nonviolently, allowing mild release of combustion gases. Debris stays mainly within the area of the fire. This debris is not expected to cause fatal wounds to personnel or be a hazardous fragment beyond 50 feet.

f. Propulsion (Type VI). A reaction whereby adequate force is produced to impart flight to the test item in its least restrained configuration as determined by the life cycle analysis.

Fig. 17

TDW • Schrobenhausen

Explosive Reaction Levels

MIL-STD-2105A (NAVY)

DaimlerChrysler Aerospace
TDW Gesellschaft für verteidigungstechnische Wirksysteme mbH

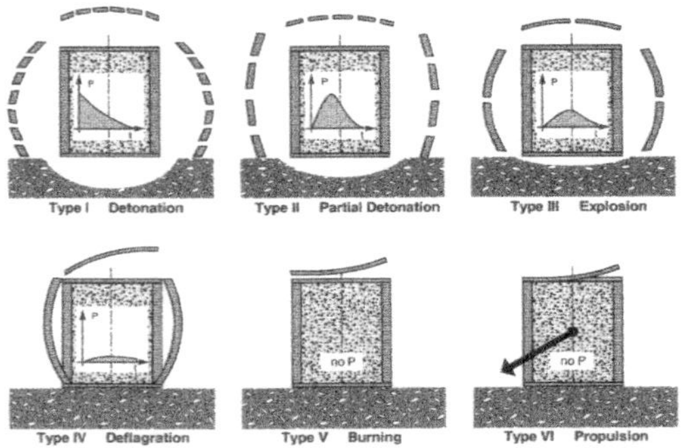

Fig. 18

TDW • Schrobenhausen

RATTAM Tests UK

DaimlerChrysler Aerospace
TDW Gesellschaft für verteidigungstechnische Wirksysteme mbH

Concern by the Royal Navy about the safety of explosive munitions mounted or stored in ships, particularly guided weapon warheads, led in the UK to an assessment programme generally known by the acronym RATTAM (Response to ATTack on AMmunition). The aim of the programme was to obtain the necessary data to advise the RN on the risk to RN ships carrying particular weapons, if those weapons were subject to terrorist attack.

It was decided to use the following weapons for the tests; they are readily available in UK service and are typical of what a terrorist might obtain:

Code	Weapon
A (i)	7.62mm AP
A (ii)	0.50 in AP
B	20mm HE
C	84mm HEAT

The weapons were fired a explosive munitions protected by a 6mm mild steel plate representing the ships side; in some tests the protective plate was not used. The 0.5 in AP has been used most in the tests.

Fig. 19

TDW • Schrobenhausen

Response Categories (RATTAM - UK)

DaimlerChrysler Aerospace
TDW Gesellschaft für verteidigungstechnische Wirksysteme mbH

Six categories of response were decided upon for the munitions under test:

(I) **No response**

(II) **Burning** - the explosive filling ignites and burns, The munition case may open up but the munition is not moved propulsively.

(III) **Deflagration** - The explosive store is ignited leading to rupture and often accompanied by the ejection of unreacted or burning explosive.

(IV) **Mild Explosion** - The explosive store is ignited leading to violent rupture. Unreacted or burning explosive may be ejected. Major pieces of the store may be thrown a considerable distance.

(V) **Severe Explosion** - The explosive store is blown apart with considerable violence. The damage from blast and fragmentation is extensive but less than that associated with detonation, Large fragments are ejected accompanied by unreacted or burning explosive.

(VI) **Detonation**

Fig. 20

TDW • Schrobenhausen

Explosive Reaction Levels
MIL-STD-2105A (NAVY)

DaimlerChrysler Aerospace
TDW Gesellschaft für verteidigungstechnische Wirksysteme mbH

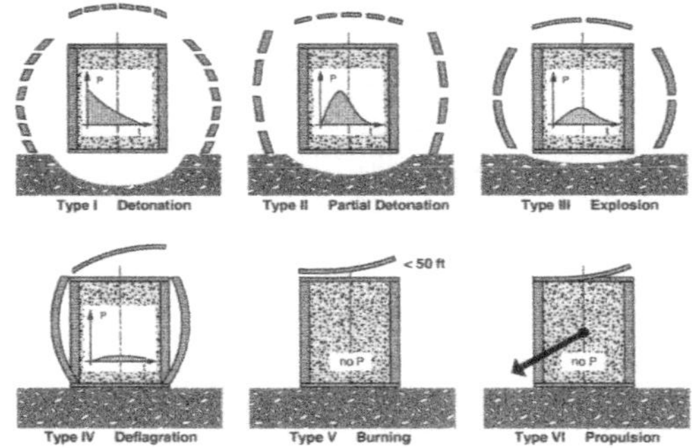

Fig. 21

TDW • Schrobenhausen

Overlay of a Fragment Threat Map onto Response Plot

DaimlerChrysler Aerospace
TDW Gesellschaft für verteidigungstechnische Wirksysteme mbH

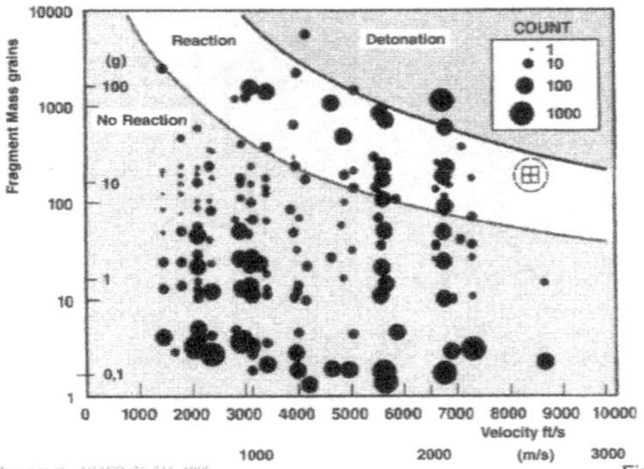

T.L. Boggs et al. AGARD-CP-511, 1992

Fig. 22

TDW • Schrobenhausen

Degradation of Example Fragments as a Function of Barrier Thickness Using Thor Equations

DaimlerChrysler Aerospace
TDW Gesellschaft für verteidigungstechnische Wirksysteme mbH

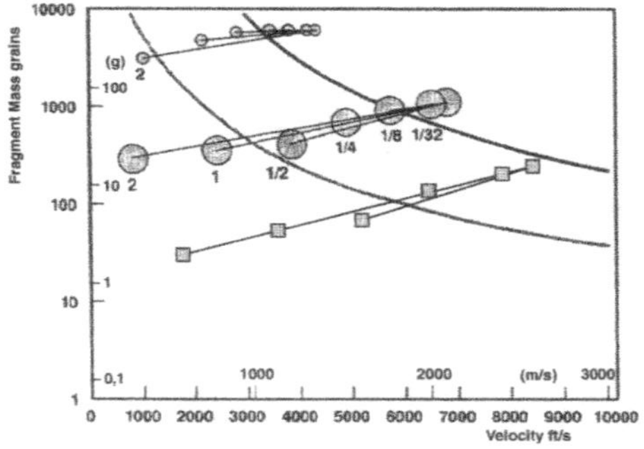

M.R. Wagenhals et al. AGARD-CP-511, 1992

Fig. 23

TDW • Schrobenhausen

Degradation of Example Fragments as a Function of Barrier Thickness Using FATE Equations

DaimlerChrysler Aerospace
TDW Gesellschaft für verteidigungstechnische Wirksysteme mbH

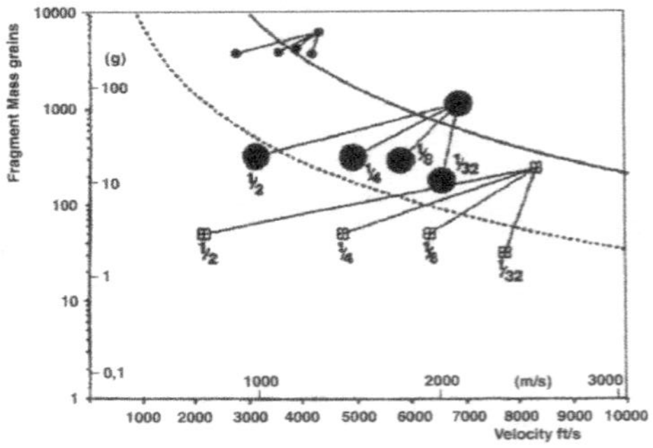

M.R. Wagenhals et al. AGARD-CP-511, 1992

Fig. 24

TDW • Schrobenhausen

Storage and Transportation

STORAGE, HANDLING AND TRANSPORTATION OF AMMUNITION AFTER REDUCTION OF 50% OF THE EXISTING IN 1990 – SECURITY INVOLVEMENT AND ENVIRONMENTAL PROBLEMS

Virgil Pompiliu Ene

Abstract

After reviewing the risk factors in storage, handling, transportation, demilitarization and destruction of conventional ammunition, the article deals with the increasing security activity in multinational operations in which the Romanian military forces take part. There are also presented the major chemical components for Romanian ammunition and their security impact on the population and environment.

At the end, the article emphasizes Romania's efforts to comply to the European and NATO standards dealing with the storage, handling, transportation, demilitarization and destruction of ammunition in safety conditions. It also refers to the action ways in case of ecological disaster in Romania.

P. Castelo Branco et al. (eds.),
Defense Industries: Science and Technology Related to Security, 257–262

STORAGE, HANDLING AND TRANSPORTATION OF AMMUNITION N THE PROCESS OF REORGANIZATION AND OWNSIZING OF THE AMMUNITION STOCKS WITHIN THE ROMANIAN ARMED FORCES - SAFETY IMPLICATIONS FOR PERSONNEL AND ENVIROMENT

LTC. ENE VIRGILIU POMPILIU
J.4 / General Staff, MOD
ROMANIA

AGENDA

- Risk factors analysis in storage, handling, transportation, demilitarization and destruction of conventional ammunition
- Increase of safety level of multinational operations in which participate Romanian units
- Factors of downsizing of the ammunition stocks
- Presentation of the major chemical components of the Romanian ammunition
- Safety impact on population and environment
- Action ways in case of ecological disaster
- Military regulations complying with the European Standards

OBJECTIVES OF THE ENVIRONMENTAL PROTECTION IN THE ROMANIAN ARMED FORCES

- to comply with national and international environmental regulations
- to implement the environmental standards of NATO member countries
- to develop an institutionalized framework within the armed forces
- to identify, assess and monitor the pollution sources in the military field
- to co-operate with similar military and civilian organizations

SAFETY LEVEL OF MULTINATIONAL OPERATIONS IN WHICH PARTICIPATE ROMANIAN UNITS

- **NATO standards**
- **UN regulations**

- **Internal military regulations are mandatory in case they are more restrictive**

Factors which influenced the 50% reducing of the 1989 level of ammunition

- Technical end-of-life
  - physical-chemical features
  - ballistic characteristics
- Downsizing and reshaping the armed forces

MAJOR CHEMICAL COMPONENTS OF THE ROMANIAN AMMUNITION

- **EXPLOSIVES**
  - primary
  - intermediary
  - secondary
- **POWDERS**
  - black
  - colloidal
  - composite
- **PYROTECHNIC COMPOSITIONS**
  - smoke
  - fire
  - light

IMPACT ON SAFETY
OF POPULATION AND ENVIRONMENT

- Increase of intoxication, nervous and skin diseases and hepatitis
- Decrease of personnel health
- Injuries, burns and deaths

IMPACT ON SAFETY
OF POPULATION AND ENVIRONMENT

(*continuation*)

- burning of vegetation and forests
- contamination of surface and depth waters
- development of a high quantity of dangerous and toxic substances
- deterioration of the ozone layer and increasing of the global heating effect
- ammunition laboratory tests
- phonic pollution of the environment
- contamination and destruction of soil and infrastructure

WAYS OF ACTION
IN CASE OF ECOLOGICAL DISASTER

- warning of the population living near the accident location
- warning of the specialized military and civilian structures
- first aid measures for limiting the effects of the ecological disaster
- delimitation, marking and securing the affected area
- going on with actions for stopping the effects of the ecological disaster
- gathering, transportation and storage measures for the contaminated materiel
- examination of the accident causes for avoiding similar ones or diminishing their environmental impact

NEW MILITARY REGULATIONS
WITH THE VIEW OF APPROACHING THE EUROPEAN STANDARDS

- **Compatibility Groups**
- **Hazard Divisions**
- **Quantity - Distance**
- **Potential explosion sites**

Possible Hazards of Propellants Connected with their Chemical Composition and Decomposition

Dr. Manfred A. Bohn

Fraunhofer-Institut für Chemische Technologie (ICT)
Postfach 1240 D-76318 Pfinztal-Berghausen

Abstract

The typical ingredients of gun propellants are nitrocellulose, blasting oils, nitroguanidine, stabilizers based on amine group activated aromatic systems, some inorganic and organic salts and plasticizers as dibutylphthalate. Some of these compounds can be toxic and / or carcinogenic or mutagenic. Doublebase rocket propellants have similiar ingredients except nitroguanidine. Composite rocket propellants contain typically a HTPB-IPDI binder, aluminium, ammonium perchlorate, plasticizers and burning catalysts. Hazards may occur from the relatively high impact and friction sensitivities and the possible toxicity of plasticizers and burning catalysts.

Nitric acid ester based propellants decompose chemically. Thereby the stabilizers are consumed. This limits the time period in which a stabilizer prevents the decomposition to be autocatalytically accelerated. The molar mass or chain length of the nitrocellulose is reduced which can lead to a malfunction in shooting. The decomposition reactions are in total exothermic and the hazard of thermal explosion is possible with ageing of propellants, with their exposure to high environmental temperatures and /or by storing propellants in non-suitable geometries and containers. Some consecutive stabilizers formed during the steady decomposition of such propellants may have increased toxicity compared to the starting compound. Burning products too may have hazardous properties.

The principles of the decomposition reactions, their mitigation, their experimental recognition and kinetic description will be discussed. This enables to assess the safe use time of propellants and assure the proper handling of ammunition in storage and in use also. The toxicity aspects of typical ingredients and burning products will be compiled and the consequences discussed. Thermal explosion hazards of storage magazines are even in the present time not yet banned as very recent examples have demonstrated. Their experimental and descriptive understanding will be shown in short.

P. Castelo Branco et al. (eds.),
Defense Industries: Science and Technology Related to Security, 263–291

Possible Hazards of Propellants Connected with their Chemical Composition and Decomposition

Dr. Manfred A. Bohn

NATO Advanced Research Workshop
'Defense Industries: Science and Technology Related Security'
'Impact of conventional munition on environment and population'
October 29 to 31, 2001, Porto, Portugal

Fraunhofer Institut Chemische Technologie

Types of hazards

general type 'toxicity'

- toxicity
- cancerogenic
- mutagenic (changes or damages on genes)
- damaging ability to embryo
- endangering to environment
  to plants, fishes, other organisms, ecological systems

general type 'accident'

- uncontrolled explosion / detonation
  > decomposition reactions can cause self ignition
  chemical stability and compatibility
- in controlled operation
  > burning products are hazardous
  > malfunction because of ageing
  chemical stability and compatibility, migration of compon.

Fraunhofer Institut Chemische Technologie

Hazards of toxicity type from ingredients

- Not sealed cartridges (artillery bags, mortar propellants)

  ingredients with toxic potential can evaporate / migrate from the cartridge

- Broken cartridges, broken rocket motors
  - contact with the propellant is possible

    evaporation and / or diffusion of toxic ingredients into the environment

- Storage of 'open' propellants in containers for any further use

  as above

- Uncontrolled open burning of propellants

Fraunhofer Institut Chemische Technologie

Gun propellants - typical ingredients

binder
nitrocellulose (NC), cellulose acetate butyrate (CAB)

blasting oils
nitroglycerine (NG)
diethylene glycol dinitrate (DEGN)

crystalline energetic material
nitroguanidine (Nigu, NQ)
hexogen (RDX)

stabilizers
DPA (diphenylamine), 2-NO_2-DPA, acardite II, ethyl centralite (EC), ethyl phenyl urethane (EPU)

plasticizer
dibuthyl phthalate (DBP), diamyl phthalate (DAP), dioctyl phthalate (DOP)

additives
potassium nitrate (KNO_3), potassium sulfate
sodium oxalate, sodium sulfate,
dinitro toluene (DNT)

Fraunhofer Institut Chemische Technologie

Rocket propellants - typical ingredients

binder
- nitrocellulose (NC) in DB RP
- HTPB-IPDI, HTPB-TDI, CTPB-IPDI, CTPB-HDI in composite RP

blasting oils
- nitroglycerine (NG)
- butanetriol trinitrate (BTTN)

stabilizers
- 2-NO_2-DPA, MNA (N-methyl-p-nitro-aniline)

crystalline material
- ammonium perchlorate (AP), aluminium (Al)
- hexogen (RDX)

plasticizer
- dioctyl adipate (DOA), diisooctyl sebacate (DOS), isodecyl pelargonate (IDP)

additives
- iron oxide (Fe_2O_3)
- lead and copper salts
- ferrocene type compounds (Catocene™, Butacene™)

EP 2001 Bohn / RP ingred

ICT
Fraunhofer Institut
Chemische Technologie

International risk marking labels

T+: very toxic

T: toxic

N: danger to environm.

Xn: less toxic, Xi: irritating

O: promot. fire

C corrosive

F: inflammable

EP 2001 Bohn / Risk marking label

ICT
Fraunhofer Institut
Chemische Technologie

Toxicity of ingredients part 1 of 2

compound		toxicity / cancer / mutag. / endang. env.
nitrocellulose	NC	none
cellulose acetate butyrate	CAB	none
nitroglycerine	NG	very toxic, environment poisson
butanetriol trinitrate	BTTN	toxic
nitroguanidine	NQ	irritating
diphenylamine	DPA	toxic, environment poisson
N-nitroso-DPA	N-NO-DPA	toxic
2-NO_2-DPA		irrtiating
N-methyl-p-nitro-aniline	MNA	toxic
N-nitroso-MNA	N-NO-MNA	carcinogenic, mutagenic, toxic
dibutyl phthalate	DBP	irritating, environment poisson
dioctyl phthalate	DOP	irritating, env. poisson, cancerog. ?
dioctyl adipate	DOA	cancerogenic ?
lead salts		toxic

Toxicity of ingredients part 2 of 2

compound	toxicity / cancer / mutag. / endang. env.
ammonium perchlorate	
HMX	less toxic
RDX	toxic
DNT	toxic, carcinogenic
TNT	toxic, environ. poisson, carcinogenic ?

Toxicity of ingredients in development

compound		toxicity / cancer / mutagenic endangering environment
Ammoniumdinitramid	ADN	less toxic
Hydraziniumnitroformate	HNF	toxic
ε-Hexanitro-hexaza-iso-wurtzitane (ε-HNIW)	ε-CL20	?
Gycidyl azide polymer	GAP	?
binder	PolyNIMMO	?
	PolyGlyn	?
plasticizer	GAP-A	?
	DANPE	?
	DNDA	less toxic

EP 2001 Bohn / Tox-Ing-dvel

Fraunhofer Institut Chemische Technologie

Operational fume main products of prop. WC 844 / US M16 rifle

compound		g / (kg of WC 844)	toxicity / cancer / endang. env.
hydrogen	H_2	15.4	highly inflammable
carbon monoxide	CO	490.4	toxic, highly inflam.
carbon dioxide	CO_2	229.0	
methane	CH_4	2.9	highly inflam.
acetylene	C_2H_2	0.026	highly inflam.
ethene	C_2H_4	0.2	cancerogenic ?, highly inflam.
ethane	C_2H_6	0.1	highly inflam.
propene	C_3H_6	0.03	highly inflam.
butane	C_4H_{10}	0.02	highly inflam.
benzene	C_6H_6	0.15	carcinogenic, mutagenic, toxic
acetonitrile	CH_3CN	0.33	toxic, inflamable
hygrogen cyanide	HCN	0.47	very toxic, environ. poisson
ammonia	NH_3	1.9	toxic, environ. poisson
nitrogen monoxide	NO	1.56	very toxic (NO_2 also)

Data from A. Snelson, a. o., ITTRI, Chicago

EP 2001 Bohn / Burnprod WC844

Fraunhofer Institut Chemische Technologie

Burning products as function of 'enclosure'

L5460 (JA2)

main burning products in the fumes as function of controlling the reactions

open burning is kinetically controlled

closed burning is thermodynamically controlled

Data from Dr. F. Volk

	experimental	experimental	thermod. calc.
load. density >	0.0013 g/cm^3	0.1 g/cm^3	0.1 g/cm^3
compound	g / (kg TLP)	g / (kg TLP)	g / (kg TLP)
H_2	5.6	11.5	
N_2	21.1	139.2	
CO	368	358	365
CO_2	159	327	318
NO	**246**	**0**	**0**
N_2O	1.7	0	0
CH_4	0.6	0.63	2.54
C_2H_4	2.2	0	0
H_2O	198	159	162
	open burning	closed burning in gun	

EP 2001 Bohn / Burnprod JA2

Fraunhofer Institut Chemische Technologie

Hazards of accident type - limited stability of propellants

Propellants based on nitric acid ester compounds are intrinsically unstable because of chemical decomposition reactions

- single base GP, NC
- double base GP, NC and blasting oils
- triple base GP, NC, blasting oils, crystalline energetic as NQ, RDX
- double base RP, NC and blasting oils
- triple base RP, NC, blasting oils, RDX

Decomposition effects thermal stability and mechanical strength of GP grains and the rocket motor charge

Resulting dangers: > self ignition by self heating

> grain failure by mechanical load → breech blow

> malfunction of ammunition by energy loss

EP 2001 Bohn / limited stability

Fraunhofer Institut Chemische Technologie

Hazards of accident type - migration of components
- crack formation by temperature load

A non-chemical cause of instability is migration of

plasticizers, burning catalysts

phlegmatizers, blasting oils

Temperature change load may cause crack formation and propagation in composite rocket propellants

With some older types of composite rocket propellants the cross-link density can change in both directions - weakening or hardening of propellant

Resulting dangers: > rocket motor explosion

> malfunction of ammunition by disturbed burning

Decomposition of NC

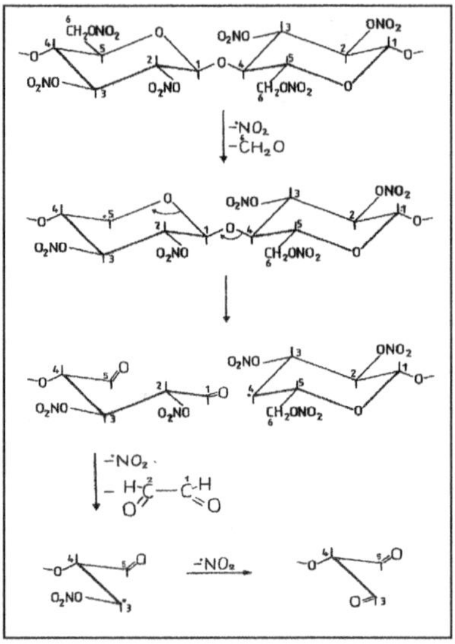

Fraunhofer Institut Chemische Technologie

Main decomposition reactions in NC and NC based propellants, part 1 of 2

type of decom-position	caused mainly by	activation energy [kJ/mol]	temp. dependent preponderance	effect in propellant	to prevent or only to slow down
homolytic splitting of the $CO\text{-}NO_2$ bond in two radicals: $\text{-C-O}\cdot + \cdot NO_2$ in NC and blasting oils NENA's polyGlyn, polyNIMMO	temperature, (thermal energy), named thermolytic bond splitting or homolysis	160-170	mostly at higher temp. > 60°C-70°C transition temp. dependent on the concentr. of the products H_2O and HNO_3 in NC or propellant	formation of NO_2 and in consecutive reactions HNO_3, HNO_2, NO, N_2O, N_2, decomposition of the NC backbone by reactions stabilizing the radical functionality, chain splitting, autocatalytic effect of NO_2, HNO_3, HCOOH	not to prevent, only to slow down by catching (chemical binding) the autocatalytically effective products
saponification of the $CONO_2$ group, in NC and blasting oils and as above	water, acids, alkalines, named hydrolytic decomp. or hydrolysis	80-100	mostly at lower temp. < 60°C-70°C	formation of HNO_3 which transforms also to HNO_2, NO, NO_2, autocatalysis, the NO_2 produces the -C-O· radical from C-OH	not to prevent, only to slow down as said above

EP 2001 Bohn / -Main decomp-NC-1

Fraunhofer Institut Chemische Technologie

Main decomposition reactions in NC and NC based propellants, part 2 of 2

type of decomposition	caused mainly by	activation energy [kJ/mol]	temp. dependent preponderance	effect in propellant	to prevent or only to slow down
chain splitting of NC chain by destruction of chain elements	thermolyt. or hydrolytically, combined with radical formation by NO_2 and O_2	80-160	according to mechanism at higher or lower temperature, mostly mixed effect	formation of CO, CO_2, H_2O, and of many low molecular organic compounds, decrease of mechan. strength	not to prevent, only to slow down as said above
chain splitting of NC by hydrolysis	water, acids	60	at lower temp.	chain shortening, formation of unstable endgroups, decrease of mechan. strength	not to prevent
backbone (structural) decomposition of NC	result of the homolysis and hydrolysis combined with radical formation by NO_2		mostly mixed effect	formation of CO, CO_2, H_2O, and of many small organic compounds, increase of the concentration of the autocatalyt. effective reactants	not to prevent, only to slow down as said above
structural decomposition of blasting oils	as with NC		mostly mixed effect	as with NC	as with NC

EP 2001 Bohn / Main-decomp-NC-2

Fraunhofer Institut Chemische Technologie

Weighing of reaction rates by investigation temperature

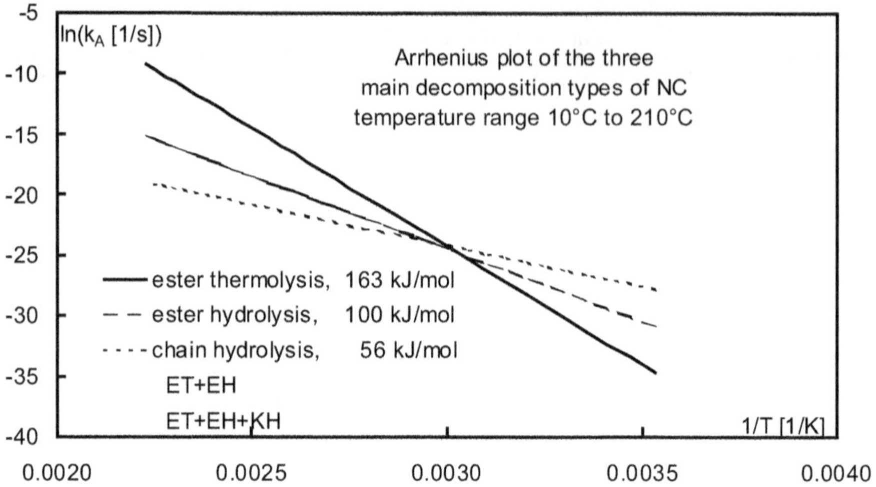

EP 2001 Bohn / Ea-Einfluß-

Fraunhofer Institut Chemische Technologie

Stabilizer consumption: activation energy as function of temp.

- decrease of stabilizer DPA in single base GP A5020 (20mm machine gun)
- change in Ea at about 65°C
- higher range is more thermo-lytical ester group decomposition
- lower range is more hydrolytical ester group decomposition

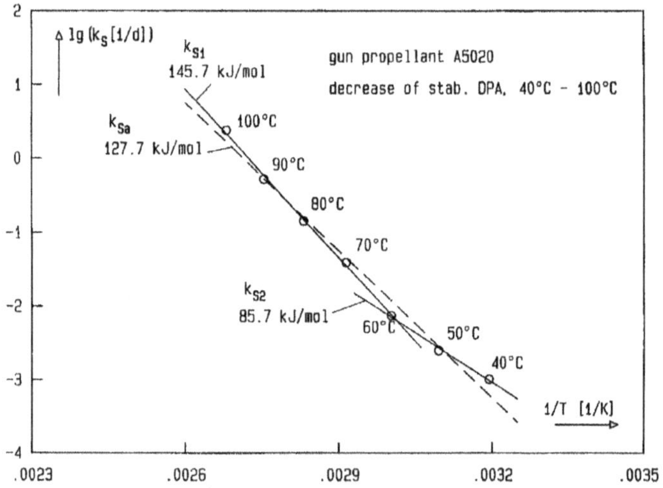

EP 2001 Bohn / S: Ea-f(T)

Fraunhofer Institut Chemische Technologie

Molar mass decrease of NC: activation energy as function of temp.

decrease of molar mass of NC in single base GP A5020 (20mm machine gun)

change in Ea at about 65°C

higher range is more thermo-lytical ester group decomposition with consecutive chain scission

lower range is more hydrolytical NC chain splitting

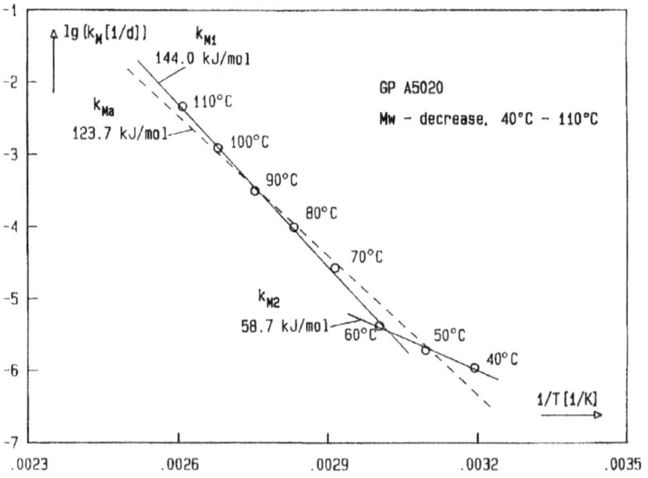

Fraunhofer Institut Chemische Technologie

Kinetic description of stabilizer consumption

reaction scheme RS I

NE	$\xrightarrow{k_{NE}}$	P + R-NE	primary decomp.
NE + P	$\xrightarrow{k_{auto}}$	2 P + R-NE	autocat. decomp.
S + P	$\xrightarrow{k_{SP}}$	P-S	stabilizer reaction

- condition for autocatalysis
  reaction rate constant $k_{auto} >> k_{NE}$
- condition for stabilization
  reaction rate constant $k_{SP} >> k_{auto}$

S must be necessary only in small amounts

a good stabilizer suppresses the autocatalytic decomposition

the primary decomposition of NE cannot be prevented

RS I is the right scheme in the case of no consecutive stabilizing products or if the primary stabilizer has a much higher reactivity as these

Fraunhofer Institut Chemische Technologie

Model 'S: exponential + linear' part 1 of 2

Combination of the reactions of first and zero order

$$\left(\frac{dS(t,T)}{dt}\right)\Bigg|_T = -k_S^1(T)\cdot S(t,T) \; - k_S^2(T)$$

$$k_S^1(T) = Z_S^1 \cdot \exp\left(-Ea_S^1/RT\right)$$
$$k_S^2(T) = Z_S^2 \cdot \exp\left(-Ea_S^2/RT\right)$$

Arrhenius behaviour

Integrated rate equation

$$S(t,T) = \left(S(0) + \frac{k_S^2(T)}{k_S^1(T)}\right)\cdot \exp\left(-k_S^1(T)\cdot t\right) - \frac{k_S^2(T)}{k_S^1(T)}$$

EP 2001 Bohn / M-S-expon+lin-1

Fraunhofer Institut Chemische Technologie

Model 'S: exponential + linear' part 2 of 2

Calculation of service times $ty_S(T)$

Definition of degree of stabilizer consumption

$$y_S = \frac{S(t,T)}{S(0)}$$

$$ty_S(T) = \frac{1}{k_S^1(T)}\cdot \ln\left(\frac{1+\dfrac{k_S^2(T)}{S(0)\cdot k_S^1(T)}}{y_S + \dfrac{k_S^2(T)}{S(0)\cdot k_S^1(T)}}\right)$$ times to reach y_S

$$t_0(T) = \frac{1}{k_S^1(T)}\cdot \ln\left(\frac{S(0)\cdot k_S^1(T)}{k_S^2(T)} + 1\right)$$ times to reach $y_S = 0$

the model has a good extrapolation ability

EP 2001 Bohn / M-S-expon+lin-2

Fraunhofer Institut Chemische Technologie

Stabilizer consumption: decrease of 2-nitro-DPA

description with model 'S: expon. + linear' in comparison with description according to first order reaction

data points marked with arrows must be omitted with first order reaction, but not with the model 'S: expon.+linear'

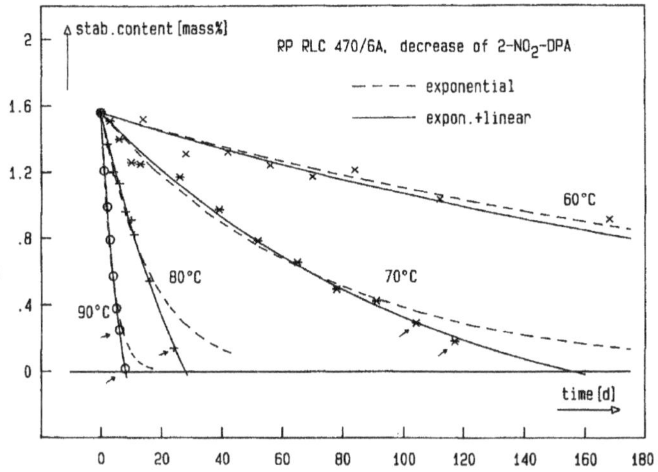

EP 2001

Bohn / Stab-RLC 470

Fraunhofer Institut Chemische Technologie

Model for stabilizer consumption - extrapolation ability

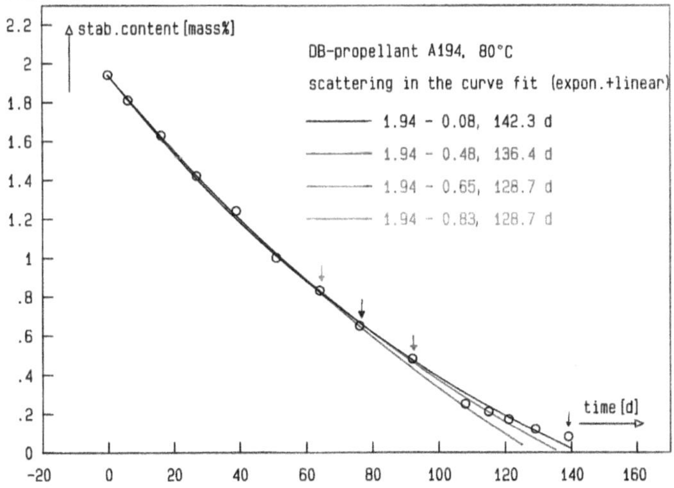

EP 2001

Bohn / S-Model-extrapol

Fraunhofer Institut Chemische Technologie

Reaction kinetic model for the chain splitting of NC - part 1 of 2

- statistical splitting of the NC chain
- splitting of the chain by decomposition of a chain element
- number of chain elements decreases according to a reaction of first order

Number of chain elements at time t and temperature T

$N(t,T) = N(0) \cdot \exp(-k_M \cdot t)$

The formulation results in:

$$\ln\left(1+\frac{m}{Mn(t,T)}\right) = \ln\left(1+\frac{m}{Mn(0)}\right) + k_M(T)\cdot t$$

EP 2001 Bohn / NC-Spalt-1

Fraunhofer Institut Chemische Technologie

Model NC-splitting part 2 of 2

with: Mn(osmosis=absolute) = An·Mn(GPC=relative)
polydispersity D = Mw / Mn

Expression used for evaluation

$$\ln\left(1+\frac{m\cdot D(t,T)}{An\cdot Mw(t,T)}\right) = \ln\left(1+\frac{m\cdot D(0)}{An\cdot Mw(0)}\right) + k_M(T)\cdot t$$

$$y_{Mn} = \frac{Mn_L}{Mn(0)} = \frac{Mn(ty_{Mn}(T))}{Mn(0)} \qquad y_{Mn} \approx y_{Mw} \quad \text{when } y_{Mn} \geq 0{,}4$$

$ty_{Mn}(T)$ is the time to reach the degree of degradation y_{Mn}

$$ty_{Mn}(T) = \frac{1}{k_M(T)}\cdot\ln\left(\frac{\frac{1}{y_{Mn}}+\frac{An\cdot Mw(0)}{m\cdot D(0)}}{1+\frac{An\cdot Mw(0)}{m\cdot D(0)}}\right)$$

the model has a very good extrapolation ability

EP 2001 Bohn / NC-Spalt-2

Fraunhofer Institut Chemische Technologie

Modelling of the molar mass decrease of NC

Modelling of NC degradation in a double base rocket propellant

on ordinate is the model expression

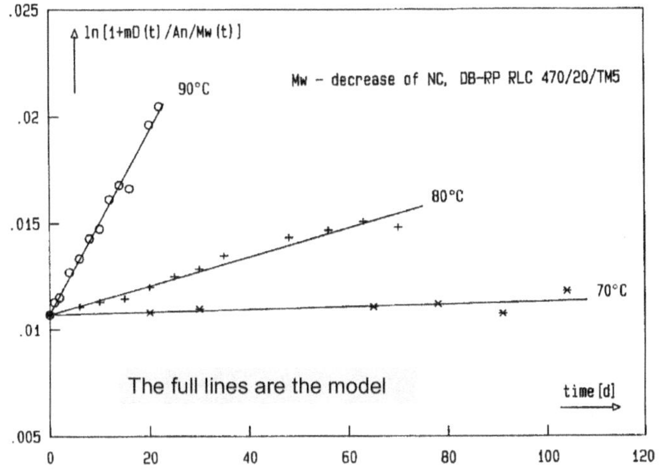

Modelling of the molar mass decrease of NC

Modelling of NC degradation in a double base rocket propellant

on ordinate is the back calculated model expression

the description of the data is very good

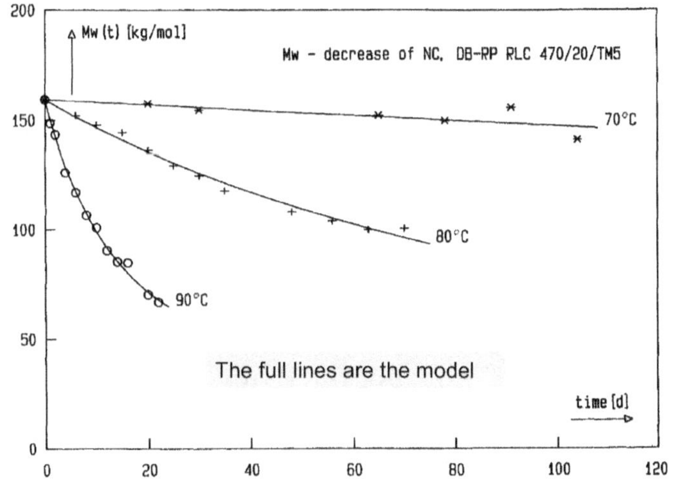

Model for degradation of molar mass of NC - extrapolation ability

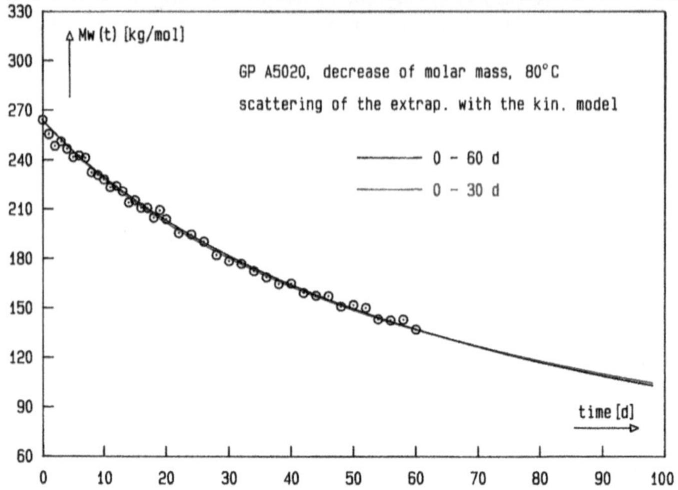

EP 2001 Bohn / Mw-Abnahme-Extrapol

Fraunhofer Institut Chemische Technologie

Molar mass decrease - brittle ductile transition

shift of the brittle - ductile transition temperature to higher values with decreasing molar mass = decreasing chain length of NC

ageing causes increase in brittle tendency

danger during firing rounds with large grain GP

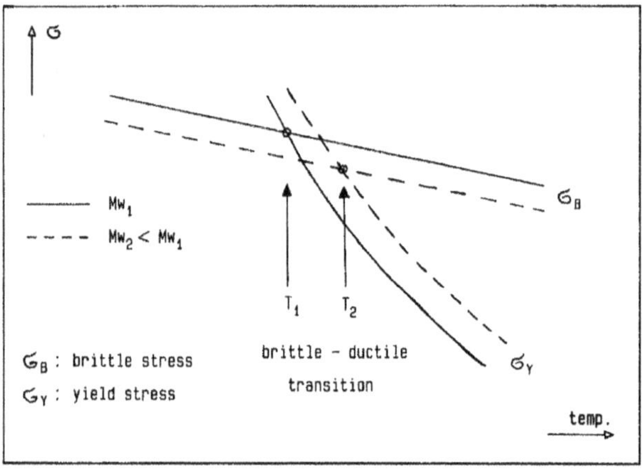

EP 2001 Bohn / brittle-ductile

Fraunhofer Institut Chemische Technologie

Example for molar mass decrease - brittle ductile transition

GP KN6540 (M30)

stress - strain curves of non-aged and aged samples as function of temperature determined at high strain rate

the brittleness increases with ageing

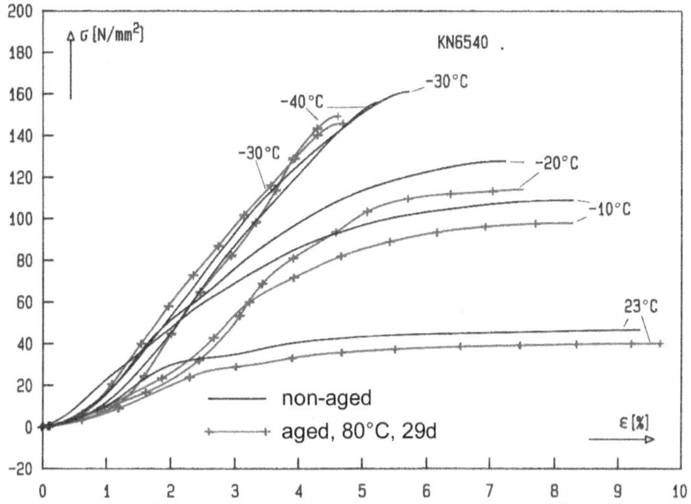

EP 2001

Bohn / KN6540-sig=f(eps)

Fraunhofer Institut Chemische Technologie

Molar mass decrease - brittle ductile transition

GP KN6540 (M30)

shift of the brittle ductile transition temperature with ageing to higher values

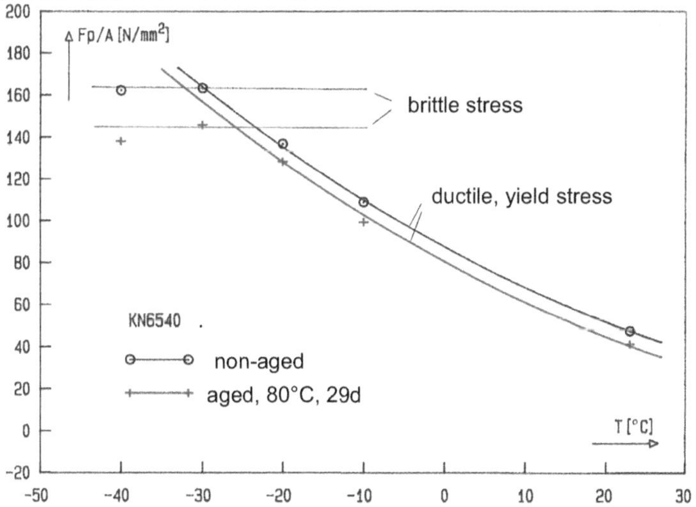

EP 2001

Bohn / brittle-ductile

Fraunhofer Institut Chemische Technologie

Description of autocatalytic mass loss of some GP

double base GP
L5460 ÷ US JA2

triple base GP
KN6540 ÷ US M30

K503: double base ball powder with 8.8 % NG

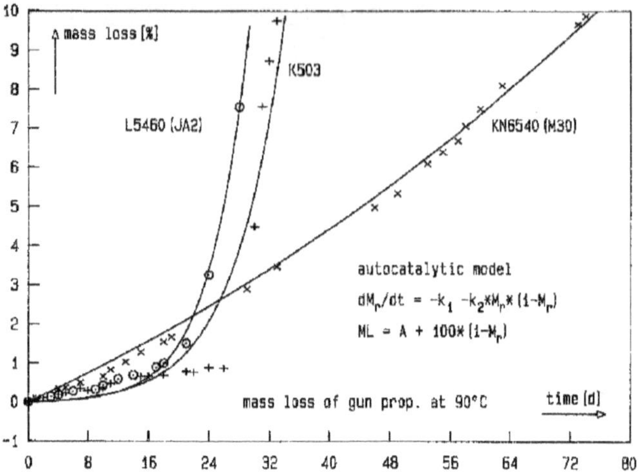

Description of mass loss - general relations

$$A \xrightarrow{k_1} B + C + S + (-\Delta H_{R,1})$$

C escapes as gas

S remains in the residue

B remains also in the residue, may be an autocat. product

Notice the difference between measured quantity M(t,T) and looked for quantity $M_A(t,T)$

$$M_r(t,T) = \frac{M(t,T)}{M(0)} = 1 - \frac{m_A - m_B - m_S}{m_A} \cdot \frac{M_A(0)}{M(0)} \cdot (1 - M_{Ar}(t,T))$$

molar masses of A, B, C, S $m_C = m_A - m_B - m_S$

$M(0) = M_A(0) + M_B(0) + M_S(0) + M_N$ M_N is a non-reacting compound

For $M_{Ar}(t,T)$ the corresponding kinetic expression has to be inserted.

$ML(t,T) = 100\% (1 - M_r(t,T))$

Description of mass loss with autocatalysis part 1 of 2

- reaction scheme

$$A \xrightarrow{k_1} B + C + S \quad (-\Delta H_{R,1}) \quad \text{primary decomp.}$$

$$A + B \xrightarrow{k_2} 2\,B + C + S \quad (-\Delta H_{R,2}) \quad \text{autocatalytic decomp.}$$

primary decomposition in a first order reaction

autocatalytic decomposition in a second order reaction

with $B(t,T) = B(0) + A(0) - A(t,T)$

- kinetic expression in molar quantities

$$\left(\frac{dA(t,T)}{dt}\right)\Bigg|_T = -k_1(T)\cdot A(t,T) - k_2(T)\cdot A(t,T)\cdot B(t,T)$$

EP 2001 Bohn /ML-autocat-1

Fraunhofer Institut Chemische Technologie

Description of mass loss with autocatalysis part 2 of 2

- transfer to masses gives

$$\left(\frac{dM_{Ar}(t,T)}{dt}\right)\Bigg|_T = -k^1_{ML}(T)\cdot M_{Ar}(t,T) - k^2_{ML}(T)\cdot M_{Ar}(t,T)\cdot(F+1-M_{Ar}(t,T))$$

$$k^1_{ML}(T) = k_1(T) \quad \text{and} \quad k^2_{ML}(T) = k_2(T)\cdot A(0) \quad \text{and} \quad F = \frac{M_B(0)}{M_A(0)}\cdot\frac{m_A}{m_B}$$

- applicable equation to determine $M_{Ar}(t,T)$ is

$$M_{Ar}(t,T) = \frac{k^1_{ML}(T) + (F+1)\cdot k^2_{ML}(T)}{k^2_{ML}(T) + \left(k^1_{ML}(T) + F\cdot k^2_{ML}(T)\right)\cdot \exp\left(\left(k^1_{ML}(T) + (F+1)\cdot k^2_{ML}(T)\right)\cdot t\right)}$$

model 'ML: first order + autocatalytic'

EP 2001 Bohn / ML-autocat-2

Fraunhofer Institut Chemische Technologie

Modelling of mass loss - example ADN

To assess new components the ageing behaviour must be investigated

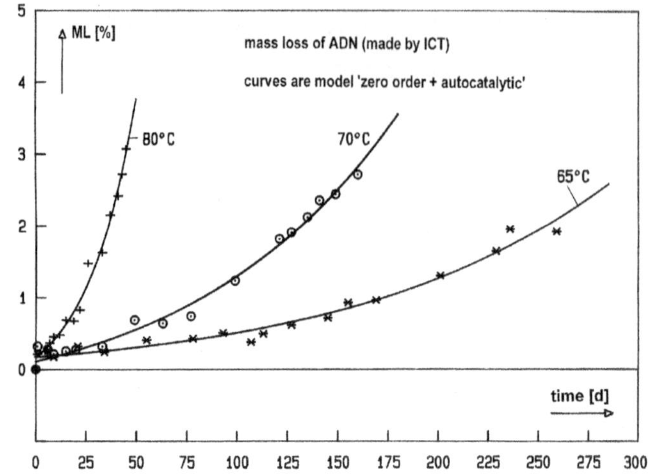

Modelling of mass loss - example ε-CL20 (ε -HNIW)

To assess new components the ageing behaviour must be investigated

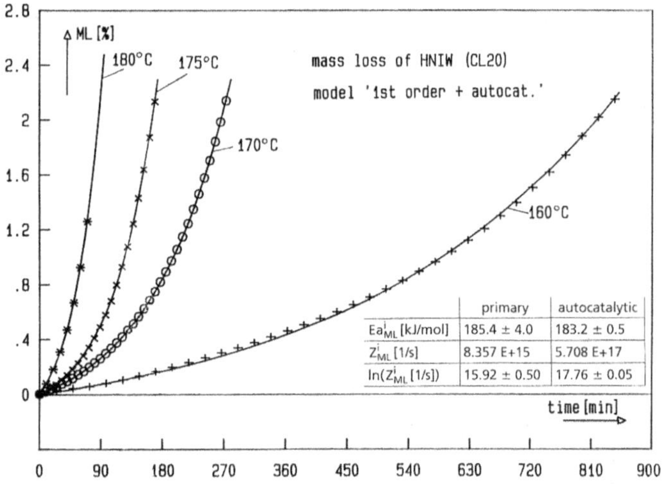

	primary	autocatalytic
Ea^i_{ML} [kJ/mol]	185.4 ± 4.0	183.2 ± 0.5
Z^i_{ML} [1/s]	8.357 E+15	5.708 E+17
$ln(Z^i_{ML}$ [1/s])	15.92 ± 0.50	17.76 ± 0.05

Modelling of mass loss - example NC

Mass loss of NC plasticized with DBP

development of the description of mass loss

an autocatalytic reaction and a consecutive first order reaction

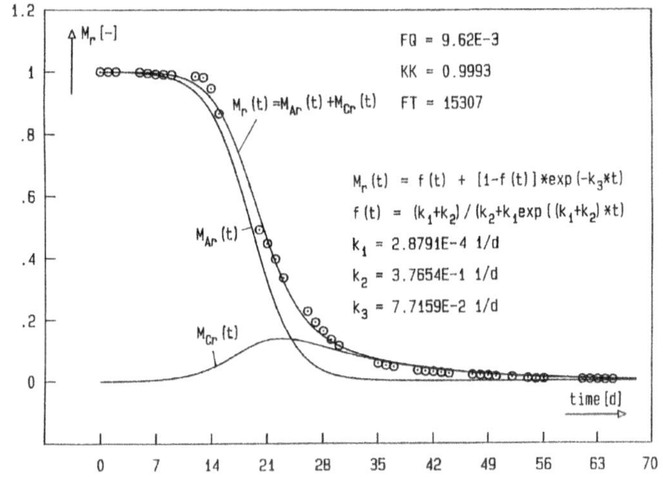

ARC: TNT-TAGN mixture

Compatibility between TAGN and TNT.

Reactivity between TNT and TAGN, recognizable by the strong reduc-tion of the onset temperature of the mixture

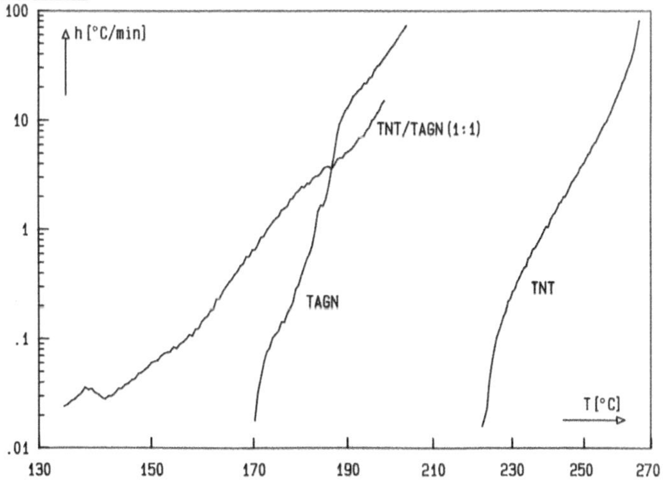

ARC: Reaction of RDX with water

Assessment of the danger a run away in the system RDX-water

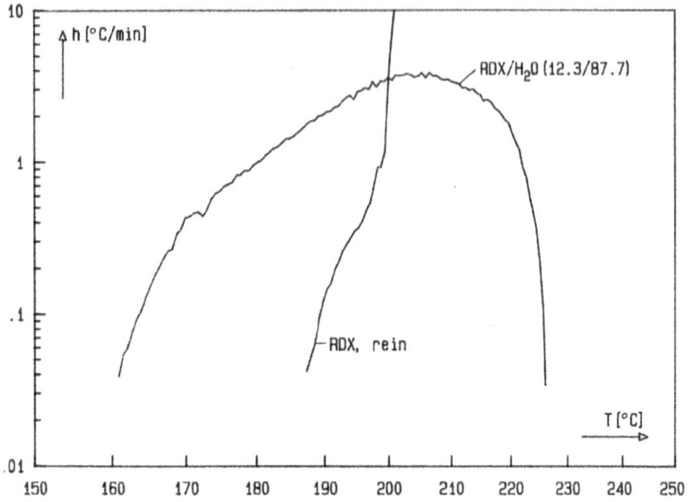

EP 2001 Bohn / RDX-water-h

Fraunhofer Institut Chemische Technologie

ARC: Reaction of RDX with water - Arrhenius parameters

Evaluation of the reaction

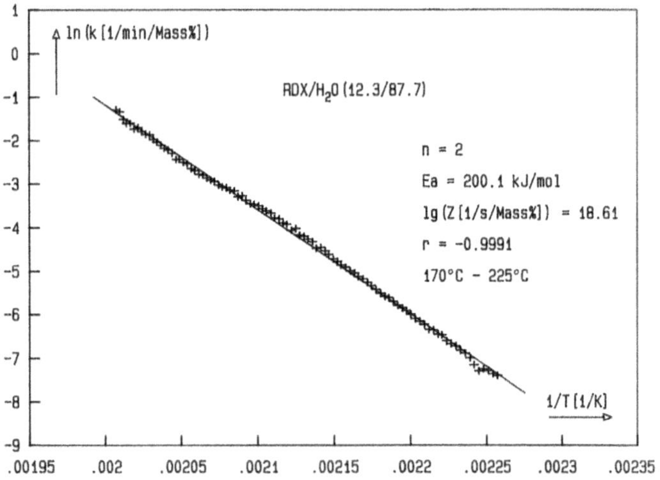

EP 2001 Bohn / RDX-Water-Ea

Fraunhofer Institut Chemische Technologie

ARC: RDX in solvents

RDX in cyclohexanon

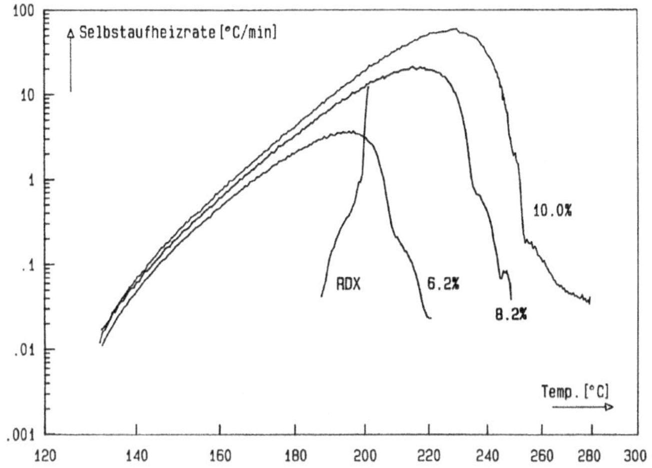

EP 2001

Bohn / RDX Cyclohex

Fraunhofer Institut Chemische Technologie

ARC: RDX in cyclohexanon

Times to deflagration (=maximum rate followed by ARC) for RDX alone

times to maximum rate for the RDX solutions

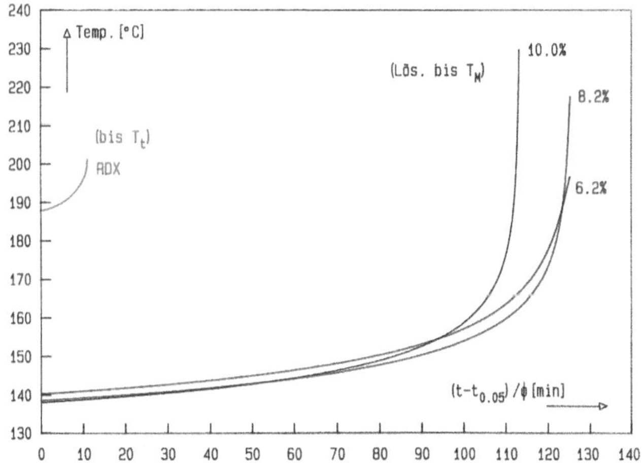

EP 2001

Bohn / RDX Cycloh-t-t

Fraunhofer Institut Chemische Technologie

ARC measuring principle

Heating up to start temperature.

Procedure:
Wait phase for equilibra-tion. Search phase, to test if a self heating exists. If no self heating then heating by heat step. A wait and search phase follow.

This procedure is repeated until an exotherm is found. Then the adiabatic mode is used and the self heating is determined.

If a self heating cannot be found the apparature switches off at reaching the preset end temperature.

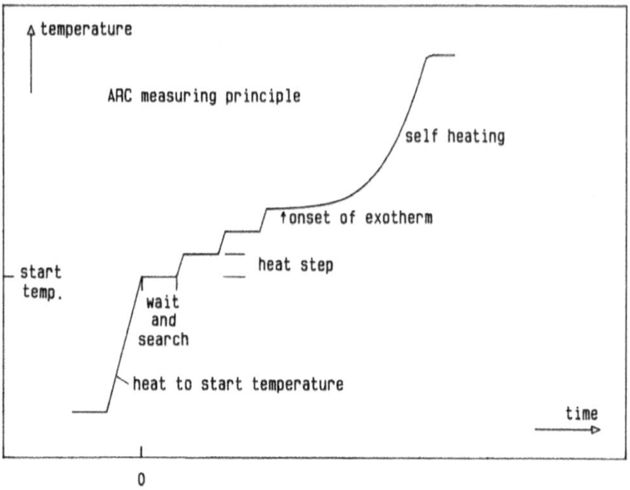

ARC: Some measuring cells

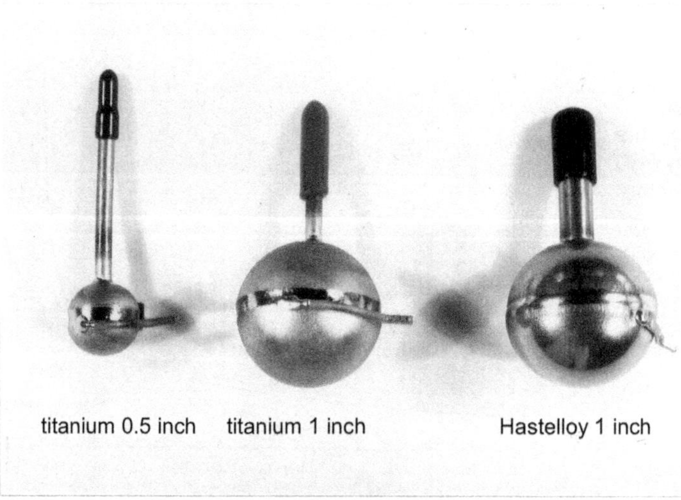

ARC: Investigation of new components

Binders and plasticizers

Determination of kinetic parameters of new components

description with kinetic models

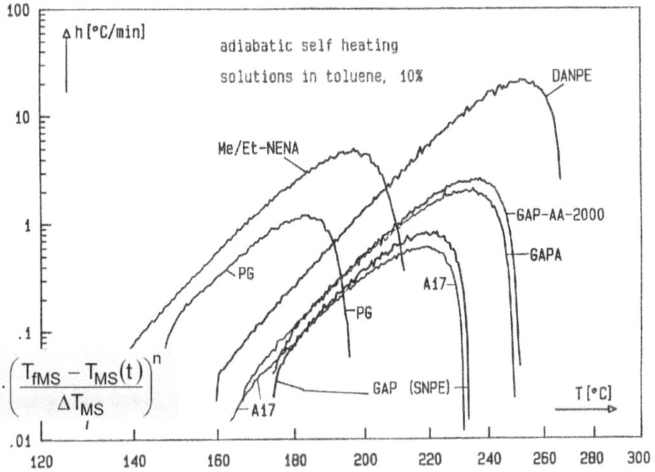

$$h(T_{MS}(t)) = k_{n,r}(T(t)) \cdot \Delta T_{MS} \cdot \left(\frac{T_{fMS} - T_{MS}(t)}{\Delta T_{MS}} \right)^n$$

EP 2001

Bohn / ARC-WM-h

Fraunhofer Institut Chemische Technologie

ARC: Adiabatic self heating of explosives - thermal cook-off-behaviour

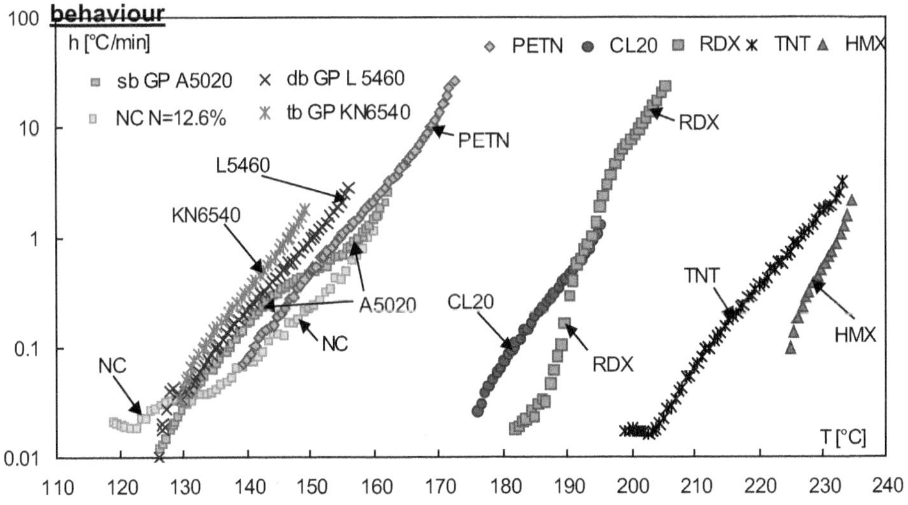

EP 2001

Bohn / ARC-TLP-SS

Fraunhofer Institut Chemische Technologie

ARC: adiabatic self heating of ADN, HNF, NC

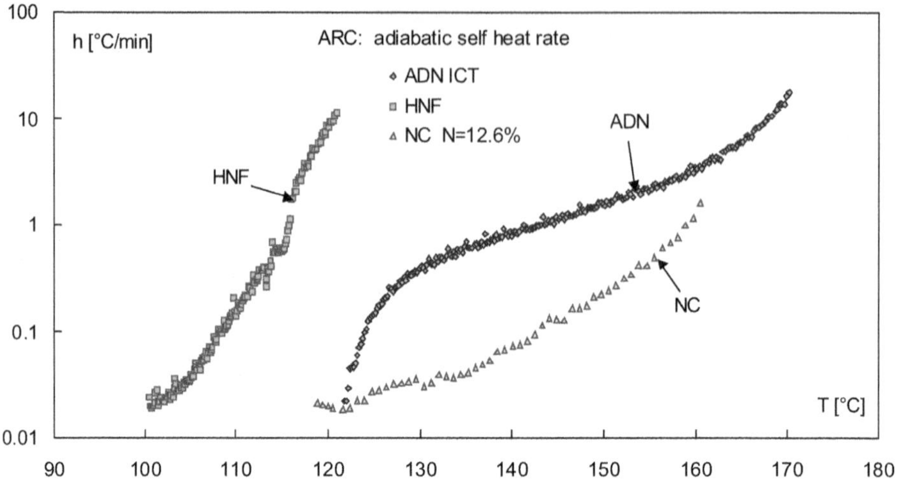

EP 2001 Bohn / ARC-ADN-HNF-NC

Fraunhofer Institut Chemische Technologie

Frank-Kamenetzkii evaluation - selfheating to thermal ignition

Calculation of the critical radius at given outer temperature

$$r_{C,FK} = \sqrt{\frac{\delta c \cdot \lambda}{\rho} \cdot \frac{RTw^2}{Ea_{Q,G}} \cdot \frac{1}{Z_{Q,G}} \cdot \exp(+Ea_{Q,G}/RTw)}$$

δc	critical Frank-Kamenetzkii parameter (FKP_C), its value is shape dependent [-]
$r_{C,FK}$	thermal critical radius [length] at the given wall temperature Tw
ρ	mass density [mass/volume]
λ	heat conductivity [energy/time/length/K]
Tw	storage temperature (wall temperature of the sample) [K]
R	general gas constant (8,3144 J/mol/K)
$Ea_{Q,G}$	activation energy of the global heat generation rate [energy/mol]
$Z_{Q,G}$	pre-exponential factor of the global heat generation rate [energy/mass/time]

FKP_C parameter for a cylinder with radius r_C and length L

$$\delta c = 2.0 + 0.843 \cdot \left(\frac{2 \cdot r_C}{L}\right)^2$$

EP 2001 Bohn / FK-eval

Fraunhofer Institut Chemische Technologie

Example of heat generation rate data used as input for the FK-evaluation

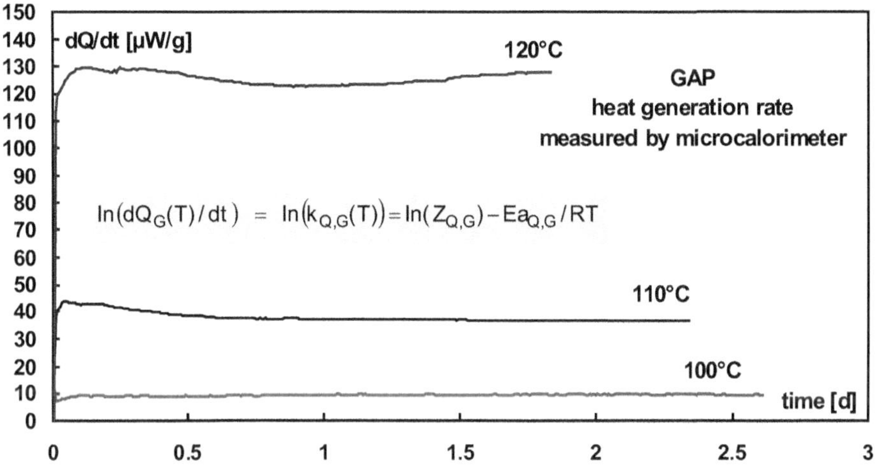

EP 2001

Bohn / GAP-WER

Fraunhofer Institut Chemische Technologie

Temperature distribution in a cylinder caused by self heating $r < r_C$

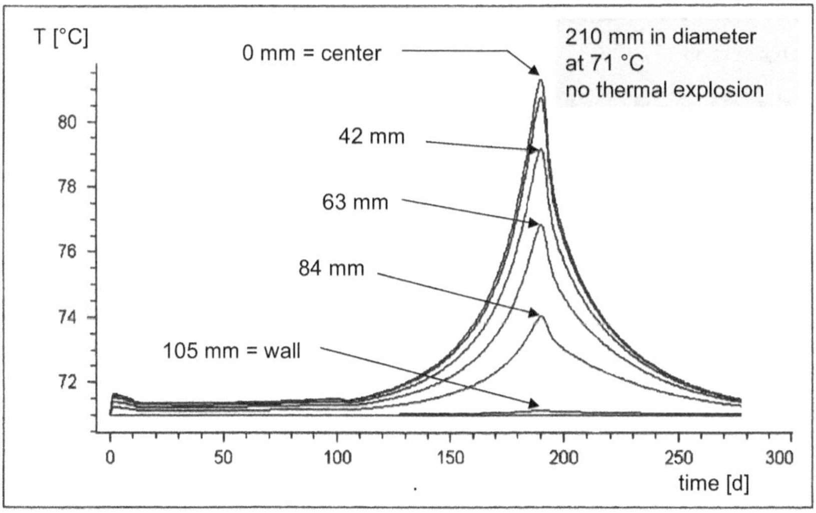

EP 2001

Bohn / SH-210mm

Fraunhofer Institut Chemische Technologie

Temperature distribution in a cylinder caused by self heating $r > r_C$

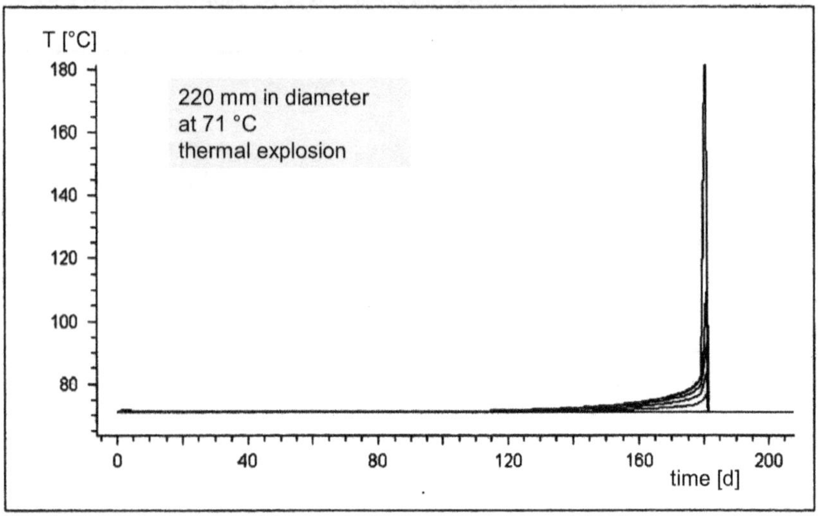

EP 2001 Bohn / SH-220mm

Fraunhofer Institut Chemische Technologie

Summary and conclusion

What is to do ?

Determination and compiling of toxicity data, including cancerogenic, mutagenic, embryo damaging and environmental endangering properties of ingredients

Investigation about the possible burning products of propellants their amounts and determining their dangerous potential by the product 'amount * endangering'

Investigation of the stability and ageing behaviour of propellants including the compatibility questions
- to prevent uncontrolled thermal explosion
- to prevent malfunction of ammunition

Development and use of kinetic modelling to quantify and assess the ageing rates with regard to safety and tolerable property changes

EP 2001 Bohn / sum conl

Fraunhofer Institut Chemische Technologie

Temperature distribution in a cylinder r < r_C

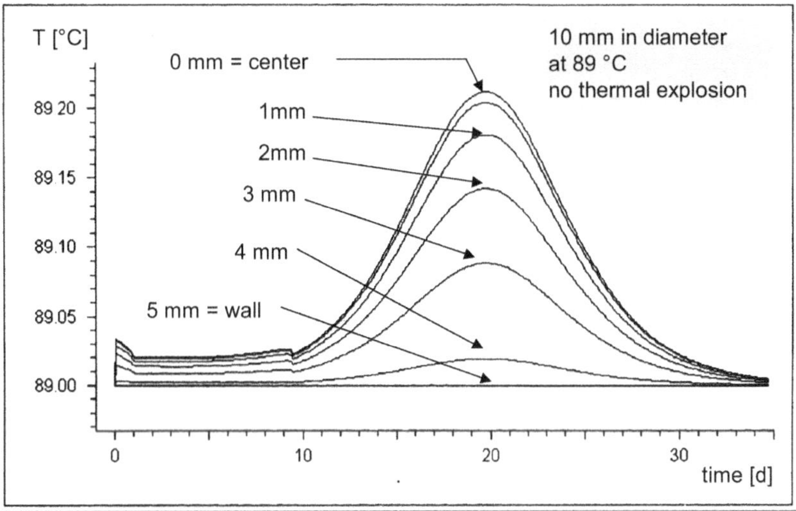

International toxicity classification

type of reception	classification		
	very toxic	toxic	less toxic
oral rat-LD_{50} [mg/kg]	< 25	25 - < 200	200 - < 2000
dermal rat / rabbit-LD_{50} [mg/kg]	< 50	50 - < 400	400 - < 2000
inhalative rat-LC_{50} [mg/(liter air, 4 h)]	< 0.5	0.5 - < 2	2 - < 20

LD_{50} lethal dose, 50% of test animals die

LC_{50} lethal concentration, 50% of test animals die

Oxidation of Unsymmetrical Dimethyl Hydrazine over Oxide and Noble Metal Catalysts. Solution of Environmental Problems of Production, Storage and Disposal of Highly Toxic Rocket Fuels

Z.R.Ismagilov[*a], M.A.Kerzentsev[a], I.Z.Ismagilov[a], V.A.Sazonov[a], V.N.Parmon[a], G.L.Elizarova[a], O.P.Pestunova[a], V.A.Shandakov[b], Yu.L.Zuev[c], V.N.Eryomin[c], N.V.Pestereva[c], F.Garin[d], H.Veringa[e]*

[a]Boreskov Institute of Catalysis, 639090, Novosibirsk, Russia,
[b]Federal Research and Production Center "ALTAI", Biysk, Russia
[c]State Rocket Center "Academician V.P.Makeyev Design Bureau", Miass, Russia
[d]Louis Pasteur University, Strasbourg, France
[e]Netherlands Energy Research Foundation ECN, The Netherlands

Introduction

Reduction and conversion of weapons production in Russia have made extremely urgent the problems of development and implementation of environmentally safe and efficient processes of disposal of rocket fuel components (RFC). One of the most pressing problems is the development of a disposal process for extremely toxic 1,1-dymethylhydrazine (unsymmetrical dimethylhydrazine -UDMH). There are no industrial facilities for UDMH treatment in Russia and CIS, as well as no reliable treatment technologies that meet economical and environmental requirements. Therefore, solution of problems of utilization of off-specification UDMH and other RFC, industrial wastes containing RFC, spills of RFC on soil, etc., is a very urgent task

Conventional thermal methods: incineration in flame and plasma method consume much energy and lead to the formation of large amounts of secondary pollutants. The need of additional purification of exhaust gas from NO_x and other pollutants complicates these methods for practical application.

Methods of processing of UDMH and other toxic fuels to useful products are not yet promoted to the state of industrial application and cannot solve the problem of UDMH utilization due to the lack of interest of enterprises to deal with such hazardous, toxic and flammable chemicals.

The problem of UDMH utilization can be solved by creation of a treatment plant based on UDMH oxidation in a fluidized catalyst bed [1, 2]. The unique feature of this method is the possibility of flameless total oxidation of organic compounds in near stoichiometric ratio with oxygen at relatively low temperatures (500-750°C), which suppresses formation of "thermal" and "fuel" nitrogen oxides. This is achieved by the use of highly active catalysts in a fluidized bed that is characterized by rather high heat transfer coefficients and thus allows the efficient removal of reaction heat. Pilot and industrial plant tests in combustion and processing of different types of organic wastes have shown high efficiency and environmental safety of this technology/1-6/.

An important advantage of the technology of catalytic fluidized bed waste destruction is small size of apparatus. A compact installation designed as a mobile unit will allow UDMH treatment at distant locations, navy bases, and thus will

P. Castelo Branco et al. (eds.),
Defense Industries: Science and Technology Related to Security, 293–313.

exclude hazardous (and sometimes impossible) transportation of rockets or rocket fuels to centralized treatment facilities.

Catalytic oxidation of UDMH.

A laboratory setup for study of UDMH catalytic destruction has been fabricated.
The setup is intended for investigation of reaction of catalytic oxidation of UDMH, and it
allows to study composition of reaction mixtures and dependencies of the reaction rate on
temperature, space velocity and UDMH concentration.

Fig.1. Principal scheme of the laboratory setup

The scheme of the setup is given in Fig. 1 and comprises:

1. System of gas supply, containing cylinders with nitrogen and air (1); devices for gas flow control and measurement – pressure reducers (2), valves (3), gas flow controllers (4) and rotameters (5).
2. System of UDMH supply. The dosage of small amounts of fuel is provided by the saturation of nitrogen flow with UDMH vapor in a saturator (6) kept at 0°C in a Dewar flask (7) with ice and water. This system ensures necessary gas tightness and safety, providing the supply of UDMH in the range of 0.1-5 g/h.
3. Flow mixer for mixing gas-vapor flow with air flow (9).
4. Catalytic reactor (11) with catalyst. We used a gradientless quartz reactor with a vibro-fluidized catalyst bed. Vibro-fluidization of the catalyst was accomplished by connecting reactor to an electromagnetic vibrator, operated with 50 hz frequency. Reactor was placed into the furnace with a fluidized bed of quartz sand (13), providing effective heat removal from the reactor and constant preset catalyst temperature.
5. Electric power supply and temperature measurement devices (14)
6. System of gas analysis (18-20) based on Gas Chromatograph Kristall-2000M. Separation and analysis of UDMH and organic reaction products was done in a Teflon column 1,5 m long, 2mm in diameter, with polymer sorbent HayeSep C modified by 10% KOH, using FID and He as carrier gas at a temperature of 155°C. Air and CO_2 were separated in a stainless steel carbon SKT column (1.5 m long, 0.3 mm diameter, carrier gas He, 30 ml/min, temperature 155°C) and detected with TCD.

Kinetic experiments were conducted at following parameters:

Catalyst temperature	20-500°C
Concentration of UDMH in the initial reaction mixture	0.1-2 mmol/l (0.2-4 vol/%)
Space velocity	5000-40000 h^{-1}
Reaction mixture flow rate	10-100 l/h
Catalyst loading	0,1-2 cm^3

The first series of studies were conducted using the $Cu_xMg_{1-x}Cr_2O_4/\gamma\text{-}Al_2O_3$ catalyst (fraction of 0,25-0,5 mm), developed for catalytic combustion in the fluidized bed. It was found that a noticeable increase of conversion to CO_2 begins at a temperature of 200°C (Fig.2), and at temperatures above 300°C practically complete UDMH oxidation into the deep oxidation products CO_2, H_2O and N_2 is observed.

Boreskov Institute of Catalysis

Solution of Environmental Problems of Production, Storage and Disposal of Rocket Fuels

1. Unsymmetrical Dimethylhydrazine
2. Large Scale Solid Propellant Charges

Z.R.Ismagilov [1], M.A.Kerzhentsev[1], V.N.Parmon [1], I.Z.Ismagilov[1], S.R.Khairulin[1], V.N.Eryomin [2], L.N.Rolin [2], G.V. Kutsenko[3], V.A.Shandakov[4], V.F. Komarov[4], G.V.Sakovich [4], L.V.Zabelin [5], F.Garin [6], H.J.Veringa[7], A.C.Heywood[8]

[1]*Boreskov Institute of Catalysis, Novosibirsk, Russia*
[2] *State Rocket Center "Academician V.P.Makeyev Design Bureau, Miass, Russia*
[3]*State Unitary Enterprise "Research and Design Institute of Polymer Materials, Perm, Russia*
[4] *Federal Research and Production Center "Altai", Biysk, Russia*
[5] *RAC, Moscow, Russia*
[6]*Louis Pasteur University, Strasbourg, France*
[7]*Netherlands Energy Research Foundation, Petten, The Netherlands*
[8] *Lawrence Livermore National Laboratory, Livermore, USA*

Boreskov Institute of Catalysis

Boreskov Institute of Catalysis

Boreskov Institute of Catalysis

Boreskov Institute of Catalysis

Types of liquid propellants used in engines of Russian missiles and carrier rockets

No	Propellant		Rocket name
	Fuel	Oxidizer	
1	Ethanol	Liquid oxygen	R-1, R-2, R-5M, R-7
2	Kerosene	Liquid oxygen	R-9, R-7, R-11, 11K25, «Sputnik», «Luna», «Vostok», «Voskhod», «Molniya», «Soyuz», «Soyuz-2», «Zenit», (1st stage) «Energia-Buran», «Energia-M», «Angara» (prospective)
3	TG-02	AK-27 I AK-27 K	R-11 «Luna-4», «Luna-14»,P-15 3 UR: 5V-11, V-1100, V-750V, 217M, 218, 5Ya-24, ICBM «Burya»
4	TM-185	AK-27 I	R-17
5	Liquid hydrogen	Liquid oxygen	«Proton M» (booster), «Angara» (in prospect), «Energia-Buran»(3rd stage)
6	Hydrazine	---	Low thrust engines: DOT-5, DOT-25, DOT-10, DOT-50
7	Synthine	Liquid oxygen	«Buran», carrier rocket «Soyuz-2»
8	Ammonia	Fluorine	«Proton» (booster)
9	UDMH	Liquid oxygen	«Kosmos» (2nd stage)
10	UDMH	AK-27 I	R-14, R-16, «Kosmos» (2nd stage.), KTDU-35
11	UDMH	Dinitrogen tetroxide N_2O_4	R-14, R-16, R-36, R-36M, UR-100, UR-100UTTKh, UR-100K, UR-100N, MR-UR100, UR-200, RS-10 (SS-11), RS-18 (SS-19), RS-20 (SS-18), R-27 (SS-N-6), R-27K (SS-NX-13), R-29 (SS-N-8), R-29R (SS-N-18), R-29RM (SS-N-23), «Luna-15», «Luna-16», «Luna-24», «Mars-2», «Mars-7», «Venera-16», «Vega-1», «Vega-2», «Proton-K», «Angara» (2nd stage) – version «Rokot», «Cyclone»

Boreskov Institute of Catalysis

Main Types of Liquid Rocket Propellants

Fuels		Oxidizers
Liquid hydrogen	H_2	Liquid O_2
Ethanol	C_2H_5OH	Liquid O_2
Kerosene	C_xH_y	Liquid O_2
Synthine	$C_{10}H_{16}$	Liquid O_2
TM-185	C_xH_y	HNO_3
Ammonia	NH_3	Liquid F_2
TG-02	$C_6H_5N(CH_3)_2$	HNO_3
UDMH	$(CH_3)_2N\text{-}NH_2$	N_2O_4, HNO_3, O_2

Boreskov Institute of Catalysis

Objective

Environmentally safe and efficient utilization of an extremely toxic and explosive rocket fuel 1,1-dimethylhydrazine:

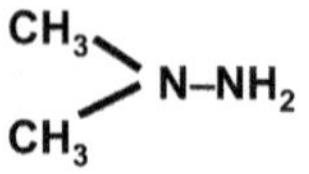

(unsymmetrical dimethylhydrazine - UDMH,technical name - "heptyl")

Method

Total catalytic oxidation of UDMH by air in a fluidized bed reactor

Boreskov Institute of Catalysis

Traditional Combustion (Chain Mechanism)

$$RH + O_2 \xrightarrow[t > 1200\ ^oC]{flame} CO_2 + H_2O + Q_1$$

$$\longrightarrow CO + Benz[\alpha]pyren$$

$$N_2 + O_2 \rightleftharpoons NO, NO_2 \quad \longleftarrow \text{thermal } NO_x$$

$$RNH + O_2 \longrightarrow NO_x + ... \quad \longleftarrow \text{fuel } NO_x$$

Heterogeneous Catalytic Oxidation (Low Temperature Flameless Combustion)

$$RH + O_2 \xrightarrow[t=400 \div 750\ ^oC]{catalyst} CO_2 + H_2O + Q_2$$

$$N_2 + O_2 \not\longrightarrow NO_x$$

$$RNH + O_2 \xrightarrow{k_1} N_2 + CO_2 + H_2O$$

$$RNH + O_2 \xrightarrow{k_2} NO_x + CO_2 + H_2O$$

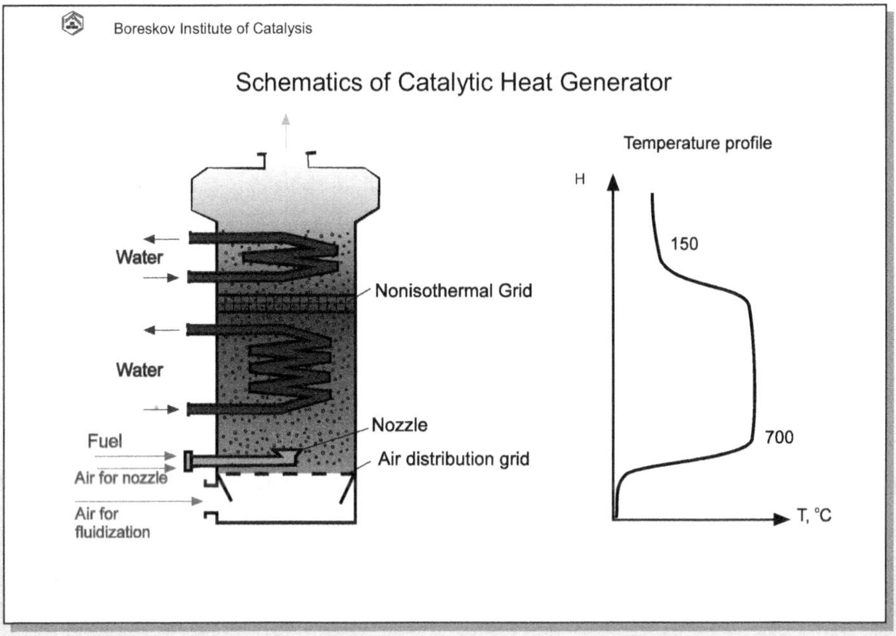

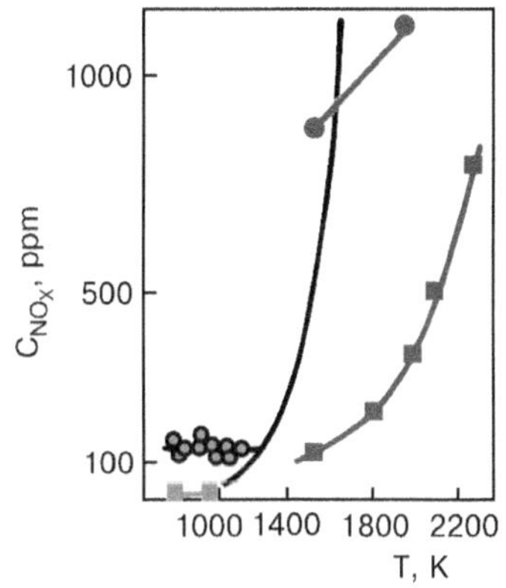

Dependance of NO_x concentration on temperature during:

- flame combustion:

(● - fuel NO_x, ■ - thermal NO_x)

- catalytic combustion in CHG

(● - fuel NO_x, ■ - thermal NO_x)

N_f= 1 wt.%. The bold line corresponds to equilibrium values of C_{NOx}

Boreskov Institute of Catalysis

List of catalysts studied for the UDMH oxidation by air

Oxide catalysts

- IC-12-73 ($Cu_xMg_{1-x}Cr_2O_4$/ γ-Al_2O_3)
- IC-12-74 (Fe_2O_3/ γ-Al_2O_3)
- IC-44 (Fe_2O_3 + Cr_2O_3 + ZnO)
- perovskites $Ca_{0,7}Sr_{0,3}FeO_{2,5(3,0)}$
- perovskites $La_{0,7}Sr_{0,3}CoO_{2,5(3,0)}$

Noble metal based catalysts

- 0,64% Pt/ γ-Al_2O_3
- 0,5% Pd/ γ-Al_2O_3
- 32,9% Ir/ γ-Al_2O_3

Catalysts of other types

- modified zeolite $Cu/(ZSM5+TiO_2+Al_2O_3)/Al_2O_3$-$SiO_2$ (further – "Cu/zeolite")
- Si_3N_4

Boreskov Institute of Catalysis

Chromatogram of the UDMH oxidation reaction products on the catalyst $Cu_xMg_{1-x}Cr_2O_4$/ γ-Al_2O_3 at 240°C
Chromatograph Kristall 2000m, detector FID, column HayeSep C

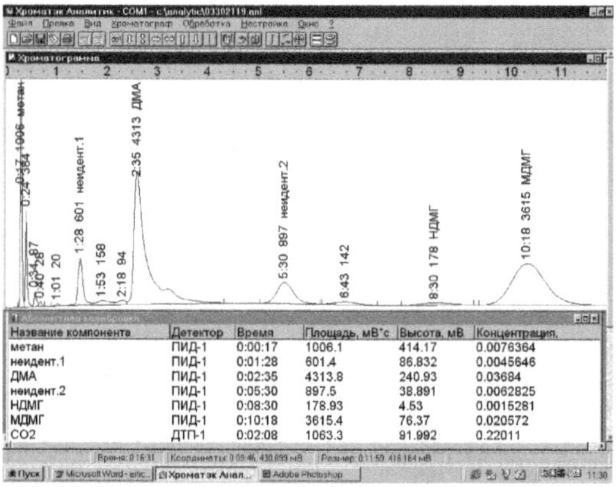

Название компонента	Детектор	Время	Площадь, мВ*с	Высота, мВ	Концентрация,
метан	ПИД-1	0:00:17	1006.1	414.17	0.0076364
неидент.1	ПИД-1	0:01:28	601.4	86.832	0.0045646
ДМА	ПИД-1	0:02:35	4313.8	240.93	0.03684
неидент.2	ПИД-1	0:05:30	897.5	38.891	0.0062825
НДМГ	ПИД-1	0:08:30	178.93	4.53	0.0015281
МДМГ	ПИД-1	0:10:18	3615.4	76.37	0.020572
CO2	ДТП-1	0:02:08	1063.3	91.992	0.22011

Main products: CO_2, H_2O, N_2, methane, dimethylamine (DMA), methylenedimethylhydrazine (MDMH) $(CH_3)_2N$–N=CH_2, N_2O

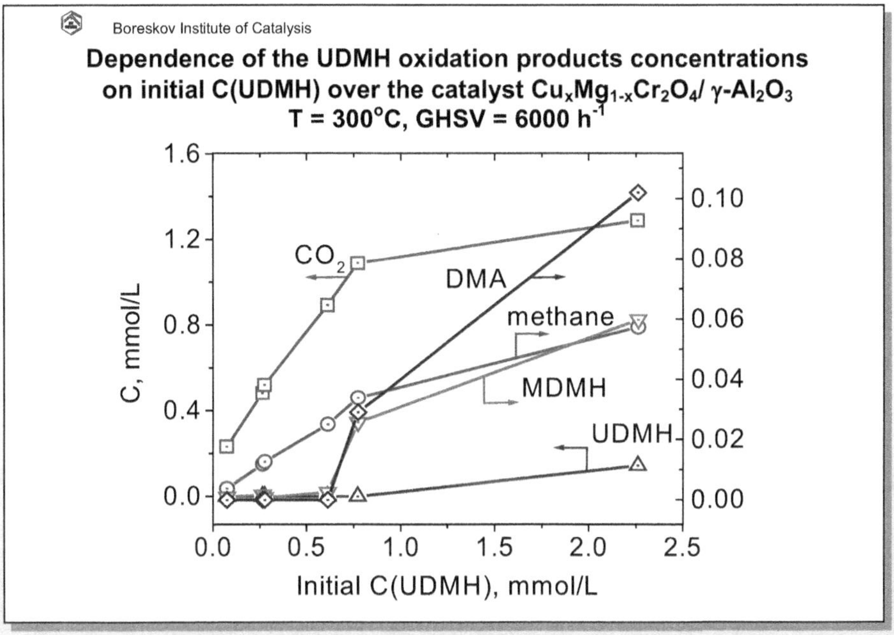
Boreskov Institute of Catalysis
Dependence of the UDMH oxidation products concentrations on initial C(UDMH) over the catalyst $Cu_xMg_{1-x}Cr_2O_4/\gamma\text{-}Al_2O_3$
T = 300°C, GHSV = 6000 h^{-1}
CO_2
DMA
methane
MDMH
UDMH
C, mmol/L
Initial C(UDMH), mmol/L
1.6
1.2
0.8
0.4
0.0
0.10
0.08
0.06
0.04
0.02
0.00
0.0
0.5
1.0
1.5
2.0
2.5

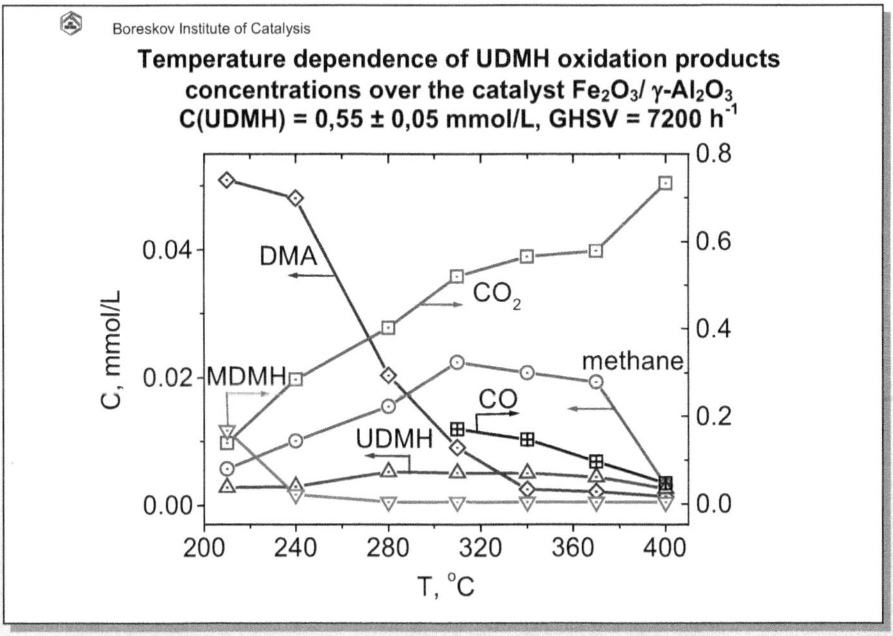
Boreskov Institute of Catalysis
Temperature dependence of UDMH oxidation products concentrations over the catalyst $Fe_2O_3/\gamma\text{-}Al_2O_3$
C(UDMH) = 0,55 ± 0,05 mmol/L, GHSV = 7200 h^{-1}
DMA
CO_2
methane
MDMH
CO
UDMH
C, mmol/L
T, °C
0.04
0.02
0.00
0.8
0.6
0.4
0.2
0.0
200
240
280
320
360
400

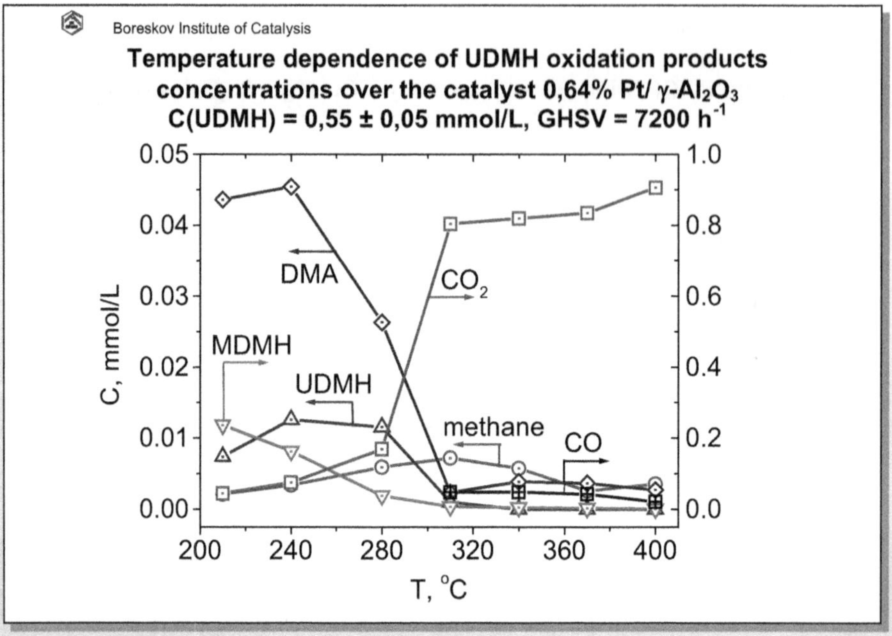
Boreskov Institute of Catalysis
Temperature dependence of UDMH oxidation products concentrations over the catalyst 0,64% Pt/ γ-Al2O3
C(UDMH) = 0,55 ± 0,05 mmol/L, GHSV = 7200 h-1
C, mmol/L
T, °C
DMA
CO2
MDMH
UDMH
methane
CO
0.05
0.04
0.03
0.02
0.01
0.00
1.0
0.8
0.6
0.4
0.2
0.0
200
240
280
320
360
400

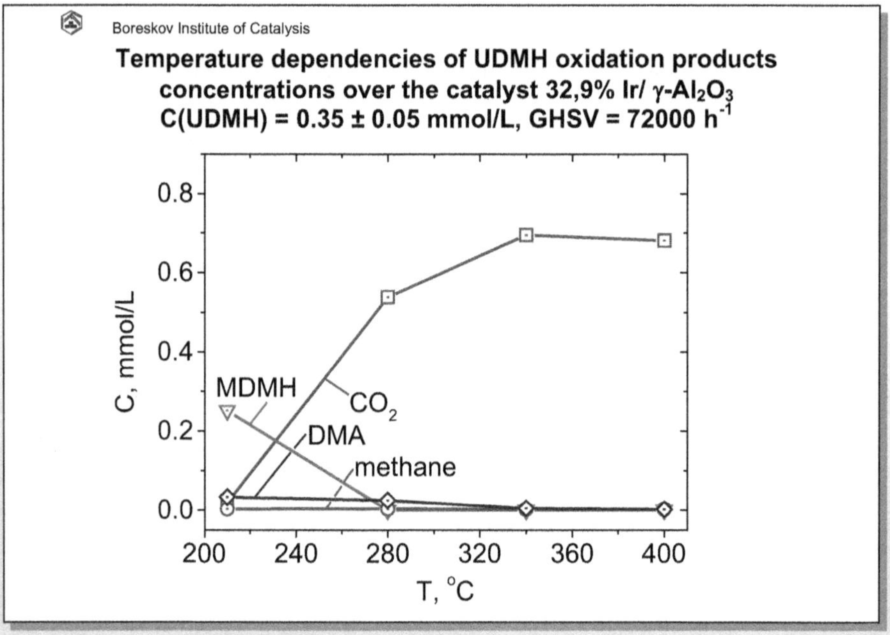
Boreskov Institute of Catalysis
Temperature dependencies of UDMH oxidation products concentrations over the catalyst 32,9% Ir/ γ-Al2O3
C(UDMH) = 0.35 ± 0.05 mmol/L, GHSV = 72000 h-1
C, mmol/L
T, °C
MDMH
CO2
DMA
methane
0.8
0.6
0.4
0.2
0.0
200
240
280
320
360
400

Boreskov Institute of Catalysis

Comparison of the temperature dependencies of CO_2 concentration upon UDMH oxidation over different catalysts
C(UDMH) = 0,55 ± 0,05 mmol/L, GHSV = 7200 h^{-1}

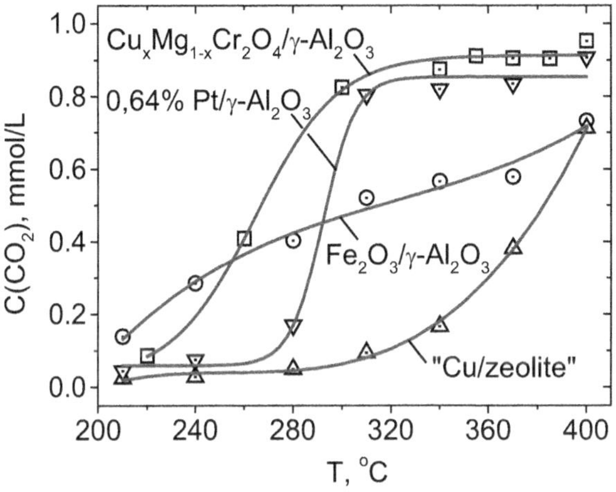

Boreskov Institute of Catalysis

Comparison of the temperature dependencies of NO concentration upon UDMH oxidation over different catalysts
C(UDMH) = 0,55 ± 0,05 mmol/L, GHSV = 7200 h^{-1}

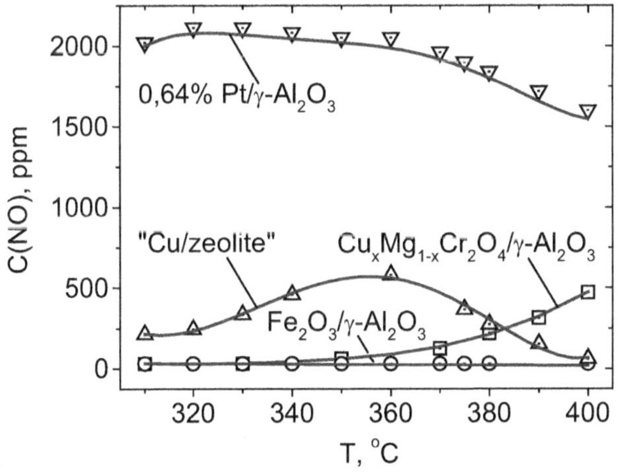

Boreskov Institute of Catalysis

Mechanism of the UDMH heterogeneous catalytic oxidation

$((CH_3)_2N\text{–}NH_2 + [\ \ ] \rightarrow [(CH_3)_2N\text{–}NH_2]_{ads}$ (1)

$[(CH_3)_2N\text{–}NH_2]_{ads} + [O]_{ads} \rightarrow [(CH_3)_2N{=}N]_{ads} + [H_2O]_{ads}$ (2)

$[(CH_3)_2N{=}N]_{ads} + [(CH_3)_2N{=}N]_{ads} \rightarrow [(CH_3)_2N\text{–}N{=}CH_2]_{ads} + [CH_3N{=}NH]_{ads}$ (3)

$[(CH_3)_2N\text{–}N{=}CH_2]_{ads} \rightarrow (CH_3)_2N\text{–}N{=}CH_2 + [\ \ ]$ (4)

$[(CH_3)_2N\text{–}N{=}CH_2]_{ads} + [O]_{ads} \rightarrow (CH_3)_2NH + CO_2 + H_2O + N_2 + NO_x$ (5)

$[CH_3N{=}NH]_{ads} \rightarrow CH_4 + N_2$ (6)

$[CH_3N{=}NH]_{ads} + [O]_{ads} \rightarrow CO_2 + H_2O + N_2 + NO_x$ (7)

Boreskov Institute of Catalysis

UDMH catalytic oxidation reactions stoichiometry

Main reaction occuring in the reactor:

$(CH_3)_2N\text{-}NH_2 + 4O_2 = N_2 + 2CO_2 + 4H_2O$ (g) $\Delta H = -1799{,}15$ kJ/mole (1)

Besides that, in the excess of air reactions of nitrogen oxide formation can occur:

$(CH_3)_2N\text{-}NH_2 + 4{,}5O_2 = N_2O + 2CO_2 + 4H_2O$ (g) $\Delta H = -1717{,}60$ kJ/mole (2)
$(CH_3)_2N\text{-}NH_2 + 5O_2 = 2NO + 2CO_2 + 4H_2O$ (g) $\Delta H = -1618{,}41$ kJ/mole (3)
$(CH_3)_2N\text{-}NH_2 + 6O_2 = 2NO_2 + 2CO_2 + 4H_2O$ (g) $\Delta H = -1731{,}37$ kJ/mole (4)

Experimental results have shown, that contribution of reaction (1) into the overall process constitutes more than 80-90 %, therefore reaction (1) was used for the material and heat balance calculations.

Boreskov Institute of Catalysis

Schematics of the pilot plant for catalytic utilization of UDMH

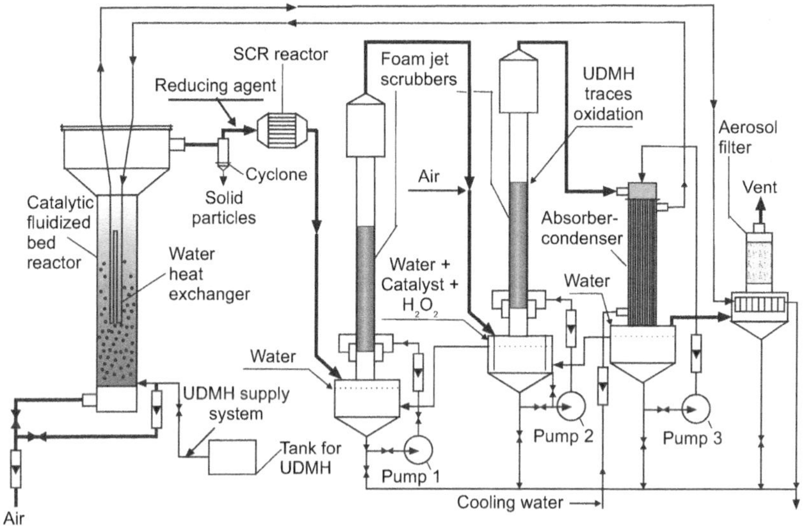

Boreskov Institute of Catalysis

Reactor with fluidized catalyst bed

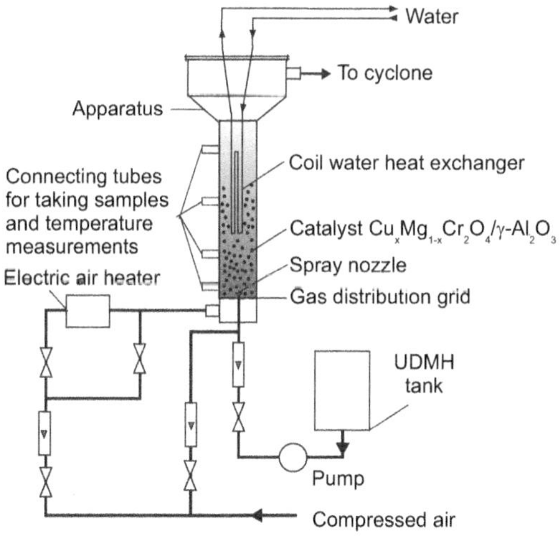

Characteristics

UDMH feed rate, kg/h	1-2
Air flow rate, Nm³/h	15-20
Gas velocity, m/s	1.2-1.9
Pressure drop, atm	0.1
Catalyst loading, kg	5-7
Bed temperature, °C	550-780
Inner diameter, m	0.12
Height, m	1.5

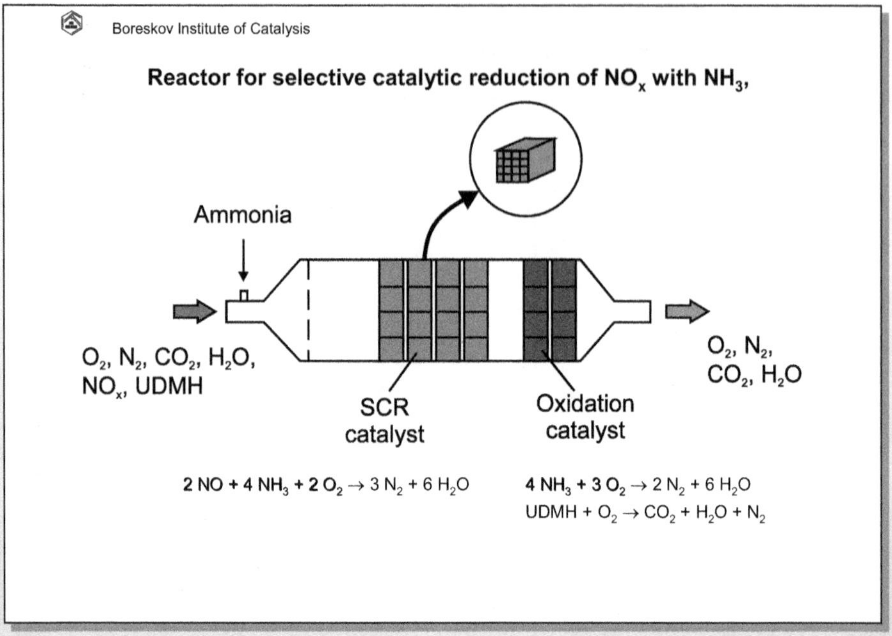
Boreskov Institute of Catalysis
Reactor for selective catalytic reduction of NO_x with NH_3,
Ammonia
O_2, N_2, CO_2, H_2O, NO_x, UDMH
SCR catalyst
Oxidation catalyst
O_2, N_2, CO_2, H_2O
$2\,NO + 4\,NH_3 + 2\,O_2 \rightarrow 3\,N_2 + 6\,H_2O$
$4\,NH_3 + 3\,O_2 \rightarrow 2\,N_2 + 6\,H_2O$
$UDMH + O_2 \rightarrow CO_2 + H_2O + N_2$

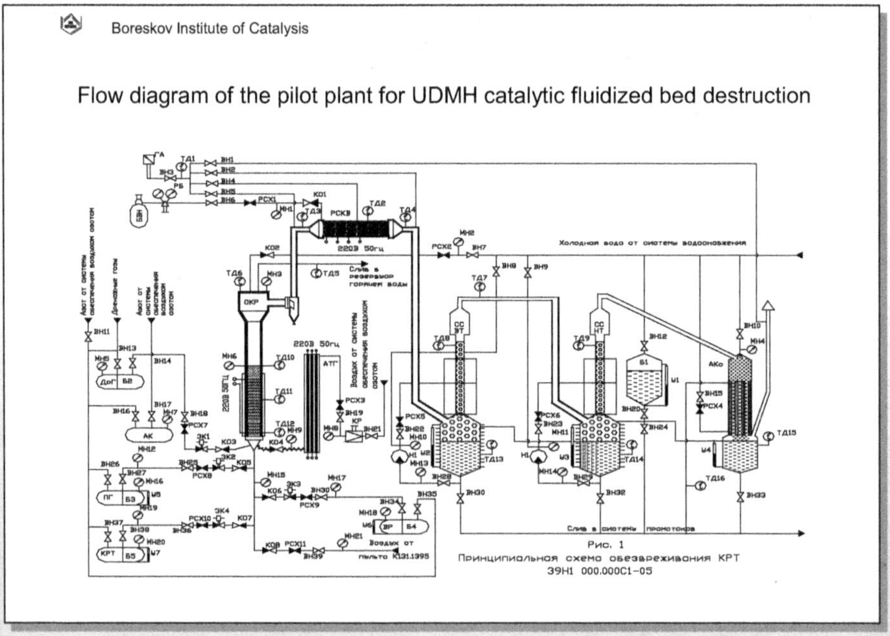
Boreskov Institute of Catalysis
Flow diagram of the pilot plant for UDMH catalytic fluidized bed destruction
Рис. 1
Принципиальная схема обезвреживания КРТ
39Н1 000.000С1-05

Boreskov Institute of Catalysis

Lay-out of the pilot plant for UDMH catalytic fluidized bed destruction

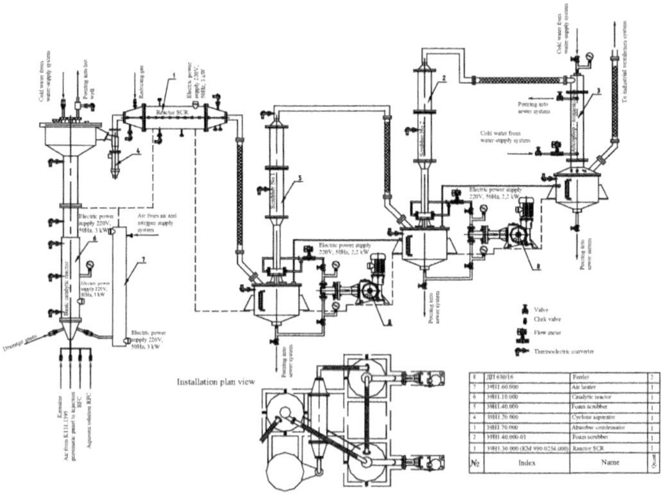

8	ДП 630/16	Feeder	2
7	39Н1.60.000	Air heater	1
6	39Н1.10.000	Catalytic reactor	1
5	39Н1.40.000	Foam scrubber	1
4	39Н1.20.000	Cyclone separator	1
3	39Н1.70.000	Absorber condensator	1
2	39Н1.40.000-01	Foam scrubber	1
1	39Н1.30.000 (КМ 990.0254.000)	Reactor SCR	1
№	Index	Name	Quant

Boreskov Institute of Catalysis

Photograph of the pilot plant for UDMH catalytic fluidized bed destruction

Boreskov Institute of Catalysis

Results of testing the main reactor with fluidized catalyst bed

No	Main air flow, m^3/h	Air flow through injectors, m^3/h	UDMH flow rate, L/h	Oxidizer excess coeff. α	T in the reactor, °C	Gas composition at the main reactor outlet		
						NO_x mg/m^3	CO mg/m^3	UDMH mg/m^3
1	16.2	1.6	1.08	2.7	630	6000	< 5	< 0.05
2	16.2	1.6	2.24	1.31	675	6000	< 5	< 0.05
3	16.2	1.6	2.56	1.14	735	6000	< 5	0.075

Results of the SCR reactor testing

No	Conc. of UDMH (init.), mg/m^3	Conc. of NO_x (init.), mg/m^3	Conc. of NH_3 (init.), g/m^3	NH_3 flow rate, m^3/h	T in the SCR reactor, °C	Gas composition at the SCR reactor outlet			
						UDMH, mg/m^3	NO_x, mg/m^3	CO, mg/m^3	NH_3, mg/m^3
6	2.14	2000	7.86	0.20	420	0.095	80	< 5	< 5

Boreskov Institute of Catalysis

Results of final complex tests of the pilot plant for catalytic UDMH destruction, February 6-8, 2001

date	time	No	V of UDMH, L/h	α	Main reactor T, °C	SCR reactor T, °C	NO_2 initial, mg/m^3	NO_2 final, mg/m^3	S (NO_x), %	UDMH before SCR, mg/m^3	UDMH after SCR, mg/m^3	UDMH in scrubber 1, mg/L	UDMH in scrubber 2, mg/L	UDMH after scrubber 2, mg/m^3	Notes
06.02	10^{23}	1	2,4	1,3	715	450	10753	269	97	1,12	1,3	0,3	0,6	0,16	
	11^{50}	2	2,0	1,6	710	450	10753	1075	90	1,7	0,2	0,34	0,3	0,64	
	14^{00}	3	2,0	1,6	710								**0,87**		14^{05}: 1 kg of catalyst added to scrubber 2
	14^{40}	4	1,5	2,2	590					-	2,7		**0,24**	0,43	
07.02	10^{00}	5										0,3	**0,16**		
	10^{40}	6	1,75	1,8	595					14,0	7,9	0,3	**0**	0,33	
	12^{00}	7	1,75	1,8	600	440	4076	1086	73	2,9	1,7	0,5	0,18	0,82	
	12^{45}	8	1,75	1,8	650	455	8152	2717	67	2,28	0,39	0,72	0,24	0,13	
	13^{10}	9	1,75	1,8	650							0,6	0,32		
	13^{30}	10	1,75	1,8	650							0,62	0,5		
	14^{00}	11	1,75	1,8	650							0,56	0,2		
08.02	10^{00}	12			700					0,87	0,98	0,3	**0,84**		Kerosene combustion 1,83 L/h, added 6 mL 30% H_2O_2
	11^{00}	13	2,1	1,5	725					0,28	0,39	1,5	**0,1**	0,15	
	11^{25}	14	2,1	1,5	725							0,26	**0,18**		
	11^{30}	15	2,0	1,6	720	465	4347	271	94	0,58	0,12	0,26	**0,26**	0,11	11^{32}: added 6 mL 30% H_2O_2
	12^{30}	16										0,5	**0,05**	< 0,05	
	13^{00}	17	2,2	1,5	725	445	9152	1086	88	0,87	0,41	0,5	0,3	0.19	
	13^{25}	18	2,2	1,5	725							0,5	0,3	< 0,05	
	14^{30}	19			700					0,13	0,13	0,2	0	< 0,05	13^{30}: switched to kerosene (1,8 L/h)

Total gas flow on the reactor with fluidized catalyst bed constituted 18 m^3/h; water temperature in scrubber 1 - 55-60°C, in scrubber 2 - 45-50°C

Boreskov Institute of Catalysis

Problem of Utilization of Solid Rocket Propellants

START-2 Treaty

410 missiles should be eliminated

Land based: RS-18 “Stiletto”, RS-20 “Satan”,

RS -22 “Scalpel” (SS -24), RS-12M “Topol”, (SS -25)

Submarine launched: RSM-52

1230 solid propellant charges (3.5-49 t)

TOTAL – 22850 t

In 1996-1997 40 missiles RSM-52 were eliminated by launching and firing, causing environment pollution: NO_x, NH_3, *HCN,* Cl_2, *HCl, CO, HC, etc*

Boreskov Institute of Catalysis

Typical composite solid propellant

Reductant: RDX (HMX, TABT) + aluminum powder

Oxidizer: NH_4ClO_4

Polymer binder (e.g., cross-linked polybutadiene)

Propellant fuels

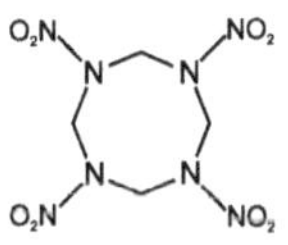

Octohydro-1,3,5,7-tetranitro-1,3,5,7-tetrazocine
OCTOGEN - (RDX)

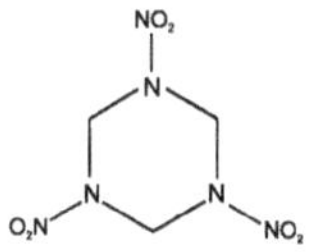

Hexahydro-1,3,5-trinitro-s-triazine
HEXOGEN - (HMX)

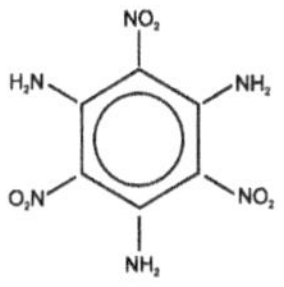

Triaminotrinitrobenzene (TATB)

Boreskov Institute of Catalysis

There are three approaches in the solution of this problem:

- burning the propellant *inside the rocket motor*,
- mechanical destruction and reprocessing of solid propellants to commercial explosives;
- chemical disruption of solid propellants based on a safe and environmentally acceptable process for chemically disrupting the propellant's polymer binder followed by extraction of the rest of the propellant's components and their recycling into commercial products.

Boreskov Institute of Catalysis

Burning the propellant directly inside the solid propellant rocket motor (SPRM)

Drawbacks:

- irregular flow of the propellant combustion products into the neutralizing chamber;
- high consumption of chemical reagents supplied for the neutralization;
- large amounts of slag residue (alumina, coke) that require further treatment;
- partial coking of the SPRM casing;
- large capital costs associated with neutralizing chamber and the multi-stage off-gas treatment system manufacturing.

Boreskov Institute of Catalysis

Formation of hazardous wastes upon combustion of solid rocket propellants
Combusted propellants amount – 17 500 tons

WASTES TYPE	WASTES AMOUNT (TONS)
Solid	
Al_2O_3	5792
NaCl	3658
Gaseous into atmosphere	
HCl	44
CO	4200
CO_2	264
H_2, H_2O, NOx	5642

L.V. Zabelin, R.V. Gafiyatullin, L.R. Guseva, "Environmental aspects of the solid propellant rocket motors (SPRM) charges utilization", Khimiya v Rossii (Chemistry in Russia), *1999, p. 4-7*, in Russian

Boreskov Institute of Catalysis

Cutting the charge by a high pressure, water jet (Thiokol, USA)

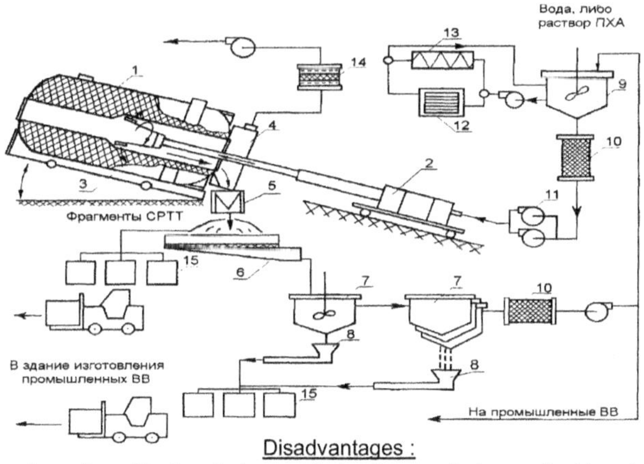

Disadvantages :

- formation of toxic substances during fragmented propellant burning;
- the environmental impact from utilizing propellant components, that produce chlorine-containing compounds, in CE.

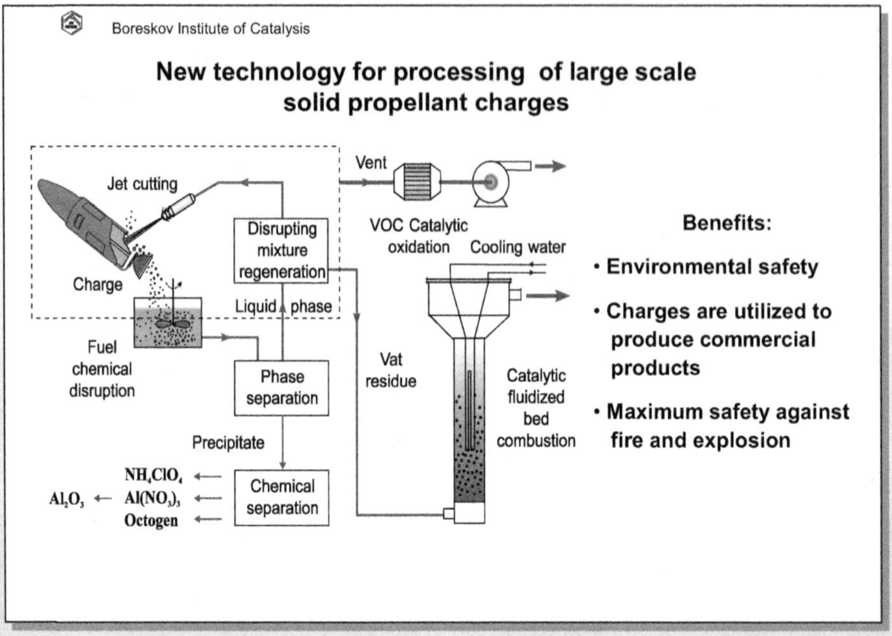
Boreskov Institute of Catalysis
New technology for processing of large scale solid propellant charges
Jet cutting
Charge
Disrupting mixture regeneration
Liquid phase
Fuel chemical disruption
Phase separation
Precipitate
NH4ClO4
Al2O3
Al(NO3)3
Octogen
Chemical separation
Vent
VOC Catalytic oxidation
Cooling water
Vat residue
Catalytic fluidized bed combustion
Benefits:
• Environmental safety
• Charges are utilized to produce commercial products
• Maximum safety against fire and explosion

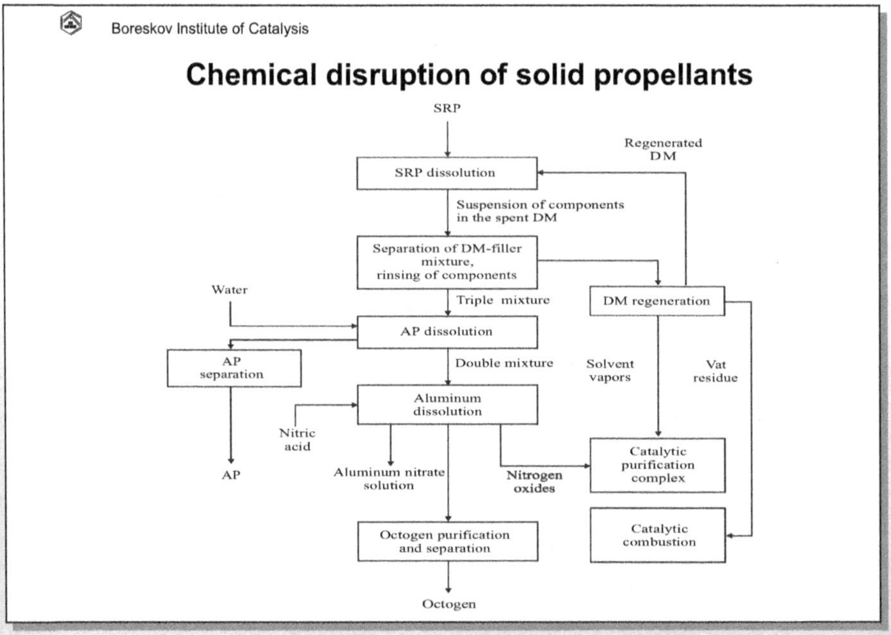
Boreskov Institute of Catalysis
Chemical disruption of solid propellants
SRP
SRP dissolution
Regenerated DM
Suspension of components in the spent DM
Separation of DM-filler mixture, rinsing of components
Triple mixture
DM regeneration
Water
AP dissolution
AP separation
Double mixture
Solvent vapors
Vat residue
Aluminum dissolution
Nitric acid
AP
Aluminum nitrate solution
Nitrogen oxides
Catalytic purification complex
Catalytic combustion
Octogen purification and separation
Octogen

Boreskov Institute of Catalysis

Scope of activities

In the course of project accomplishment, the following five main tasks will be solved:

- Fundamental studies of physico-chemical principles of SRP chemical disruption.
- Development and testing of the SRP chemical disruption process in the experimental bench-scale installation to be created in the frame of the project at FRPC "Altai".
- Analysis and studies of modern techniques for large-scale SRP charge fragmentation and formulation of recommendations for the selection of SRP samples fragmentation method and conditions for its realization.
- Preparation of an input data set for the design of experimental-industrial plant for processing charges of up to 3.5 tons.
- Development of a block diagram for scaling the utilization process to charges with masses of up to 50 tons.

Boreskov Institute of Catalysis

The authors gratefully acknowledge

the financial support of the projects # 959 and #1632

by the International Science and Technology Center

Demilitarization and Recycling

INVESTIGATIONS OF RISKS CONNECTED TO SEA DUMPED MUNITIONS

Drs N.H.A. van Ham

TNO Prins Maurits Laboratory
P.O.Box 45 2280 AA Rijswijk
Netherlands

This presentation is based on the Netherlands investigations on conventional munitions, dumped in the North Sea and the Oosterschelde. It will give a general outline of the toxic components present in the munitions. The expected scenario's to describe how these toxic components can leak into the environment. A description of the possible reactions of these toxic components will be given, as well as chemical decomposition or biological degradation from the components

P. Castelo Branco et al. (eds.),
Defense Industries: Science and Technology Related to Security, 317–333.

Investigations of Risks connected to Sea Dumped Munitions

- Components in munitions
- Toxicity of munition components
- Reactions of seawater with munitions
- Reaction of munition components with seawater
- Monitoring of dumping sites
- Mitigation of environmental risks from dumping sites
- Conclusions

Investigations of risks connected to sea dumped munitions May 2001

Investigations of risks connected to sea dumped munitions May 2001

Investigations of risks connected to sea dumped munitions May 2001

Munition Components

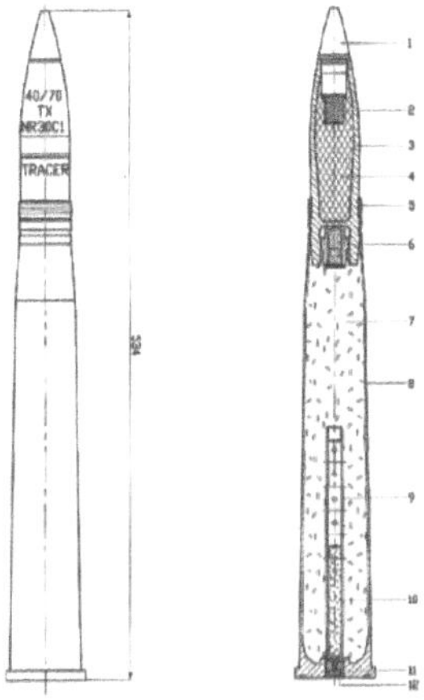

- Metal parts 70 %
- Explosives 11 %
- Propellants 16 %
- Pyrotechnics 2,8 %

Metals used in Munitions

• Iron / Steel	• 76 %
• Lead	• 12 %
• Brass	• 9 %
• Aluminium	• 3 %

Explosives used in Munitions

• Trinitrotoluene (TNT)	• 90 %
• Hexogene (RDX)	• 5 %
• Tetryl (CE)	• 3 %
• Pentaerytryttetranitrate (PETN)	• 2 %

Components in Propellants

• Nitro-cellulose	• 70 %
• Nitroglycerine	• 15 %
• Diphenylamine	• 1 %
• Dinitrotoluene	• 10 %
• Potasiumsulfaat	• 1 %
• Dibuthylphtalate	• 3 %

Pyrotechnic Composition

• Zinc oxide	• 35 %
• Barium nitrate	• 5 %
• Strontium nitrate	• 3 %
• Aluminium	• 11 %
• Lead	• 1 %
• Hexachloro-ethane	• 35 %

Weight of Metals In 1000 kg Munition

Component	Weight kg
Steel	532
Lead	84
Brass	63
Aluminium	25

Weight of Propellant Components in 1000 kg Munition

Nitro-glycerine	25
Nitro-cellulose	112
Diphenylamine	1,6
Dinitrotoluene	**15,7**
Dibutylphtalate	4,7

Weight of High Explosives in 1000 kg Munition

Trinitrotoluene	**99**
RDX	5,5
Tetryl	3,3
PETN	2,2

Weight of Pyrotechnic Components in 1000 kg Munition

Hexachloroethane	9,8
Zinkoxide	9,8
Bariumnitrate	1,4
Strontiumnitrate	0,84
White Phosphorus	1,0

Investigations of risks connected to sea dumped munitions May 2001

Corrosion of Munitions by Sea Water

- **Quality of steel, %carbon**
- **% Oxygen in sea water**
- **Temperature of sea water**
- **Velocity of sea water current**
- **Contact corrosion, steel / copper**

Investigations of risks connected to sea dumped munitions May 2001

Investigation of Dumpsite

- Side Scan Sonar
- Remote Operated Vehicle (ROV)

Investigations of risks connected to sea dumped munitions May 2001

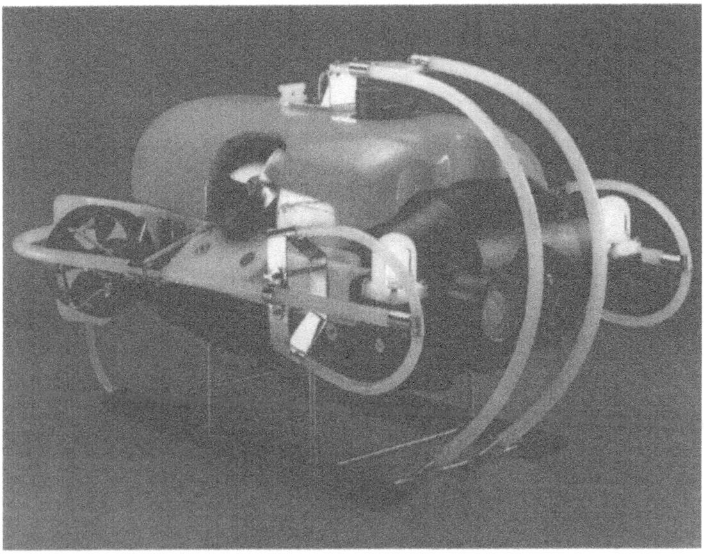

Investigations of risks connected to sea dumped munitions May 2001

29

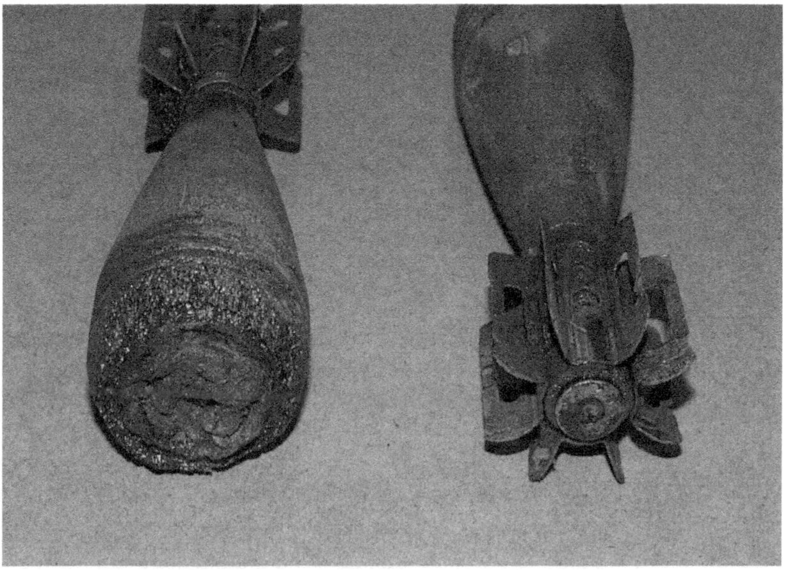

Investigations of risks connected to sea dumped munitions May 2001

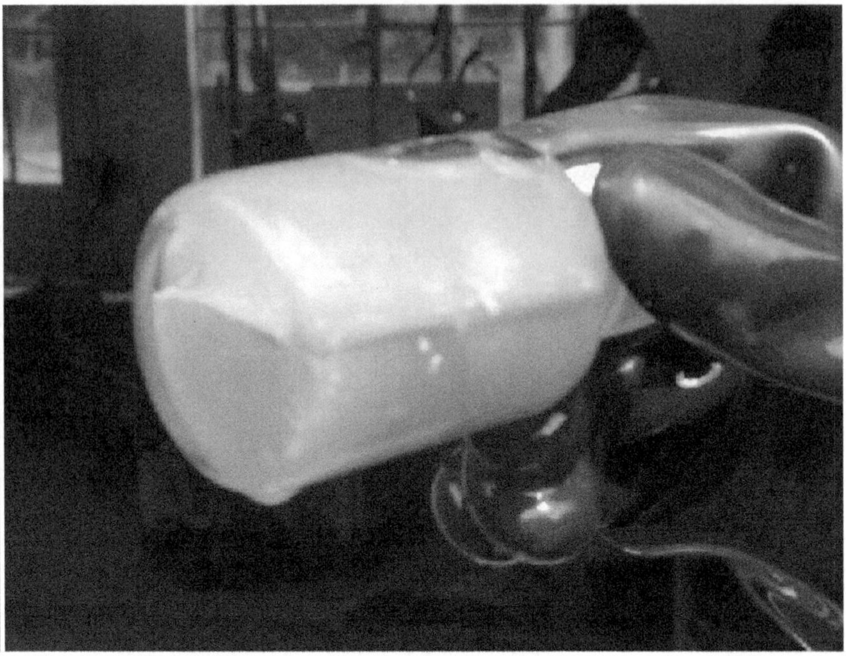

Concentration / Time Profile

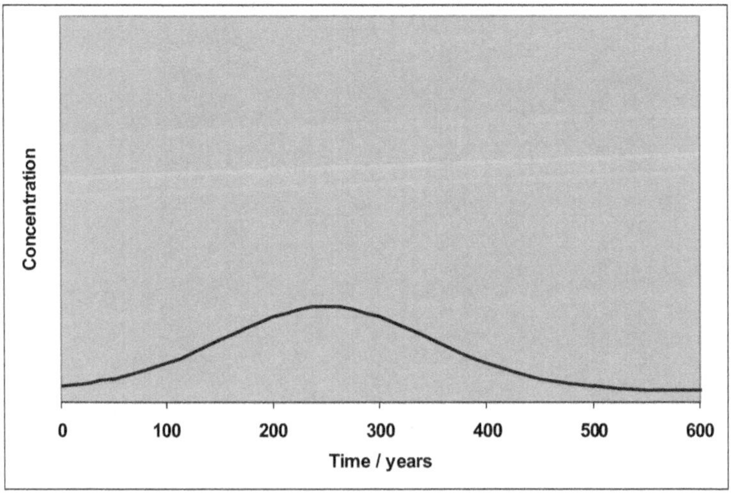

Monitoring of Dumping Sites

- **Water samples**
- **Soil samples**
- **20 Samples per square km**
- **Regular sample taking**

Investigations of risks connected to sea dumped munitions May 2001

Water Sample Apparatus

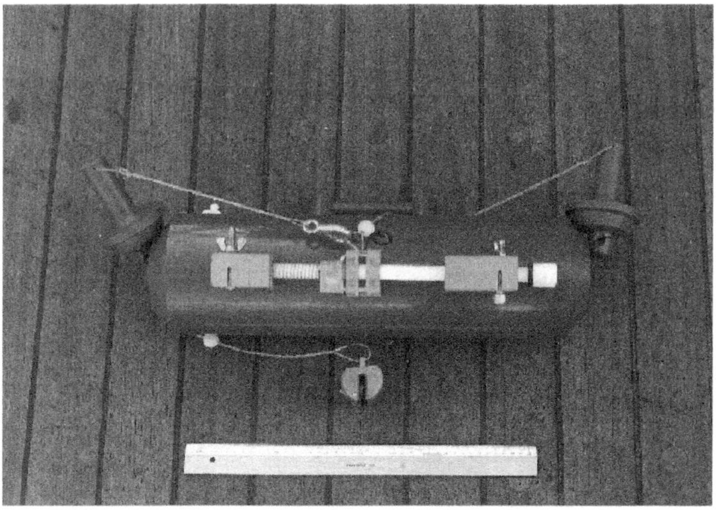

Investigations of risks connected to sea dumped munitions May 2001

Van Veen Happer

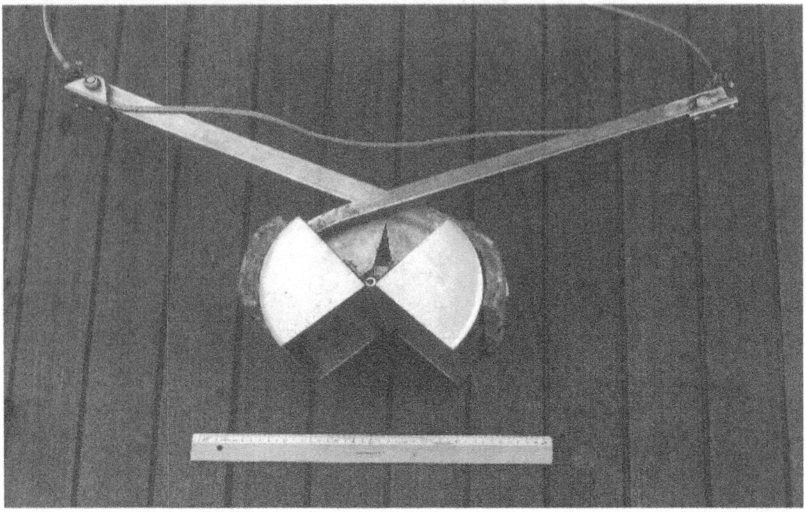

Natural Degradation

- Most munition components have low solubility in water, resulting in low concentrations
- In sea environment, munition components are diluted with very large quantities of water
- Chemical decomposition
- Chemical reactions
- Biological degradation

Conclusions 1

- Sea dumped munitions are subject to degradation
- Due to degradation the hazardous chemicals in the munitions will leak into the environment
- Hazardous chemicals can be explosives, propellants or pyrotechnics

Conclusions 2

- When there is sufficient dilution, natural degradation might be acceptable (no immediate actions needed)
- When the concentrations in and around the dump site might create a human risk, recovery or isolation of the munitions might be considered
- Dump sites will create a long lasting environmental burden. A strict monitor plan has to be followed to protect the population from health hazards

Investigations of risks connected to sea dumped munitions May 2001

Demilitarisation and related environmental research on munitions at FOI

Tomas Carlsson, M.Sc. (Chem. Eng.)

Swedish Defence Research Agency, FOI
Weapons and Protection Division
Grindsjön Research Centre, SE-147 25 Tumba, Sweden

The environmental constraints on all activities in the society are continuously sharpened. Conventional munitions have an impact on the environment during all stages of their life cycle: production, use and demilitarisation. FOI investigates the environmental impact of munitions at all these stages. Research and development are directed towards technologies and products that will reduce emissions. The presentation will cover an overview of some projects at FOI.

Today, Sweden and other countries are demilitarising conventional munitions at a high pace. The stockpile is to be reduced and all AP mines are being phased out. In Sweden more than 90% of the demilitarised explosives are recycled for commercial applications. Nevertheless, recycling processes, e.g. the TNT melt-out process, generate wastewater and hazardous waste. As this waste usually is contaminated with explosives, it has to be disposed of on-site. FOI has developed a process that uses fluidised bed incineration to treat wastewater, waste explosives and explosive-contaminated combustible wastes. In addition to that, FOI has also developed a process to recycle smoke agent ingredients. The next step in the development of demilitarisation is to be able to re-classify energetic materials as "virgin" materials in order to be able to re-use them in new munitions. This would significantly reduce both environmental impact and the life-cycle cost of conventional munitions.

Ammonium perchlorate-based propellants generate HCl during firing. In many propellants, e.g. some double-base propellants and also ammonium perchlorate-based ones, lead compounds are added to the composition for several reasons. FOI has made an inventory of the lead content in propellant compositions in munitions of the Swedish Armed Forces. New propellants should avoid chlorine and heavy metals. ADN, a new propellant, for which an efficient synthesis process was developed at FOI, does not contain any chlorine, and during production no environmentally hazardous solvents have to be used either. Other green energetics from FOI are FOX-7 and FOX-12, which, like ADN, already have been transferred to industrial production.

FOI has also made an inventory on potential environmental hazards from flare munitions. Dioxins have been detected in the smoke from PVC-based flares. Toxicological studies are performed on explosives-contaminated soils and sediments, as well as new energetics. The fate of explosive residues in the environment is studied both from the perspective of detecting buried landmines, and to determine harmful migration of explosives to drinking water and sensitive wildlife.

P. Castelo Branco et al. (eds.),
Defense Industries: Science and Technology Related to Security, 335–347.

Demilitarisation and related environmental research on munitions at FOI

Tomas Carlsson, M.Sc. (Chem. Eng.)

Swedish Defence Research Agency, FOI
Weapons and Protection Division
Grindsjön Research Centre, SE-147 25 Tumba, Sweden

Outline

Demilitarisation

Environmental impact analysis

Green energetics

Future research

Demilitarisation flowsheet

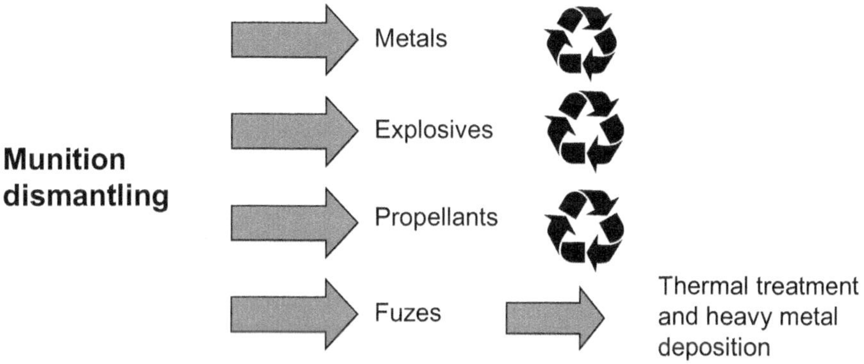

FOI Weapons and Protection October 2001

Residual

Explosives-contaminated waste

Packing material safety concerns

Pyrotechnics

FOI Weapons and Protection October 2001

Incineration of explosives-contaminated waste challenges

How do you assure that the explosives do not explode?

How do you assure safe feeding?

How do you assure safe feed preparation?

How do you meet environmental requirements in terms of destruction efficiency and NO_x removal?

Fluidised bed incineration concept

FOI Weapons and Protection October 2001

Fluidised-bed incineration

High heat capacity

High heat transfer

Good mixing

Proven technology

Non-detonable slurry feed

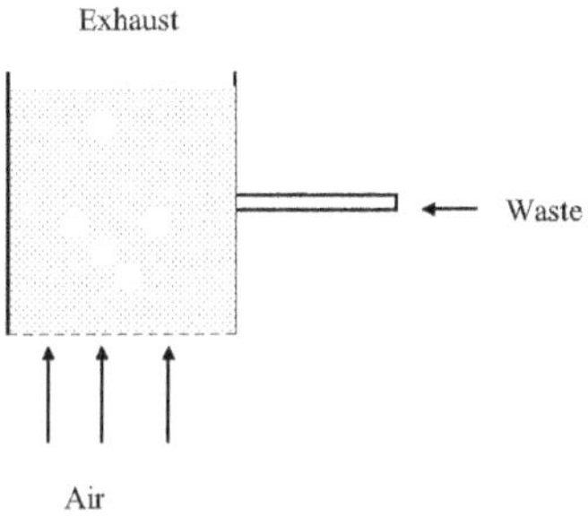

FOI Weapons and Protection October 2001

Grinding tests

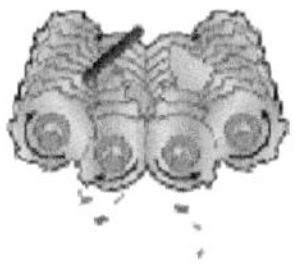

Shredder

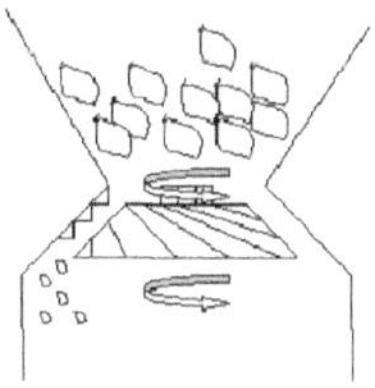

Colloid mill

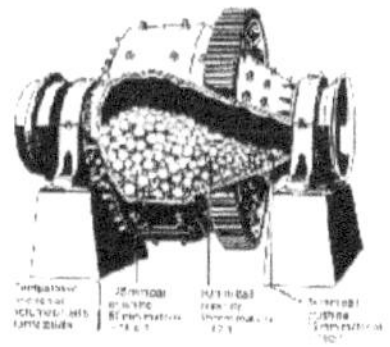

Ball mill

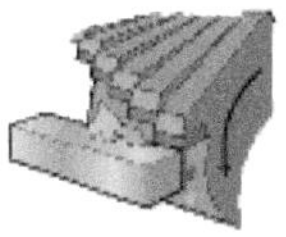

Cutting mill

FOI Weapons and Protection October 2001

Test materials

Explosives (TNT, RDX ...)

Propellants (double-base, AP-HTPB)

Textiles (e.g. filters)

Plastics

Rubber

Carboard cases

Slurry sedimentation tests (46 days)

Explosive	Stabilizer	Observation
10% fine TNT	1.5% Keltrol T	No sedimentation
18% fine TNT		No sedimentation
30% coarse TNT	1.0% Keltrol T	No sedimentation
30% coarse TNT	0.5% Keltrol	Slow sedimention at the top
50% coarse TNT		Not pumpable

Fine: <10μm, Coarse: 1-2 mm

BAM Steel tube detonation test

Composition	No microspheres	Microspheres (Scotchlite 25) ,10 volym-%
15% TNT, 1 % Keltrol T	-	No detonation
20% TNT, 1 % Keltrol T	-	No detonation
25% TNT, 1 % Keltrol T	-	No detonation
30% TNT, 1 % Keltrol T	No detonation	Detonation
40% TNT, 1 % Keltrol T	Detonation	

Incineration trials

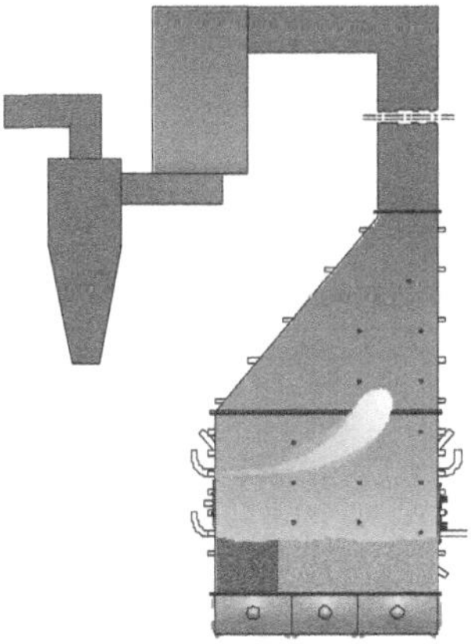

Fuel:

- Nitromethane slurry (25%,20-80 kg/h)
- Textiles
- Aluminium powder
- KP-HTPB
- Pellets (115 kg/h)

Slurry injection

Emissions

		No 2B	No 4	No 5	No 6
Slurry	[%]	14	25	34	42
Effect	[kW]	580	560	595	590
Air	$[m^3_n/h]$	713	736	747	748
Rec. Exhaust	$[m^3_n/h]$	100	100	100	0
λ	[-]	1.35	1.45	1.40	1.46
SR_1	[-]	0.55	0.60	0.61	0.63
O_2	[%]	5.5	6.5	6.0	6.6
Normaliserat till 6% O_2					
CO	[ppm]	159	287	402	1089
NO	[ppm]	374	437	564	608

NOx reduction by ammonia injection (SNCR)

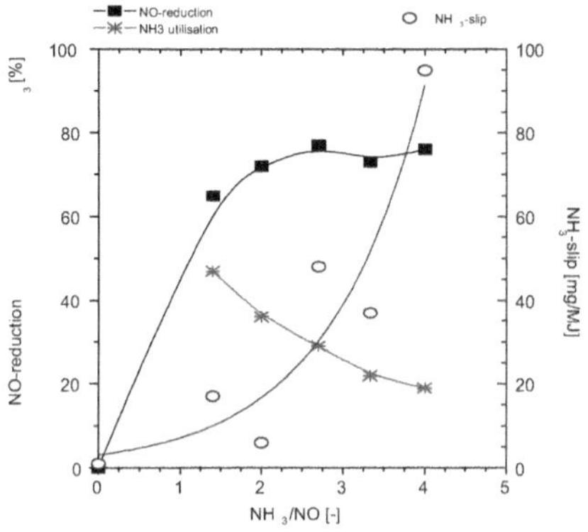

Cost analysis

25 kg/h dry TNT-slurry, 80% H2O

Co-firing with recycled wood-ships

Slurry preparation unit

Emission control

Heat recovery 250 kW

Investment cost: 20 MSEK (US$20M)

Destruction of smoke compositions

Superstructure flowsheet

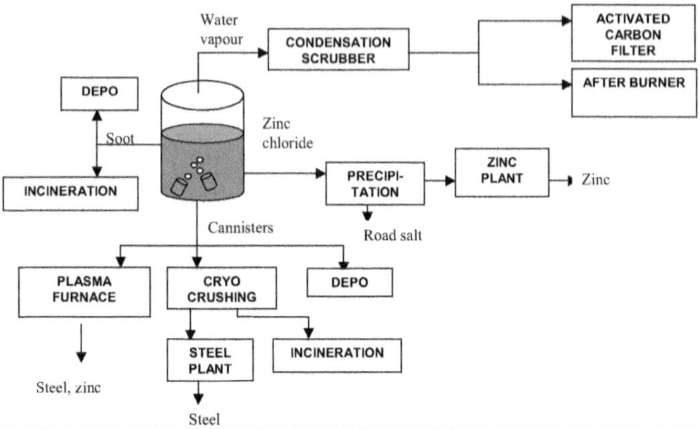

FOI Weapons and Protection October 2001

Environmental impact analysis of munitions

Inventory of metal additives in propellants

Measurement of emissions from flare ammunition

Fate of explosives in buried/dumped ammuntion

Toxicity of explosives

Inventory of metal additives in propellants

Propellant	*Metals*	*Additive function*	*Percentage of objects with additive*	*Average addtion*
Double-base propellant in rocket engines	**Lead, copper, chromium**	**Catalyzer**	**100%**	**2-10 kg lead per ton**
Gun propellants (all calibers)	**lead, tin, barium**	**Lubrication, Cupper removal**	**25%**	**2-7 kg lead per ton**

Measurement of emissions from flare ammunition

Emissions	*Distress lights*	*Distress rockets*
Dust (g/kg)	435	310
NOx (g/kg)	10,4	0,4
VOC (g/kg)	0,27	1,1
HCl (g/kg)	1,9	<0,5
Chlorobenzenes (µg/kg)	92	17
Chlorophenols (µg/kg)	84	6,7
PCB (µg/kg)	0,58	1,9
Dioxin (PCDD/PCDF) (µg/kg)	3,6	<3,6

Environmental Impact Analyses

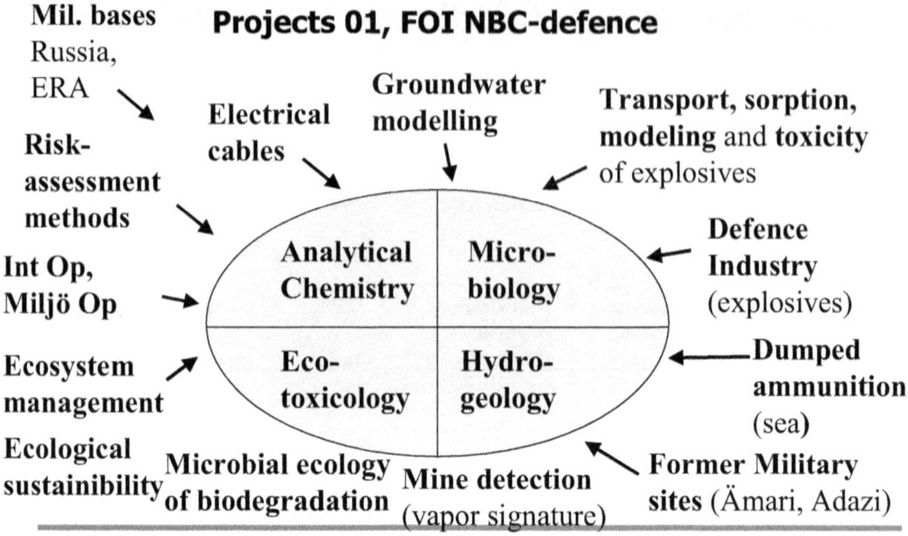

FOI Weapons and Protection October 2001

Green energetics

- ADN - Chlorine-free oxidiser
- FOX-7, FOX-12

FOI Weapons and Protection October 2001

Future environmental research at FOI

- Recycle and reclassification of high-valued explosives
- Life-cycle analysis of environmental impact of muntions
- International operations and environment

DETERMINATION OF THE COMPOSITION OF UNKNOWN OBJECTS BY MEANS OF NUCLEAR NON-DESTRUCTIVE TECHNIQUES

A. KUZNETSOV
V.G. Khlopin Radium Institute
2-j Murinski pr., 28, 194021 St Petersburg, Russia

1. Introduction and short overview of unknown object identification by means of non-destructive methods.

The main existing methods of non-intrusive analysis, their advantages, disadvantages, and areas of their applications are listed in the table below.

	Method	Applications	Advantages	Disadvantages
1	Determination of the chemical content of vapors by chromatographic analysis, electronic nose, trained dogs, etc.	Search for narcotics, explosives and other dangerous substances	Sensitive to small quantities	Unable to identify hermetically packed substances, needs frequent "cleaning" to maintain sensitivity
2	Obtaining images of objects with pulsed radar, thermal, acoustic, X-ray and other methods	Search for density anomalies in constructions, soil, etc.	Very fast	Unable to identify the anomaly, high probability of false alarms
	Nuclear methods			
3	Detection of presence of a characteristic chemical element (e.g. nitrogen) by characteristic gamma-rays from thermal neutron capture reactions (**TNA**)	Search for explosives and other dangerous chemicals	Simplicity and relatively low cost	High radioactive background and heavy shielding for personnel protection, high false alarm rate
4	Determination of relative concentrations of light elements by secondary gamma-rays from inelastic scattering of fast neutrons (**FNA**)	Search for narcotics, explosives; analysis of oil-containing layers	Determines elemental content; high probability of identification	Unable to determine spatial structure of the object, low effect-to-noise ratio
5	Nuclear magnetic (quadruple) resonance (**NQR**)	Search for explosives and narcotics	High probability of identification for some substances	Needs strong electromagnetic fields, narrow field of application
6	Elemental analysis and localization of hidden objects using Associated Particle Sealed Tube Neutron Generator (**APSTNG**)	Search for explosives, narcotics and other hidden objects, waste monitoring	Determines elemental content, high probability of identification, fast analysis, localization	Difficult to obtain high enough flux of neutrons due to limitation of the count rate of corresponding associated particles

P. Castelo Branco et al. (eds.),
Defense Industries: Science and Technology Related to Security, 349–356.

At present there are several devices that use nuclear (**TNA** and **FNA**) techniques:

- **PINS** – Portable Isotopic Neutron-Spectroscopy Chemical Assay System – that was developed jointly by ORTEC and INEEL (USA). It uses TNA and FNA to identify the content of sealed containers. The device uses ^{252}Cf spontaneous neutron source and gamma-detector based on cooled high purity germanium (HPGe) crystal.
- **PELAN** – Portable ELementary Analysis with Neutrons – developed at the Kentucky University, USA, for detection of explosives and narcotics. It uses TNA and PFNA methods (pulsed fast neutron analysis). The device uses neutron generator produced by the MF Physics, USA, and gamma-detector based on BGO crystal.
- **NIGAS** – device that was developed by the Bruker Saxonia Analytik, Germany, for the identification of the content of ammunition, and that uses TNA and PFNA techniques. The device uses D-D neutron generator ING-07D developed and built in VNIIA, Moscow, Russia, and cooled gamma-detector on the basis of HPGe crystal.

1.1.1 What is the idea of the neutron non-destructive analysis?

Inspected object is irradiated with neutrons from the neutron source. Neutrons scatter or are captured on nuclei inside the object and produce gamma-rays; each chemical element has its own characteristic gamma-ray signature. Gamma-ray spectrum is measured and then decomposed into contributions from different chemical elements.

Result: chemical composition of the object in known.

By measuring the ratio of concentrations of chemical elements of the object and using, for example, Daltiz triangle plot method to identify explosives, narcotics and common goods (see Fig. 1), explosives or their components can be detected.

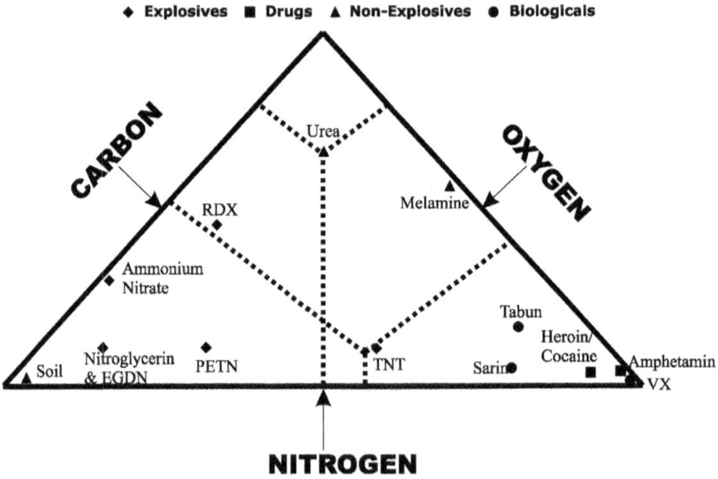

Fig. 1 Daltiz Plot of common explosives, drugs, non-explosive, and biological agents

1.1.2 What are the problems?

Very high gamma background (from reactions on surrounding materials, from the neutron source, cosmic radiation) effectively increases the time that is needed to identify a small object (typical identification times – several minutes).

1.1.3 What do we propose to improve the existing technique?
We detect particles that accompany neutron emission from the neutron source and get the emission time for each neutron. Gamma-ray spectra are measured within narrow time window (several nanoseconds); all gamma-rays produced by neutrons in the object are accepted, while the background gamma-rays are rejected (they do not fit into the time window).
Result: the background is suppressed by more than an order of magnitude, and the identification time is reduced in the same proportion.

2. Technique for identification of explosives based on the timed neutron source.

For identification of unknown objects we use a technique based on irradiation with neutrons from timed isotopic source with subsequent determination of relative concentrations of light elements (carbon, oxygen, nitrogen, hydrogen, etc) in the inspected object.

The main idea for modernization of the technique based on irradiation with neutrons from radioactive sources (for example, ^{252}Cf) is to simultaneously measure secondary gamma-radiation from the object and accompanying (charged) particles. The latter allows using «marked» neutrons (i.e. determining, from which reaction the given neutron originates), and carrying out detection of secondary gamma-radiation in narrow (nanosecond) time intervals, thus considerably improving the background conditions.

The accompanying particles (fission fragments of ^{252}Cf) are detected by a specially developed ionization chamber with ^{252}Cf source inside. A small chamber with a 2 μg ^{252}Cf source (~5×10^6 neutrons per second) is shown on Figure 2.

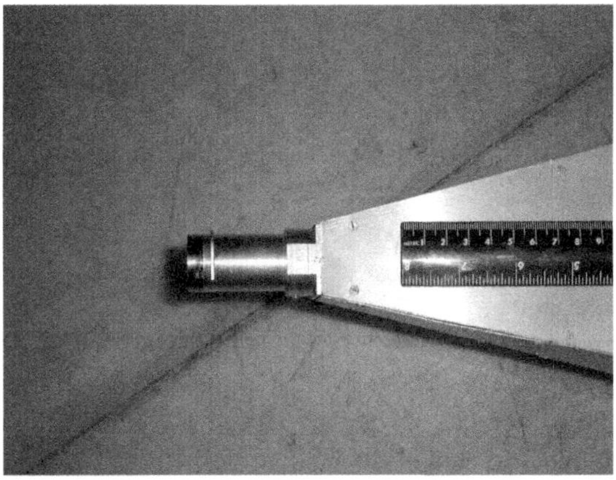

Figure 2. Ionization chamber with 2μg ^{252}Cf source and preamplifier coupled to a pyramidal shielding, consisting from layers of iron, polythene and lead.

Gamma-rays are detected by a NaI(Tl) crystal with dimensions: diameter 150mm × 100mm coupled to a photo multiplier. The detector is placed on a mobile platform (see Figure 3). The mobile installation is serviced by an electronic block, which was specially developed for use in various temperature and weather conditions. It includes programmable microprocessor, PC-connection, spectrometer with possibility to use nanosecond windows of gamma detection, and a battery power supply. Hardware and software allow one to contineously monitor count rates of the ionization chamber, gamma-detector, as well as the rate of coincidences. Temperature effects on the spectrometer are corrected on-line by altering the amplification in spectroscopic channels.

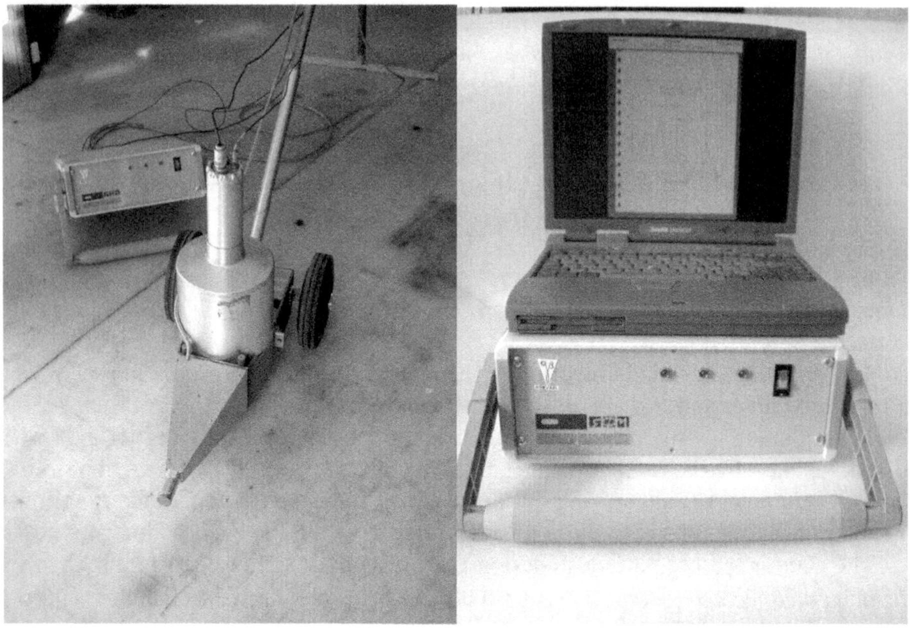

Figure 3. A prototype of the mobile device for detection of explosive substances.

Measurements of spectra of secondary gamma-rays were carried out using objects placed either in open air, or under a layer of a soil of various thickness, composition and humidity. The following types of soil were used: fertilized soil with humidity 20 % by weight, and sand with humidity 0.4 % by weight. Measurements for the following samples were done:

- TNT phantom (with relative concentration of carbon, oxygen, nitrogen, and hydrogen equal to those of the real TNT) of cylindrical shape weighting 500g and 700g.
- Damp root of a tree - cylinder with weight about 700 g.
- Steel cylinder with weight 700 g.
- Stone weighting about 900 g.

The view of the experimental area is shown on Figure 4.

Figure 4. Geometry used in experiments with the prototype of mobile device for explosives detection with 2 μg ^{252}Cf timed neutron source.

Samples were located at distances from 2cm to 5cm below the surface of the soil. Time of measurements varied from 5 to 20 minutes depending on the depth and on the required probability of identification.

Response functions of the existing gamma-detector to gamma-rays from reactions of neutrons with main chemical elements and their compositions were simulated in the computer code based on Monte-Carlo technique. Geometry of the neutron source, gamma-detector and construction materials were taken into account. Some of the response functions were measured experimentally for comparison with experiment. In all cases, experimental and simulated response functions coincide within uncertainty limits.

We are currently using a decision taking procedure that relies on principal component analysis with Mahalonobis distances (PCA/MD). This method allows one to obtain probability that the unknown spectrum belongs to the given group of spectra from the training set. The main advantage of the method as compared to the least-squares fitting procedure is that it accounts for common variations of spectra at the given energy as well as correlations between different energy points of the spectrum (i.e. co-variation). In other words, practically all the information contained in the spectrum is taken into account.

The PCA/MD method of identification requires several measurements of the background conditions at different locations around the inspected object. These measurements, together with results of the Monte-Carlo calculations, allow one to build a library of training spectra, which are then used by the system for recognizing the unknown substances by their gamma-spectrum.

The analysis of experimental spectra shows, that this technique allows us to identify explosive substances weighting around 500 g within 5-15 minutes (depending on the depth and type of soil) with probability 99 % at 1 % of false alarms. The results of identification are not sensitive to presence/absence of metal, and can reliably distinguish explosives from organic substances with similar elemental composition (tree root, melamine, etc).

3. Technique of identification of explosive substances using neutron generator with nanosecond timing based on detection of the accompanying α-particles.

The most important disadvantages of using the timed isotopic neutron source are: a) the need to use heavy shielding against radiation during storage and transportation of the equipment; b) relatively low flux of fast neutrons, which is limited by the radiation safety rules. To achieve better characteristics of the current prototype one has to use a different type of timed neutron sources, for example a neutron generator with nanosecond timing. Unlike isotopic source, a portable neutron generator can be switched off for storage and transportation. Detection of α-particles, which accompany neutron emission in T+D reaction, allows one to collimate neutron flux on the suspected location of the hidden object and to identify it through the analysis of spectrum of secondary gamma-rays.

We have carried out model calculations of the parameters of the device that uses a portable neutron generator with total neutron flux 10^8 neutrons/second with and without the system for detection of the accompanying particles.

The sectioned detector of the accompanying α-particles should possess the following characteristics:

- Have intrinsic detection efficiency ~100% for α-particles;
- Have sub-nanosecond time resolution;
- Must be capable of working at high count-rates (up to 10^7 particles/second).

To meet the above requirements, we constructed a semiconductor detector and tested it at the NG-400 neutron generator of the Applied Physics Laboratory at Radium Institute. The detector was exposed to an integrated doze of neutrons and α-particles 10^{13}, which corresponds to about 500 hours of operation of the portable neutron generator. The detector remained operational throughout the irradiation.

The prospective characteristics of the prototype of the device for explosives detection equipped by neutron generator with nanosecond timing:

- Identification of the hidden object weighting 100 g within 10 seconds with probability >99% and probability of false alarms <1%.
- Simultaneously inspected area 50cm × 50cm, position resolution within this area – 10cm × 10cm.
- Time of preparation of the device - around 5 minutes. Automated controls and the decision-taking procedure ("single button" approach) will not demand special knowledge from the personnel.
- Mobility: weight of the device - not more than 40 kg, dimensions: 50×40×30 $см^3$.
- The installation will comply with the existing radiating safety rules and will not cause damage to health of people, goods and vehicles.
- Low maintenance cost – low power consumption ~20 Watt, estimated lifetime without replacement of the neutron source ~ 2000 hours.

4. Further development of the prototype of the mobile device for detection and identification of explosive substances.

Further development of the existing prototype of the mobile devise for explosives detection will be done along the following lines:

- Laboratory and field tests of the prototype of mobile installation for detection of explosive substances with 2μg ^{252}Cf timed neutron source.
- Inclusion of the portable neutron generator with position-sensitive detector of accompanying particles (α-particles) into the existing prototype of the mobile device.

5. Conclusion.

Prototype of a mobile device for non-intrusive elementary analysis based on timed neutron source was built and tested for explosives' identification. It can be used as a confirmation sensor, when an "anomaly" is detected by other "fast" localization sensor, for example the one based on electromagnetic UHF waves. This prototype can also identify any other dangerous hidden object, even sealed or packed in metallic shell, by non-destructive elementary analysis. To increase the area of application of the device, a portable neutron generator with associated particle detection will be used as a neutron source. It will allow us to reduce time of identification down to several seconds and to obtain 3D image of the inspected area in terms of elemental content for the purposes of localization and identification of the hidden object. This work is under way now.

References

Kuznetsov A.V., (1999), "Development of a Spontaneous Portable Neutron Source with a Built-in System for Registration of Accompanying Particles for Background Reduction" // In IAEA Research Coordination Meeting "Application of Nuclear Techniques to Anti-personnel Landmines Identification", 23-26 November 1999, RBI, Zagreb, Croatia.

Kuznetsov A.V., (2000), "A Spontaneous Portable Neutron Source with a Built-in System for Registration of Accompanying Particles for Background Reduction"// In Proc. of The International Conference "Explosives and Drug Detection Techniques", 18-24 June, 2000, Crete, Greece.

Kuznetsov A.V., (2001), "A Prototype of Mobile Device with Isotopic Timed Neutron Source"// In Proc. of IAEA First Project Co-ordination Meeting on Regional Technical Co-operation Project RER/1/005 "Field Testing and Use of Pulsed Neutron Generator for Demining", 12-14 February, 2001, Vienna

Kuznetsov A.V., (2001), "Nuclear sensor for explosives detection based on timed radioactive source" // In Proc. of the Second Australian-American Joint Conference on the Technologies of Mine Countermeasures conference, 27-29 March, 2001, Sydney, Australia.

Kuznetsov A.V., (2001), “A prototype of mobile device for explosives and landmines identification based on timed radioactive source”// In Proc. of the 7th International Conference on Applications of Nuclear Techniques "Nuclear and Atomic Industrial & Analytical Applications", 17-23 June, 2001, Crete, Greece.

Kuznetsov A.V., (2001), “Concept of a combined mobile device for explosives and landmines identification based on timed neutron source and electromagnetic UHF waves”// to be published in Proc. of NATO ARW #977941 “DETECTION OF EXPLOSIVES AND LAND MINES: METHODS AND FIELD EXPERIENCE”, 11th-14th of September, St.-Petersburg, Russia

REUSE OF DEMILITARIZED AND/OR EXCESS ENERGETIC MATERIALS AS INGREDIENTS IN COMMERCIAL EXPLOSIVES

Oldrich Machacek
Universal Tech Corporation, USA

SUMMARY

This paper outlines the extensive work, which has been accomplished by Universal Tech Corporation (UTeC) in the area of recycling energetic materials, that have been derived from either demilitarization efforts or manufacturing excess. The following information is broken into two sections: Rocket Motor Propellants and Conventional Ammunition Propellants.

ROCKET MOTOR PROPELLANTS

UTeC received its initial exposure to recycling opportunities of Large Rocket Motor propellants through a contact with United Technologies, Chemical Systems Division (CSD), in early 1992. CSD and UTeC entered a joint venture, which involved investigating the development of a commercial explosive formulation containing a Class 1.3 composite propellant from the Minuteman III Stage III Rocket Motor. This propellant was a typical Ammonium Perchlorate (AP)/Aluminum/Rubber Binder type propellant. During 1992, UTeC was successful in developing a large diameter packaged commercial explosive product, containing up to 30% of the downsized composite propellant. The propellant was introduced into standard wategel slurry.

The production of a watergel slurry incorporates a basic two step process. First, a Mother Solution, or liquid phase (which normally constitutes 30-60% of the final product), is produced. The second step constitutes the blending of the Mother Solution with any additional dry nitrate salts and gelling agents. The blending process is utilized to homogenize the mixture and to entrain air into the mixture. Air, which is an essential sensitizer to the explosive, is entrained until the required density of the product is achieved. When ready to package, the mixture has the consistency of thick oatmeal, which is easily pumped into packages. The necessary equipment consists of temperature controlled storage tanks, a mixing chamber and packaging equipment.

In order to incorporate the composite propellant, on a large scale, an additional step had to be added. This step was the downsizing of the propellant into a useful and manageable dimension, both for production and detonation requirements. This was found to be approximately 1-inch (25-mm) pieces or smaller. The method chosen to accomplish this task was a counter-rotating disk shredder. For safety reasons, the shredding process was conducted under a liquid (Mother Solution) and was conducted as an unattended operation.

As a result of a successful development program, UTeC, in cooperation with CSD, designed a facility dedicated to the processing of demilled or excess rocket motor propellants, and their subsequent use as an energetic ingredient in a commercial explosive product. This facility (located near Columbus, Kansas, USA) known as the "PRUF" Plant, houses a propellant shredder as well as the final Blasting Agent explosive product mixing and packaging equipment. The PRUF Plant was constructed in 1995 and was partially re-engineered in 1997.

P. Castelo Branco et al. (eds.),
Defense Industries: Science and Technology Related to Security, 357–360.

In the PRUF Plant's production process, approximately 3,500 lbs. (1,590 Kg.) of composite propellant (usually in a block configuration weighing no more than 20 Kg. each) were placed onto a belt conveyor. The propellant was then remotely fed into the awaiting disk shredder, which was flooded with Mother Solution. The shredder downsized the propellant into irregularly shaped pieces, with an average maximum cross section of about ¾-inch (19-mm). The shredded propellant was continuously taken away from the shredder by an auger system and collected in a hopper. When the appropriate amount of propellant had been shredded and transferred to the mixer, the other ingredients (Ammonium Nitrate, additional Mother Solution, Guar Gum, etc.) were added, and the final Class 1.5 Blasting Agent slurry was made. The slurry was then packaged into 4-inch (120-mm) to 6-inch (152-mm) diameter plastic shot bags, for use in commercial blasting applications.

From the plant's re-startup in March, 1998, to January 2000, the PRUF Plant processed approximately 1.8 million lbs. (818,000 Kg.) of Class 1.3 composite propellants and produced approximately 7.8 million lbs. (3,545,000 Kg.) of finished Class 1.5 Blasting Agent explosives, known as SLX 600-20. The daily production rate was approximately 4,500 lbs. (2,050 Kg.) of propellant processed, or approximately 19,500 lbs. (8,860 Kg.) of finished Blasting Agent.

SLX 600-20 has a calculated Weight Strength of 972 cal/gm, a density of 1.22 g/cc and a VOD in the 4.3 to 4.9 km/sec range. The product is marketed and sold by Slurry Explosive Corporation (SEC) as a packaged booster sensitive explosive, in cartridge diameters of 4 inches (100 mm) and larger. Some of the propellants that have been processed through the PRUF Plant are from the following projects: Titan, Orbus, Minuteman and Hawk. UTeC is currently searching for additional sources of these types of propellants, which may be suitable for use in the PRUF Plant process.

CONVENTIONAL AMMUNITION PROPELLANTS

The practice of using surplus smokeless powder propellants as ingredients in commercial explosive products in the United States dates back to the 1920's. More recently, they, along with other types of surplus military high explosives, were used as the primary sensitizers in the watergel slurry type blasting agents being developed and sold by the major commercial explosive companies during the late 1950's and early 1960's. Based upon the availability of these surplus smokeless powder propellants, their usage by U.S. explosive manufacturers has varied significantly up to the present date. Also, changes in U.S. regulations governing the transportation and storage of commercial explosives has altered the types of explosive products that can utilize propellants as energetic and/or sensitizing ingredients.

In the last six to ten years, a more strict interpretation and enforcement of the U.S.'s environmental laws, governing the disposition of hazardous waste streams and energetic materials, has revitalized the efforts of many U.S. commercial explosive manufacturers to incorporate surplus military explosives as ingredients in their commercial explosive products. Because of heavy restrictions being placed upon the often-used practices of opening burning and open detonation of these hazardous waste materials, in many instances, the military has been forced to use more costly alternative disposal solutions, such as contained incineration. For these reasons, the U.S. military has encouraged and, sometimes partially funded, the development of technologies based upon the recycling of waste and/or surplus energetic materials into commercial explosive products. The recycling of such energetic materials as useful ingredients in a commercial explosive does not require the material to be treated as a hazardous waste, according to the U.S. environmental regulations. Furthermore, significant advances in the U.S. military's tactical weapons arsenal has rendered many of its conventional weapon systems obsolete, thereby generating large quantities of similar energetic materials as surplus and in need of disposal. These stockpiles are usually large enough to serve as a reliable and consistent source of raw materials for the development of a commercial explosive product.

UTeC, as well as several other U.S. explosive companies, including Orica USA Inc., and Dyno Nobel, have produced explosive products containing propellants of this type. UTeC has developed two different packaged Blasting Agent product types using conventional smokeless powders. SLX 600-20A/20B can be described as watergel slurry packaged blasting agent products using single base (SLX 600-20A) or double base (SLX 600-20B) smokeless powders as an energetic ingredient. Slurran 430 can be described as a triple base smokeless powder sensitized watergel slurry packaged Blasting Agent. A more detailed description of these smokeless powder containing Blasting Agent explosive products are given in the following sections.

SLX 600-20A

This product is currently being made at the PRUF Plant facility. It contains 25-28% single base smokeless powder as an energetic ingredient and virtually the same additional water-soluble fuel and oxidizer ingredients as SLX 600-20. The product is manufactured using the same mixing and packaging equipment as that used for the production of the SLX 600-20 product. One distinct advantage of using the smokeless powder grains over the composite propellant is that the smokeless powder grains do not need to be downsized. Without the downsizing step, the finished product production rate is nearly tripled. The plant currently consumes approximately 280,000 lbs. (127,000 Kg.) of smokeless powder propellant, and produces approximately 1 million lbs. (454,000 Kg.) of finished product per month. Naturally, since the downsizing step has been removed, only certain smokeless powder propellant grain sizes and shapes can be used. The propellant grains can vary in diameter from 7 mm to 12 mm and in length from 12 mm to 30 mm. The propellant types that have been used in this product are M1, M6, LKL and SPD. SLX 600-20A has a calculated Weight Strength of 795 cal/gm, a density of 1.25 g/cc and an unconfined VOD in the 4.5 to 5.2 km/sec range. The product is marketed and sold as a packaged booster sensitive explosive, in cartridge diameters of 4 inches (100 mm) and larger.

SLX 600-20B

SLX 600-20B is the same type of watergel slurry blasting agent as SLX 600-20A, except that SLX 600-20B contains 25-28% double base smokeless powder propellant, as an energetic ingredient. The ammonium nitrate and hexamine nitrate based watergel slurry contains whole propellant grains. These cylindrical grains usually contain seven perforations and have an average particle size of 7 to 12 mm diameter X 12 to 26 mm length. The propellant types that have been used in the product are M26 and T28. SLX 600-20B has a calculated Weight Strength of 827 cal/gm, a density of 1.25 g/cc and an unconfined VOD in the 4.5 to 5.2 km/sec range. The product is marketed and sold as a packaged booster sensitive explosive, in cartridge diameters of 4 inches (100 mm) and larger.

Slurran 430

Slurran 430 is a smokeless powder sensitized watergel slurry packaged Blasting Agent. The Slurran 430 product differs from the previously described packaged propellant based explosives, in that it contains a significantly higher percentage of smokeless powder propellant. The Slurran 430 product contains about 60% whole grain triple base smokeless powder propellant. The propellant types that have been used in this product are M30 and M6 (single base).

It was found that relatively large grain smokeless powder propellants could be made to readily detonate in their original grain configurations, when surrounded by a high-density liquid medium. During the development work, it was determined that the higher the density of this liquid medium, the easier it was to make the propellant detonate, and the smaller the resultant explosive's critical diameter.

The production process is quite simple. It involves first filling a polywoven shotbag with the required volume or weight of propellant grains. Then, a mixture of a delayed gelling solution and an oxidizer salt solution is poured over the column of propellant until all the grains are submerged. Then the cartridge is clipped shut and loaded into a box or onto a truck for transport. The gelling agents eventually thicken the product's liquid phase to the extent that it resembles gelatin. Because of the high percentage of propellant grains in the final explosive product, the cartridges are rigid. We currently consume approximately 420,000 lbs. (190,000 Kg.) of propellant and produces approximately 700,000 lbs. (318,000 Kg.) of finished explosive product per month. Slurran 430 has a Weight Strength of 885 cal/gm, a density of 1.50 g/cc and an unconfined VOD in the 6.0 to 6.5 km/sec range. The product is marketed and sold as a packaged booster sensitive explosive, in cartridge diameters of 3 inches (100 mm) and larger.

The use of composite and conventional ammunition propellants as ingredients in commercial explosives is a viable method of their disposal. These techniques not only allow for the beneficial use of the surplus energetic materials as an effective ingredient in commercial explosive products, but they also solve many environmental and economic problems associated with the propellant's potential destruction as a hazardous waste.

Dismantling of conventional munitions and reuse of energetic materials for industrial purposes

Andrzej Maranda, Jerzy Nowaczewski
Andrzej Papliński

Military University of Technology
00 - 908 Warszawa, ul. Kaliskiego 2, Poland

ABSTRACT

The investigations carried out in Military University of Technology on recovery and utilisation of explosives obtained from withdrawn munitions are reported. Commercial reapplication of recovered explosives for rock blasting in quarries is considered.

Detonation characteristics of granulated TNT obtained during the process of melting down TNT-filled charges were investigated. Also performance of explosive mixtures obtained by introduction of nitrocellulose propellant into slurry and ANFO was examined.

Positive results were obtained by applying granular TNT and phlegmatized hexogene for rock blasting. Also an application of crumbled or bunched nitro-cellulose and nitro-glycerine propellants in quarrying was confirmed.

TNT and RDX boosters as well as other work charges acquired from dismantled items were developed.

Analyses as well as results of experimental registrations of detonation characteristics of explosive compositions of TNT and a component of positive oxygen balance (ammonium nitrate) are also presented.

INTRODUCTION

The problems of removal and/or possible utilisation of explosives from obsolete and unserviceable munitions become an important question for technical and environmental reasons [1 – 4]. The progress in development of new energetic materials and warheads technologies results in producing of growing stockpile of projectiles and propelling assemblies, replaced by smaller in number but more effective items.

It is appraised that the explosive shares of about of 5 % of total cost of the munition unit. At first, the coloured metals were of interest to remove from withdrawn munitions. Up to the end of 80-ties a common way of reclaimed explosives destruction was an open burning. Lately, quite remarkable efforts are undertaken to replace this rather costly way by technologies that ensure smaller environment degradation and offers some positive economical outcome [5 – 9].

P. Castelo Branco et al. (eds.),
Defense Industries: Science and Technology Related to Security, 361–372.

The investigations on recovery and utilisation of explosives from withdrawn munitions in the Military University of Technology are carried out for nearly ten years. The main effort is focused on appointment and examination of possible commercial reuse of disposed high explosives and propellants. An application of reclaimed explosives in surface mining industry has been extensively examined.

The work charges designed for rock blasting in quarries were elaborated. Several kinds of boosters acquired from dismantled munition units were adapted to boost charges of less sensitive explosives. Also charges for fragmentation of oversized blocks were developed. Several explosive items produced from dismantled munitions had been certified by the Mining Head Office for use in surface mining exploitation.

1. TNT Recovery and investigations

Up to now, the TNT constitutes a relatively large part of explosives contained in disposed munitions. Due to that, quite extensive investigations have been made to explore the possibility of reuse of reclaimed TNT. As TNT is a strong explosive owing relatively low shock susceptibility and well thermal stability it offers some advantages as a potential commercial explosive.

Granulation of TNT. The sketch of the process line for TNT recovery by the method of melting down is presented in Fig. 1. An item that contains TNT is subjected to a stream of preheated water vapour. Melted TNT flows down to a tank with whirling water where granules of prescribed diameter range are formed. Granulated product is centrifuged and in a semidry form is packed into polyethylene bags.

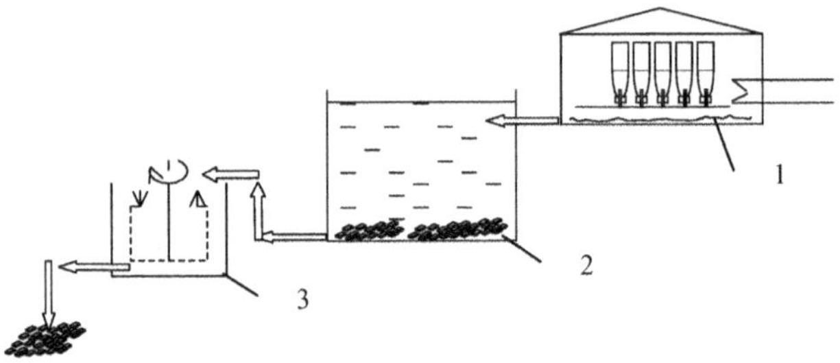

Fig.1. Sketch of the process line for melting down and granulation of TNT:
1 – the furnace, 2 – granulator, 3 – centrifuge

The obtained TNT granules are of about of 1 – 3 mm in size. 30 % of granules (by mass) has a diameter of over of 3 mm and nearly 25 % of the material is of particle sizes of below 1 mm.

Detonation parameters of granulated TNT. Detonation parameter measurements were performed for dry TNT and water-flooded TNT. Detonation velocity was measured by short-circuit gauges, the investigated explosive charges were placed in steel and polyvinylidene tubes. Tested charges were initiated by 20 g hexogene boosters. Critical detonation diameter was measured by the method of cone charge. The results of measurements are set up in Tables 1 and 2.

Table 1. Detonation velocity (D) and critical detonation diameter (d_{cr}) of granulated TNT

Fraction [mm]	Average density [g/cm^3]	D [m/s]	d_{cr} [mm]
0,2 – 0,6	0,91	4300	10
0,6 – 1,02	0,87	4200	14
1,02 – 1,2	0,82	2870	19
1,2 – 1,5	0,80	1910	24
1,5 – 2,0	0,79	1540	30
full granulate	0,80	1950	17

Inner/outer diameter of PCV tubes used in this round of experiments was of 36/40 mm.

The obtained results confirm that the critical detonation diameter of granulated TNT is far below the cross section of bore – holes employed by rock hewing. On account of higher explosion energy of TNT in comparison with typical mining explosives, some reduction of drilling costs can be attained, by applying bores of diminished diameters.

As in many cases blasting operations are performed in water-saturated rocks, the investigations of detonation parameters of flooded TNT charges were also carried out.

The measured detonation velocities are set up in the Table 2.

Table 2. Detonation velocity of dry and flooded TNT

Fraction	D [m/s]	
[mm]	Dry TNT	Flooded TNT
0,6 – 1,02	4300	4650
1,02 – 1,2	4160	4850
1,2 – 1,5	4100	5360
1,5 – 2,0	3950	5210
2,0 – 5,0	3850	4790
full granulate	4400	4840

The referred results were obtained in measurements performed in steel tubes of 36/42 mm.

Due to increased mass density, the detonation velocity of flooded TNT is even higher that in dry conditions.

2. COMMERCIAL EMPLOYMENT OF RECOVERED EXPLOSIVES

Application of granulated TNT. Several rounds of tests were performed to examine the applicability of granulated TNT for rock blasting in quarries. Exploitation tests carried out in rocks of diversified firmness like limestone (Wierzbica), porphyry (Zalas) and melaphyry (Czarny Bór) have confirmed that TNT can be applied as mining explosive in quite various exploitation conditions. Effectiveness of TNT charges in both dry and flooded bore-holes were confirmed.

Employment of granulated TNT was the first commercial application of explosive obtained from dismounted munitions. In subsequent years further investigations were performed on other explosives and blasting agents.

TNT and RDX boosters. Due to high density, of about of 1,60 g/cm3, TNT charges are an reliable device for initiation of work charges of saletrols and saletrots. After control tests the Mining Head Office has admitted for use in surface mining of four types of TNT boosters which differs in mass, diameter and longitude. Annual spending of TNT boosters reach up of several dozen of thousands that do confirm that this product has acquired an acceptation in surface mining practice.

Also hexogene boosters became an effective tool for initiation of less sensitive commercial explosives. Some hexogene charges form a series of types with mass of 20 g to 150 g.

Application of reclaimed hexogene. In several items also the phlegmatized hexogene is present. There are hexogene pressed with wax, aluminium powder or moulded hexogene/TNT mixtures. Explosive is washed out from the shell by mean of the intense water stream. For TNT/hexogene mixtures, in the case of high content of fusible component, granulate can be obtained, by applying the melting down procedure. In result of fulfilled tests, the Mining Head Office has admitted for use in surface mining phlegmatized hexogene (A-IX-1),

Applications of reclaimed explosives in blast charges. Several investigations had also been done to render a use of reclaimed explosives in work charges of various kinds. In mining, shaped charges obtained from dismounted warheads are an effective agent in breaking down of oversized blocks that sometimes emerge into rubbled rocks. Also TNT and RDX charges (boosters) are used by fragmentation of oversized blocks. Furthermore, reclaimed explosives are effectively used in demolition of obsolete buildings, industrial constructions, etc.

3. APPLICATION OF DISPOSED PROPELLANTS

Smokeless nitro-cellulose powders are less stable explosives than high explosives like TNT and RDX. Nitro-cellulose powders posses a relatively long chemical stability that is attained by addition of stabilising agents that prevent a chemical decomposition. The attained chemical stability period is of about of 40 - 50 years. However, physical stability is more difficult to pertain for such a long period of time. Powder grains undergo structural changes that results in alteration of ballistic characteristics of the material. Than, being still capable to explode, nitro-cellulose powders are withdrawn from exploitation.

At present, rather limited possibility or reapplication of disposed propellants is recognised. Due to difficulties in retirement of stabilising components, the possibility of reuse of nitro-cellulose for e.g. lacquer production is restricted.

As it seems, the only economically reasonable reuse of reclaimed propellants is to adopt them for an application as the working agent in rock breaking down in mining.

Grained powders. In the first phase of performed investigations, the reuse of grained propellants was examined. Explosion energy of nitro-cellulose powders is comparable while the volume of gaseous products per unitary mass of explosive exceeds that of TNT and RDX. To provide an effective application in rock blasting, a reliable regime of detonation initiation of powder charges should be ensured.

Investigations of detonative properties of charges based on nitro-cellulose powders (NCP) have been carried out. Detonation velocity of sole NCP was measured as well as mixtures of NCP with ammonium nitrate. ANFO containing NCP as well as slurry explosives in which TNT was replaced by NCP were also examined. Experiment particulars and detonation initiation were the same as described above, measurements were carried out in steel tubes of 46/56 mm.

Several results are comprised in Table 3.

Table 3. Chemical composition, mass density and detonation velocity of explosives containing NCP

Explosive No:	1	2	3	4	5	6
Components [% of mass]						
NCP	100	80	60	40	40	30
Ammonium nitrate		20	40	60	58	53
Oil					2	
Water						14
Carbamide						2
Thickening agent						1
Explosive parameters						
Density [g/cm3]	0,95	0,97	0,98	0,99	0,99	1,43
Detonation velocity [m/s]	4040	3760	3600	2930	3420	5280

As can be seen, nitrocellulose powder and its mixtures with ammonium nitrate have higher detonation parameters than flaky TNT. Nearly the same detonation velocity attains ANFO containing of 40 % of NCP. Among analyzed slurries the highest detonation parameters were obtained for the explosive containing 30 % of NCP and of 14 % of water (Table 3).

Exploitation tests have confirmed usability of NCP and by now a steadily demand on this material is sustained. In particular, NCP is used by blasting of rocks with high consistency and also in flooded ledges. Explosives based on NCP are placed in lower part of water filled boreholes while the upper part of the bore is completed by ANFO charge. To detonate of such a combined charge, boosters of total mass of about 300 g are applied.

Bunched nitroglycerine and nitrocellulose propellants. Afterwards, application of nitroglycerine-based propellants was considered. Nitroglycerine propellants are less sensitive to detonation and do not detonate in flooded medium. As many of nitroglycerine propellants being at disposal appear to be in bunched form, several investigations on filling blasting bores by batches of bunched propellants were undertaken. At series of experiments positive results have been obtained by applying bunches of nitroglycerine propellant enclosed in polyethylene bags placed in bores of diameter of 160 mm. Also, to attain a higher fulfillment degree, filling of blasting bores by loose bunches of nitroglycerine powder was considered. Positive results were obtained by shooting in bores with diameter of 100 mm.
Possible employment of bunches of nitrocellulose propellant were also performed. Its possible use in flooded boreholes has been confirmed.

Crumbled nitroglycerine propellants. As the filling of blasting bores by bunched propellant is laborious, the attempts were made to crumble tubular nitroglycerine propellant into pieces of length of about of 20 mm. This has enabled more proficient filling of boreholes. Up to now, more than 300 ton of crumbled nitroglycerine propellant was employed in surface mining. However, it can be used in dry bores only.

By blast production, as an efficient method, combined charges are applied. Lower part of blasting bore is filled with nitrocellulose-based propellant and the upper part - with nitroglycerine propellant. Mutually, according to accessibility, batches of bunched propellants are also applied.

4. TNT – BASED EXPLOSIVE COMPOSITIONS

As TNT has the negative oxygen balance, its detonation products contain many species of low oxidation degree. Then, relatively huge TNT charges applied by rock blasting produce remarkable amounts of injuriously compounds. This causes environment abridgements on broad use of TNT in field shooting. In regard to this several attempts has been made to elaborate explosive compositions based on TNT more suitable for commercial applications. In MUT the investigations about producing of TNT/AN (ammonium nitrate) explosive mixtures are performed.

There are many additives that do moderate the low detonative properties of AN as well as increase their explosion energy (high energetic explosives, oils, aluminium particles etc.). In view of positive oxygen balance of AN one of important additives can be TNT [10]. Added to AN, TNT reveals a very positive influence on the properties of so obtained charges. In mixture with AN increase oxidation degree of carbon, which is present in TNT in excess. Also the atomic composition of AN reacts more effectively (full exhaust of oxygen occur). This results in remarkable growth of the heat of explosion.

An important feature that can be observed by analysis of AN/TNT mixtures is that in a certain region of compositions, detonation parameters of mixture do exceed the detonative characteristics of each individual component. Detonation characteristics of AN/TNT mixtures were analyzed in the full range of proportions. Obtained results are set up in the Table 4.

Table 4. Detonation characteristics of TNT/AN explosive mixtures

ρ_{0MW} = 1,00 g/cm^3

TNT / NH_4NO_3	D	p_{C-J}	T_{C-J}	Σy_i	C(s)	$Qv+\varepsilon_{0MW}$
%	mm/μs	GPa	K	mol/kg	mol/kg	MJ/kg
100 / -	5240	7,25	3720	32,26	13,64	4,47
80 / 20	5334	7,48	3151	35,18	8,55	4,33
60 / 40	5391	7,64	3306	38,01	3,75	4,21
40 / 60	5398	7,54	3141	40,57		4,15
30 / 70	5327	7,37	3196	40,68		4,32
20 / 80	5188	7,05	3187	40, 08		4,42
10 / 90	4765	5,77	2466	41,89		3,14
- / 100	4146	4,15	1686	43,73		1,88

Σy_i – total sum of gaseous products

The detonation parameters were estimated by means of thermochemical calculations [11,12], with BKW-EOS used to describe nonideal behavior of detonation products. Carbon (soot) properties were modeled by Cowan EOS [13]. The Hobbes - Baer calibration of BKW parameters ([14]) was applied. Values of BKW constants α, β, κ and θ were of 0.5, 0.174, 11.85 and 5160, respectively.

As can be seen from presented results, maximum of detonation velocity (D) is attained at higher TNT content than maximum of detonation pressure (p_{C-J}). It's due to growing amount of gaseous products of detonation (Σy_i) by AN increase. Attained values of detonation velocity of mixtures containing of 20 – 30 % of TNT are higher than detonation velocity of the sole TNT. However, explosive mixtures containing of 20 – 30 % of TNT produce remarkably greater amount of gaseous detonation products that TNT. Moreover, the heat liberated in chemical reactions in detonation ($Qv+\varepsilon_{0MW}$) is relatively high, remarkably exceeds 4 MJ/kg. High explosion energy with sufficient amount of gaseous products build up an advantageous combination of improved ability to rock breaking down and fragmentation.

Therefore we can conclude that addition of dozen of percent of TNT to ammonium nitrate can produce an explosive of good exploitation attributes.

Due to the constitutional properties of ammonium nitrate, explosive mixtures based on AN reveal significant influence of physical structure (particle shape and sizing) and chemical composition on detonation characteristics like detonation velocity and critical detonation diameter. Pure ammonium nitrate posses very large critical detonation diameter. At density of 0,85 g/cm3 critical detonation diameter of over 300 mm was observed [15]. Addition of TNT results in modification of detonation zone width what influence on both detonation velocity and critical detonation diameter. An experimental material has been collected to elucidate the detonation properties of AN/TNT heterogeneous systems.

The results of nonideal detonation velocity measurements for AN/TNT mixtures containing of 20 %, 40 % and 10 % of TNT are presented in Fig. 2a – c.
Investigated mixtures were prepared from commercially available crystalline ammonium nitrate of grains size less than 0,8 mm was used. Two kinds of TNT were used: fine grained TNT (size below 0,8 mm) and coarse TNT ($\delta = 1,2 \div 2,5$ mm).
Detonation velocity was measured for charges confined in polyvinyl chloride tubes with inner diameter of 37, 46, 78, or 100 mm. Wall thickness of the tubes was 5 % of inner diameter and their length was of 500 mm. Detonation was initiated by use of 30 g explosive charge of plastic explosive. The run - off distance of detonation wave was of 150 mm, detonation velocity was measured at three bases, each of 80 mm long. By experiments with tubes of 100 mm, two bases of 80 mm were used and the run-off distance was of 130 mm.
The mixtures of AN with coarse TNT have density of 0,90 g/cm^3, while AN mixtures with fine grained TNT were at density of 1,01 and 0,90 g/cm^3.

The obtained values of non-ideal detonation velocity of investigated explosive mixtures were depicted versus reciprocal charge diameter (1/*d*).
As can be seen from the Fig. 2a – c, the experiments confirm applicability of the linear formula of Eyring [16]

$$D = D_i\left(1 - \frac{a}{d}\right) \tag{4.1}$$

where D_i is an extrapolated detonation velocity to infinite diameter and a is a constant.

Inferring from the value of the parameter *a* detonation zone length of investigated explosives can be appraised.

As can be seen from the Fig. 2a, dimensions of TNT particles influence on detonation velocity. By smaller TNT grains (grain size $\delta \leq 0,4$ mm), when pores between AN particles are partly filled by TNT powder the attained detonation velocities are higher and the influence of charge diameter on detonation velocity becomes moderated.

For the mixture containing of 40 % of TNT (Fig. 2b), less intense influence of charge diameter upon the detonation velocity can be observed.

However, for 90/10 AN/TNT mixture (Fig. 2c), rapid drop of detonation velocity with decrease of charge diameter occurs.

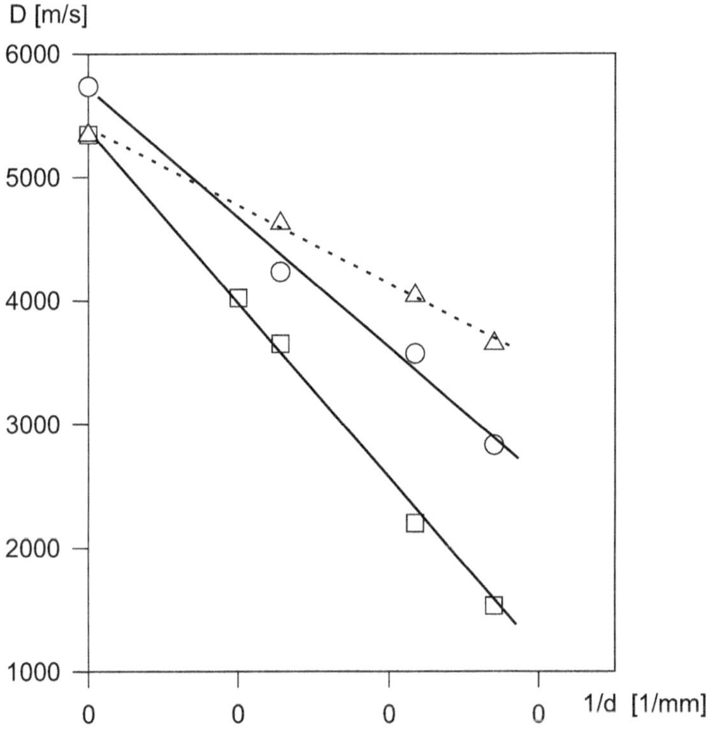

Fig. 2a. Dependence of detonation velocity on reciprocal charge diameter for AN/TNT 80/20

□ - coarse TNT, δ = 1,2 - 2,5 mm; ρ_{0exp} = 0,90 g/cm^3

○ - fine-grained TNT, $\delta < 0,8$ mm; ρ_{0exp} = 1,01 g/cm^3

Δ - fine TNT, $\delta < 0,4$ mm; ρ_{0exp} = 0,90 g/cm^3

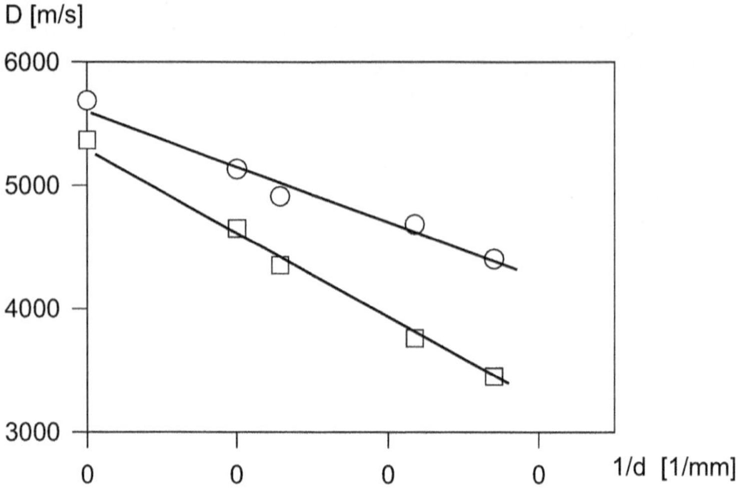

Fig. 2b. Diameter - effect curves for AN/TNT 60/40

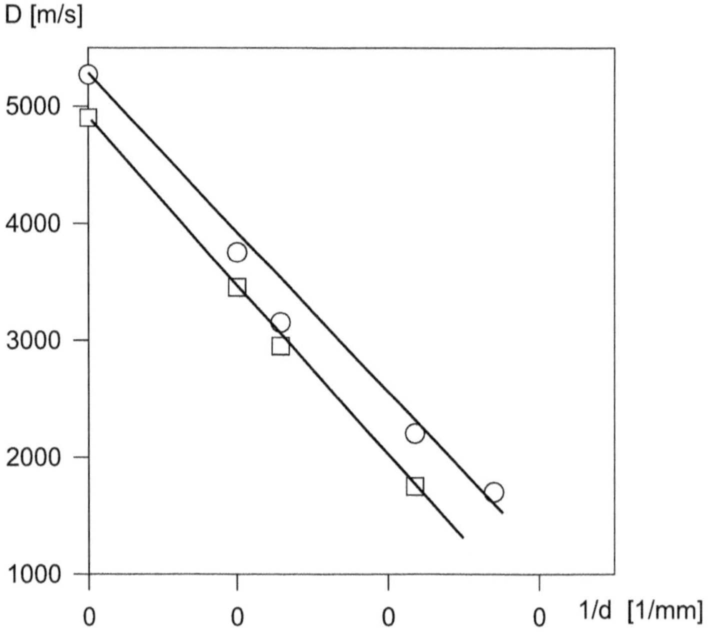

Fig. 2c. Diameter - effect curves for AN/TNT 90/10

According to obtained results one can observe that TNT/AN explosive mixtures do detonate and preserve high values of detonation velocity for charges of diameters of about several dozen of millimeters. As they possess relatively high explosion energy, their commercial employment offers some positive advantages.

FINAL CONCLUSIONS

The investigations carried out in the Military University of Technology about recovery and utilisation of explosives from withdrawn munitions are reported.
Reuse of TNT, RDX as well as nitro-cellulose and nitro-glycerine propellants was considered.

•) Detonation characteristics of granulated TNT obtained in the recovery process were investigated.
•) Positive results were obtained by applying granular TNT for rock blasting.
•) TNT and RDX boosters as well as other additional charges acquired from dismantled items were developed.
•) Application of crumbled or bunched nitro-cellulose and nitro-glycerine propellants in quarrying was carried out.
•) Investigations of detonation characteristics of AN/TNT explosive compositions were performed.

The paper was completed as the part of the KBN Project 0 T00A 034 19.

REFERENCES

[1] G.Bunte, V.Gröbel, T.Härdle, V.Weiser, *Druckabhängigkeit der Schastoffemissionen beim abbrand eines Doublebase-festtreibstoffes (DB-FTS)*, Proc. 23rd International Annual Conference of ICT, Waste Management of Energetic Materials and Polymers, Karlsruhe, June-July 1992, 49:1 - 6.

[2] M.Schneider, *Perspektiven Möghlicher alternativen zur conventionellen vernichtung von Explosivstoffen – Biologishe Behandlung und Wiedergewinnung durch chemischen Behandlung*, ibid., 7:1 -14.

[3] V.A.Tartiakovsky, S.A.Shevelev, M.D.Dutov, A.Kh.Shaknes, A.L.Rusanov, L.G.Komarova, A.M.Andrievsky, *Problems of trotyl (TNT) processing into condensation monomers, polymers and dyes*, Proc. 26th International Annual Conference of ICT, Pyrotechnics. Basic Principles, Technology, Application, Karlsruhe, July 1995, 43:1- 11.

[4] P.Wanniger, *Conversion of high explosives*, Proc. 1st International Autumn Seminar on Propellants, Explosives and Pyrotechnics, Theory and Practice of Energetic Materials, Beijing, October 1994, 392 - 402.

[5] J.Campos, R.Fernandes, F.Matos, A.Pires, *Abrasive water jet cutting of munitions and explosive incineration in a fluidised bed*, Conversion Concepts for Commercial Applications and Disposal Technologies of Energetic Systems, ed. H.Krause, NATO ASI Series no 14, Kluwer Academic Publishers, Dortrecht 1997, 227 - 245.

[6] H.H.Krause, *Advanced disposal of recycling technologies of energetic materials*, ibid., 91 - 97.

[7] N.H.A.Van Ham, *Munitions disposal as a total concept*, ibid., 47 - 54.

[8] A.Maranda, M.Miszczak, J.Nowaczewski, *Application on the explosives withdrawn from military stocks in the coal mining industry*, Proc. 24th International Annual Conference of ICT. Energetic Materials – Intensitivity and Enviromental Awareness, Karlsruhe, June – July, 1993, 72-1 – 72-13.

[9] S.Cudziło, A.Maranda, J.Nowaczewski, W.Trzciński, R.Morawa, *The analysis of demilitarized explosive application for the initiation of industrial explosives*. Proc. Fifth International Symposium on Explosive Technology. Pretoria, October 1994, 199 - 209.

[10] M.A. Cook, E. B. Mayfield, W.S. Partridge, *Reaction rates of anmmonium nitrate in detonation*, J. Phys. Chemistry, 59, No 8, 1955, pp. 675.

[11] A. Papliński, *About evaluation of chemical equilibrium composition by the steepest descent method*, 14-th International Symposium on Combustion Processes, Czestochowa, 26-29.09.1995.

[12] A.Papliński, *About evaluation of HE detonation and explosion parameters by use of BKW equation of state*, 7e Congrès International de Pyrotechnie EUROPYRO 99, Brest, Francja, 7 – 11.06.1999, pp. 478 - 486.

[13] Ch. L. Mader, *Numerical modeling of detonations,* Univ. of California Press, Berkeley,1979.

[14] M. L. Hobbs, M. R. Baer, *Nonideal thermoequilibrium calculations using a large product species data base*, Shock Waves, 2, 1992, pp. 177.

[15] A. Miyake, A. C. van der Steen, H. H. Kodde, *Detonation velocity and pressure of the non-ideal explosive ammonium nitrate*, 9th Symp. (International) on Detonation, Portland, Oregon, 1989, pp. 560.

[16] H. Eyring, R. E. Powell, G. H. Duffy, R. B. Parlin, *The stability of detonation*, Chem. Rev., 45, 1940, pp. 69

ENVIRONMENT FRIENDLY DESTRUCTION OF ENERGETIC MATERIALS AND MUNITION ON THE TERRITORY OF GEORGIA

Prof. Dr. Avtandil Dolidze
Institute of Physical and Organic Chemistry of the Georgian Academy of Sciences
5, Jikia str. Tbilisi 380086 Georgia

After abandoning the territory of Georgia, the Soviet Army left behind approximately 30 places stocked with explosive and energetic materials, different munitions, military polygons, etc. At present the rest of the Russian military bases are leaving the territory. The Black Sea and the rivers close to these bases are in danger of pollution. Moreover, there is a danger of soil and atmosphere pollution as well. Local population worries about comparative increase of various diseases and child mortality. The greatest part of the territory (former military) should be used for peaceful purposes and a special program for rehabilitation of these territories should be worked out.

The Chemical-radiation Governmental Commission under the supervision of the Minister of Environmental and Natural Resources Protection of Georgia has been established. The Georgian population worries about the information concerning revealing radiation sources of military-medical purposes. During the latest several years, 100 sources of various activity and capacity have been revealed. Calibrating components and Cesium mostly dominated among the sources. Generally, as usual, the revealed sources are being kept in the Nuclear Reactor Department of the Institute of Physics of the Academy of Sciences and Institute of Radiology.

Military materials mostly are equipped with trotile and trotile-hexogen mixture, artillery powder and other explosive substances. The conditions of keeping these substances do not satisfy the demands of safety techniques, though the work for improvement of preservation conditions have been provided. The warehouses are in bad condition, roofs leek and rain reaches the military materials, in addition, anti-fire protection is not provided. The greatest part of bombs and rockets are located on open plateaus, that is why the materials become useless and places of corrosion could be found.

P. Castelo Branco et al. (eds.),
Defense Industries: Science and Technology Related to Security, 373–380.

As a result, about 40 thousand tons of explosive substances and 10 thousand tons of gunpowder are technically useless and should be liquidated. The problems of their processing are under study. The most acceptable is processing of trotile-containing military materials and industrial explosive substances, the demand for which is quite high. According to our information, the Ministry of Defense of Georgia is inclined to establish a joint venture (with 'USTIN' L.L.C., Ukraine), for processing trotile- and hexogen-containing explosive substances in order to utilize them in commercial goods. Gunpowder is supposed to be processed locally.

Different situation is with residual components of liquid rocket fuel from the air-defense rocket system C-75. In 1996 the Ministry of Defense of Georgia asked the Head of the Republic to take urgent measures to render the components of liquid rocket fuel harmless. According to the data of the Georgian Ministry of Defense these components – more than 1000 tons, have been dislocated in former Soviet rocket bases in West and East Georgia. In West Georgia they are disposed along the Black Sea perimeter: in Angisa, Supsa, Chaladidi, Anaklia and State governed warehouse of strategic reserve at military airport in Meria, Ozurgeti region, 5-7 km far from the Black Sea. In East Georgia rocket bases have been dislocated around Tbilisi: in Yagluji, Soganlugi, Digomi, Jorjiashvili and Rustavi. The Georgian Ministry of Defense managed to collect fuel resources of West Georgia in Chaladidi and East Georgia in Soganlugi. Moreover, fuel resources were left in Maria, as well. That batch of components of liquid rocket fuel was brought to Georgia in 1988.

Oxidizer 'MELANGE' – AK-20k

1. Mass part of nitrogen acid, %, no more than	73.0
2. Mass part of nitrous oxides (recalculated on N_2O_4), %	17.5-22.5
3. Mass part of water, %	1.2-2.4
4. Mass part of HF, %	0.5-0.75
5. Mass part of H_3PO_4, %	1.0-1.3
6. Mass part of mechanical impurities, %, no more than	0.02
7. Density at 20°C, g/cm^3	1.57-1.60

Fuel 'SAMIN' – TG-02

1. Mass part of isomeric xylidines, %	50±2
2. Mass part of di- and triethilamines, %	50±2

3. Mass part of diethilamine, %, no more than	1.6
4. Mass part of water, %, no more than	0.5
5. Mass amount of mechanical impurities, %, no more than	0.01
6. Density at 20°C, g/cm^3	0.845±0.01

This fuel with oxidizer based on nitric acid forms self-igniting mixtures and is used as basic as well as starting fuel.

The oxidizer MELANGE - dark-orange colored liquid is a strong poisoning substance. Due to very low boiling temperature of nitrous quad-oxide (+22°C) Melange easily evaporates, its vapors are poisonous.

The fuel SAMIN – yellowish-brownish oily liquid with specific smell is very poisonous while breathing vapors and it destroys skin texture after contact.

The 600 tons of these components were stocked at military airport of Meria and in case of their sudden junction (provocation or terrorist act) self-ignition with explosion could take place. The explosion wave could cover the territory with the radius of 20-25 km and the poisonous vapors could move considerably far along the wind direction. The area surrounding the airport of Meria is a highly urbanized region with hard social conditions and there is a serious danger of further ecological complications in the Black Sea coast region. The situation is aggravated by the presence of Baku-Supsa oil pipeline terminal about 15-16 km far from Meria where there are tanks with up to 1 million tons in total of crude oil.

The Institute of Physical and Organic Chemistry was assigned as a provider of works on utilization of liquid rocket components by presidential decree no. 401 of August 21, 1997. The Ministry of Environmental and Natural Resources Protection of Georgia had to provide financial control. According to the agreement between the Ministry and the Institute, Prof. Avtandil Dolidze was assigned to be a coordinator of these works. For utilization of more than 1000 tons of components about 600 thousand GEL (equal to 300 thousand US dollars) have been apportioned. It was about twice cheaper than what foreign specialists demanded for the same works. Unfortunately, only 30 thousand GEL (or 15 thousand US dollars) was granted due to financial problems in 1998. In collaboration with Azerbaijanian colleagues, the components of rocket fuel were separately collected in non-standard aluminum tanks.

Melange was placed in Soganlugi while Samin - on the territory of Institute, opposite side of Tbilisi. That excluded the danger of possible explosion. The year 1998 passed, but there was still no financing. 56 tons of Samin were stocked on the territory of the Institute. We were looking for a solution.

According to the Soviet Army instructions Samin was usually neutralized by Boric acid solution, later mixed with kerosene and burned in open furnaces. Melange used to be neutralized by caustic sodium solution and obtained aqua-soluble salts were liquidated. Naturally we couldn't just burn the mixture of Samin and Kerosene outdoors within the city boundaries or subtropics zone near the Black Sea.

Initially we tried to work out on easier, ecologically less dangerous and cheap scheme for utilization of the components. The first model was neutralization of Samin by highly diluted Melange that means mutual neutralization in diluted solutions. Then the obtained salts were mixed with natural zeolite, turf and soil for further burial of the mixture in the ground, without any significant damages to the environment. But, unfortunately, this scheme was not fulfilled. Social factor has an important role in this kind of work. The local population is very suspicious about these activities and demands detailed information. I met twice the representatives of Meria population and different political organizations of Ozurgeti trying to explain them the situation. The representatives firmly rejected the idea of burial of neutralization products in the ground and requested that the components (especially Samin) should be taken out of the region.

Thus, during the winter period of 1998-1999 we organized fractional distillation of Samin on the territory of the Institute. Due to financial shortage hand made distillation apparatus from 200 liters iron tanks had to be used. A column with nozzle from porcelain mussel was welded to the iron tank. The tank was correspondingly isolated and a ribbon heater of explosion-proof construction was coiled in the lower part. This apparatus was used for neutralization of Samin in quantities of 150 liters. Initially we used two distillation apparatus, later five and ultimately eight. The temperature and water supply to the refrigerators were controlled. Three fractions were separated: up to 110°C consisting mostly of triethilamine, transition fraction and fraction of isomeric xylidines at 190-225°C. The transition fraction made only several percent of initial capacity. Fractions of triethilamines and isomeric xylidines were distillated in practically equal quantities. Scientific researches were carried out in Lab conditions: from the xylidines fraction the xycaine type preparation (anaesthetic materials) was obtained:

$$3CH_2ClCOOH + PCl_3 \rightarrow 3CH_2ClCOCl + P(OH)_2$$

$$(CH_3)_2C_6H_3NH_2 + ClCH_2COCL \xrightarrow[-HCl]{} (CH_3)_2C_6H_3NHCOCH_2Cl \xrightarrow[-HCl]{HN(C_2H_5)_2}$$

$$(CH_3)_2C_6H_3NHCOCH_2N(C_2H_5)_2 \xrightarrow{+HCl} (CH_3)_2C_6H_3NHCOCH_2N(C_2H_5)_2{\cdot}HCl$$

In experiments on animals it was shown that these mixtures have the effect similar to the one of xycaine (lydocaine). The synthesis of substances against pests was conducted:

$$SOCl_2 + HCON(CH_3)_2 + (CH_3)_2C_6H_3NH_2 \rightarrow (CH_3)_2C_6H_3N{=}CHN(CH_3)_2$$

$$CH_3C_6H_4SO_2Cl + HCONHCH_3 + (CH_3)_2C_6H_3NH_2 \rightarrow (CH_3)_3C_6H_3N{=}CHNHCH_3$$

0.5%-emulssion of the substances is effective against stable populations of pests. Xylidines were also used for obtaining carbamate compounds containing N-aryle fragment, which are characterized by Juvenoidic activity, similar to the one of phenoxicarbe. $(CH_3)_3C_6H_3N{=}CHCH_2NHOOC_2H_5$ and product of its hydrogenation $(CH_3)_3C_6H_3NHCH_2CH_2NHOOC_2H_5$ was synthesized. They are preventing metamorphosis of some harmful pests.

Xylidine fraction 190-225°C was re-rectified to obtain of xylidine mixtures with boiling temperature 215-225°C for getting stabilizers of hydrogen peroxide solutions. By interaction of obtained xylidines with ice vinegar acid the mixture of acetylxylidines isomers was formed. As a result of its re-crystallization from water crystalline substance with melting temperature 128-129°C was obtained. Acetylanilide is used as a stabilizer for medical purpose hydrogen peroxide. It is well known that quantity of stabilizer should not exceed 500mg/l. Comparison of stabilizing features of acetylanilide and mixtures of acetylxylidines showed practically equal results during long-term tests. Thus, acetylxylidines could be recommended as stabilizers for 3-35% solutions of hydrogen peroxide.

The noted examples of xylidine fraction distillation were tested in Lab conditions but it was necessary to perform high tonnage test as well. Anti-detonating compositions based on isomeric xylidines with octane number 106 were prepared for this reason. It's known from the literature that mixtures of xylidines were used in United States up to 1958 as anti-detonating additives to aviation petrol. Georgian crude oil has paraphine-naftenic nature and after direct distillation the petrol fraction (so called nafta) has octane number – 52-56. Additional technological processes are necessary for obtaining basic technological petrol with octant number 66. Addition of 2-3 volume % of anti-detonating compositions, based on xylidines, brings about 4-8 point increase of octane number. Thus, the greatest part of processed Samin was used for increasing of octant number of low-octant petrol.

Triethilamine was used in Lab conditions as a solvent in multi-stage synthesis of sex-pheromone of Californian beetle. The triethilamine fraction was passed to the Georgian Institute of Plant Protection for tests in high tonnage application. According to the conclusion of the Institute, the triethilamine fraction without dilution could be used as contact herbicide. 100% effectiveness was obtained after several days, though it could not be used for agricultural purposes due to its unpleasant smell. For that period the Georgian Railway Department stopped the import of herbicides for railway line treatment due to financial shortage. Treatment of most polluted parts of railroad net by triethilamine was performed in less populated places using special railway carriage – a herbicide splashier. Thus the triethilaime fraction became useful, as well. It should be mentioned that the Institute processed only 230 tons of Samin. That was total amount of Samin in Georgia though military personnel indicated twice as much. It appeared that during 8-10 years the local population used Samin for increasing of mazute and furnace fuel volumes, and also for home lamps. Certainly, it happened during the civil war and later economical degradation period. Thus processing of 230 tons of Samin practically excluded possibility of explosion and only Melange was to be neutralized.

First of all, total amount of Melange, more than 700 tons, was offered to Rustavi Chemical Plant AZOT. Ammonium Celitra is produced there from nitric acid (concentration 58%). The chemical plant refused to undertake this work due to financial problems (the processing of nitric acid is 4 times cheaper than transportation of Melange in special tanks from the place 320 km far from Rustavi) and necessity of making changes in technological process. Moreover, the noted quantity of Melange was not enough for even a single shift. That is why we worked out a special Melange utilization scheme for subtropical zone because the greatest amount of Melange was located in Meria. The following scheme was elaborated together with

Anaseuli Institute of Subtropical Economy (city of Ozurgeti): diluted Melange was neutralized by aqueous solution of slack lime and so called Norwegian celitra was obtained. It was mixed in defined portions with local turf forming organic-mineral fertilizers for red soils of subtropical zone. The corresponding open area tests conducted by this Institute showed efficiency of the noted method. Though the project was not fulfilled again due to financial problems.

In the spring 2000 in Chaladidi, near Poti, unconditioned tank was damaged and partial splash of Melange took place. We received partial financial support for liquidation of post-accidental damages after the protests of local population. We urgently moved there a titanium alloy tank of $100m^3$ volume. Melange in portions of 3 litres was transferred from the damaged tank in a new one, by means of stainless steel manual pumps. This allowed overcoming destruction of damaged aluminum tank and avoiding overflow of more than 50 tons of Melange.

There are some additional problems in this region. The local population, supported the former President of the State and therefore didn't trust us, as representatives of the new government. We were obliged to read lectures several times in order to explain them our goodwill. We advised them to leave the territory of neutralization for safety reasons but the next day the population representatives asked us to allow them to monitor the work. To calm them down we invited the local fire brigades for providing water-pumping works and widely demonstrated non-toxic nature of solutions obtained after neutralization by washing our hands and feet in them. Unfortunately, some problems still remained. After pumping Melange into the partitioned tank there still remained a small quantity of acid, which was impossible to transfer by means of mechanical pumps. That is why we decided to flood this small quantity of Melange by large amount of water in the tank of $100m^3$ volume. At that moment a cloud of nitrous oxides came out from the top of the tank and the wind blew it to the village. In an hour the cloud disappeared, but that gave rise to some protests again. The next time we didn't try to flood residual Melange, but added water by portions in time intervals and everything went well. The following method was used for speeding-up dilution and neutralization processes: we were taking out the mixture from the bottom of the tank through one nozzle and using the pump were transferring it to another nozzle placed above the mixture. That method guaranteed effective mixing and regulated output of nitrous oxides.

At present 450 tons of Melange is left in Meria, close to the Black Sea, while 237 tons in Chaladidi and Soganlugi is already neutralized according to the above described scheme. Melange in Meria is stocked in three tanks of $100m^3$ volume. Six tank covers out of nine are missing and as a result all the tanks are open. Poisonous vapors of nitrous oxides evaporate free that is especially well observed in summer. There were several requests from the local population to the State of Georgia to finance the neutralization works but unfortunately we received no money during 2001.

Moreover, this year we had to perform work on fuel neutralization from eight rockets C-200 left over by Russian Army in Vaziani. The fuel components, which remained in rockets, were pumped out separately. First, the fuel and then the oxidizer were pressed out from the rockets by water. The military personnel utilized the rockets as scrap metal after they were cooled and emptied. As far as we know similar rockets are still left in Chaladidi, as well.

Thus, the neutralization works of liquid rocket fuel components were performed just partially due to financial shortage.

Some observations and conclusions

[dstl]

NATO Workshop

-Some observations & Conclusions

Adam Cumming

Workshop

- We need a ***Cost Effective*** approach to our Common Problems
  - Disposal and Clean Up - we have different environments but the same issues
  - Risk Reduction & Management
    - for future needs and legacy problems
- Some Technology Exists
  - more mature in some areas
  - Not available to all
  - ***What is missing? What MUST we have to do the job?***

[dstl] 02 May 2004

Dstl is part of the Ministry of Defence

P. Castelo Branco et al. (eds.),
Defense Industries: Science and Technology Related to Security, 383–384.

Workshop 2

- The Issues require:
  - **EFFECTIVE** assessment of risk **which means QUANTIFICATION** of risk to allow decisions to be made on acceptability of processes and approach
  - Inclusion in *WHOLE LIFE ASSESSMENT & COSTING* - This make some technology affordable!!
  - A collaborative approach, with shared experience and shared studies/assessment - see earlier!
  - ***Need to develop SUFFICIENT understanding of processes.***

[dstl] 11 September 2004

Dstl is part of the Ministry of Defence

Workshop 3

- Capability Gaps
  - We need to put the pieces together for risk quantification tasks
  - We need to put the pieces together for applicable technology solutions
  - Is there sufficient general data to build the necessary assessment tools?
  - Can we reduce this to a useable and effective Quantitative method or methods?
  - Can we develop generally acceptable methods for cost offset through recycling for both existing and future munitions?

[dstl] 11 September 2004

Dstl is part of the Ministry of Defence

List of Directors and Key Speakers

Name	Anschrift	Tel.Nr.	FaxNr.	e-mail
Schubert, Hiltmar Prof. Dr.	Fraunhofer-Institut für Chemische Technologie Joseph-von-Fraunhofer-Str. 7 76327 Pfinztal GERMANY	0049 721 4640-112	0049 721 4640-111	schu@ict.fhg.de
Campos, José Prof.	Laboratory of Energetics and Detonics - Mechanical Engineering Depart. -Faculty of Sciences and Technology - Polo II - University of Coimbra 3030 COIMBRA - PORTUGAL	+ 351 239 790700/+ 351 91 7638948	+ 351 230 790701	ledapc@mail.telepac.pt
Castelo Branco, Pelágio General	DANOTEC - Av de Moscavide - Apartado 18 - 1885-998 MOSCAVIDE - PORTUGAL	+ 351 21 9457490	+351 21 9457499	pcbranco@danotec.pt
Teodorescu, Catalina	Acquisition Policies and Industrial relations Directorate, Department for Armament, Romania Ministry of National Defence, 9-11 Drumul Taberei St., 77305 Bucharest, ROMANIA	+4013151779	+4013127460	teodorec@ro.pims.org
Bohn, Manfred Dr.	Fraunhofer-Institut für Chemische Technologie Joseph-von-Fraunhofer-Str. 7, 76327 Pfinztal GERMANY	0049 721 4640-162	0049 721 4640-111	bo@ict.fhg.de
Held, Manfred Prof. Dr.	TDW Gesellschaft für verteidigungstechnische Wirksysteme mbH Postfach 1340 86523 Schrobenhausen GERMANY	+49 8252 996345	+49 8252 996126	manfred.held@tdw.lfk.dasa.de
Krone, Uwe Dr.	Nico Pyrotechnik, Hanns-Jürgen Diederichs GmbH & Co. KG, P.O.Box 12 27, 22943 Trittau, GERMANY	+49 4154 805 181	+49 4154 2451	u.krone@nico-pyro.de
Wanninger, Paul Dr.	Rheinmetall W&M GmbH, Heinrich-Erhardt-Straße 2, 29345 Unterlüss GERMANY	+49 5827 806-129	495827806832	Paul.Wanninger@rheinmetall-wm.com
Wild, Richard	Diehl Munitionssysteme GmbH&Co.KG, Werk Maasberg, Karl-Diehl-Straße 1, 66620 Nonnweiler GERMANY	+49 6873-70-569	+49 6873-70-644	Richard.Wild@diehl-mun.de
Van Hamm, Nico	TNO Prins Maurits Laboratory, P.O. Box 45, 2280 AA Rijswijk NETHERLANDS	+31 15 284 3588	+31 15 284 3963	ham@pml.tno.nl
Paplinski, Andrezej Dr.	Military University of Technology, 00-908 Warszawa, ul. Kaliskiego 2, POLAND	48 22 6837746	48 22 6839041	apap@wul.wat.waw.pl
Gois, José Carlos Prof.	Laboratory of Energetics and Detonics - Mechanical Engineering Depart. -Faculty of Sciences and Technology - Polo II - University of Coimbra 3030 COIMBRA - PORTUGAL	+ 351 239 790700	+ 351 230 790701	jose.gois@dem.uc.pt
Mendes, Ricardo Prof.	Laboratory of Energetics and Detonics - Mechanical Engineering Depart. -Faculty of Sciences and Technology - Polo II - University of Coimbra 3030 COIMBRA - PORTUGAL	+ 351 239 790700	+ 351 230 790701	ricardo.mendes@dem.uc.pt

Portugal, Anónio Prof.	Laboratory of Energetics and Detonics - Chemical Engineering Depart. -Faculty of Sciences and Technology - Polo II - University of Coimbra 3030 COIMBRA - PORTUGAL	+ 351 239 798741	+351 239 798795/798703	atp@eq.uc.pt
Coman, Adrian	Military Technical Academy Romania	4013355763	4013354660/198	adricoman@m.t.a.ro
Ene, Virgil	LOG. DIV (J4), General Staff, ROM. MOD	4014107423	4014104040/1307	yahoo@enemaria2001
Neagu, Constantin	C.M. ROMARM - SA		401 4117495	
Ismagilov, Zinfer R. Prof.	Head of Department of Environmental Catalysis Boreskov Institute of Catalysis SB RAS Pr.AK. Lavrentieva, 5 Novosibirsk 630090 RUSSIA	007 3832 341219	007 3832 397352	zri@catalysis.nsk.su
Kuprinyonok, Victor M.	Piskaryovsky Pr. 19-50 Saint-Petersburg Leningrad Region RUSSIA	812 259 4877	812 315 1159	technolog@nevsky.net
Kuznetsov, Andrey Dr.	Head of Applied Physics Laboratory V.G.Khlopin Radium Institute 194021, 2 Murinski pr., 28 St. Petersburg RUSSIA	007 812 2470173	007 812 2478095	ISTC1050@ATOM.NW.RU
Mikhaylov, Anatoly L. Dr.	Director of Institute of Physics of Explosion Russian Federal Nuclear Center - VNIIEF	831 30 4 51 89	831 30 4 59 58	root@gdd.vniief.ru
Soloviev, V.	Bauman Moscow State Technical University Power Engineering Research Institute 5, 2-nd Baumanskaya Moscow, 107005 RUSSIA	007 095 263-6974	007 095 267-98-93	voron@mx.bmstu.ru
Vasil'ev Anatoly A.	Lavrentyev Institute of Hydrodynamics SB RAS 630090 Novosibirsk RUSSIA	???	???	gasdet@hydro.nsc.ru
Vasipenko, Vadim G. Dr.	37, Mira prosp. Sarov, Nizhny Novgood RUSSIA	831 30 4 51 89	831 30 4 59 58	root@gdd.vniief.ru
Voronetsky, Andrei V. Prof. Dr.	Bauman Moscow State Technical University Power Engineering Research Institute 5, 2-nd Baumanskaya Moscow, 107005 RUSSIA	007 095 263-6974	007 095 267-98-93	voron@mx.bmstu.ru
Carlsson, Tomas	Weapons and Protection, Swedish Defence Research Agency, FOI, Grindsjön Research Centre, SE-14725 Tumba SWEDEN	+46 8 5550 3941	+46 8 5550 3949	toc@foi.se
Cumming, Adam S. Dr.	Defence Research Agency (DERA), Fort Halstead, X2, Sevenoaks, Kent TN14 7BP, UNITED KINGDOM	0044-195951-4754	???	ascumming@compuserve.com
Machacek, Oldrich Dr.	UTC Universal Tech Coporation 5925 Forest Lane Dallas / Texas 75230 USA	001 972-490-9906	001 972-490-9908	utecoldm@airmail.net
Stern, Alfred	NAVESA Indian Head, Research & Technology Dept., Bldg.600, 101 Strauss Avenue, Indian Head, MD 20640-5035, USA	001-301-744-4419	001-301-744-4419	Sternag@ih.navy.mil

GPSR Compliance

The European Union's (EU) General Product Safety Regulation (GPSR) is a set of rules that requires consumer products to be safe and our obligations to ensure this.

If you have any concerns about our products, you can contact us on ProductSafety@springernature.com

In case Publisher is established outside the EU, the EU authorized representative is:

Springer Nature Customer Service Center GmbH
Europaplatz 3
69115 Heidelberg, Germany

Batch number: 10809439

Printed by Printforce, the Netherlands